Engineering Mathematics

Volume III

Engineering Mathematics

Volume III

Chidanand S. Mujawar

Professor of Mathematics
S.G. Balekundri Institute of Technology, Belgaum
KLE's College of Engineering and Technology, Belgaum
(Former Head, Department of Mathematics)
&
Visiting Professor, P.G. Department V.T.U., Belgaum

TECHSAR™

NEW DELHI

Published and Printed by

TechSar Pvt. Ltd.
4435-36/7, Ansari Road, Daryaganj
New Delhi-110 002 (India)
E-mail: tech_info@techsarworld.com
Website: www.techsarworld.com

ISBN: 978-93-81141-36-6

Reprint 2025

Dedicated
to
His Holiness

Shri Siddharam Swamiji
Rudrakshimath, Naganur, Belgaum

"The master key to a successful life is accepting the circumstances as they exist and trying to mould them in the desired way"

FOREWORD

Engineering Mathematics-III authored by Prof. Chidanand S. Mujawar is primarily meant for graduate students of Engineering & Technology. The presentation is simple, rigorous and compatible with the curriculum of Engineering Mathematics.

Presentation is detailed with examples and notations used are standard. The chapters are largely independent covering Fourier Series, Fourier Transform, Numerical Analysis, Applications of Partial Differential Equations and Z-Transform.

The author has been an outstanding teacher for over 38 years. He has taught for 28 years in K.L.E. Society's College of Engineering and Technology affiliated to Visvesvaraya Technological University, Belgaum. He deserves our praise and encouragement for accomplishing this task. I wish the users good luck and happy reading.

Dr. Sharanabasava C. Pilli
Principal, K.L.E. Society's
College of Engineering and Technology,
Belgaum

PREFACE

The first and second volumes **Engineering Mathematics-Part-I & II** have been very well received by the teachers and students of all the engineering colleges. The rapid sale of the first-volume bears testimony to the overwhelming response. I thank them all for their appreciation.

The present book has been written according to the new syllabus of the university as suggested by many teachers of various engineering colleges. The book has been divided into eight units covering the topics, Fourier series, Fourier transforms, Applications to Partial Differential equations. Series solutions of Bessel's and Lagendre's differential equations, Numerical Analysis Optimization, Curve fitting, Difference equations and Z-Transforms. Each topic is treated in a systematic and logical manner. An outstanding and distinguishing feature of the book is the large number of typical solved examples followed by well graded question banks for each topic. Many examples and problems have been selected from recent papers of various engineering examinations.

I take this opportunity to thank my colleagues for noticing the errors and proper suggestions for the improvement of this book.

I will gratefully accept from the readers any corrections/criticism and suggestions for the improvement of the book.

Chidanand S. Mujawar

SYLLABUS

Engineering Mathematics - III

Code: 10 MAT 31	IA Marks: 25
Hrs/Week: 04	Exam Hrs.: 03
Total Hrs: 52	Exam Marks: 100

Part – A

Unit-I: Fourier Series

Convergence and divergence of infinite series of positive terms, definition and illustrative examples*.

Periodic functions, Dirichlet's conditions, Fourier series of periodic functions of period 2π and arbitrary period, half range Fourier series. Complex form of Fourier Series. Practical harmonic analysis. [7 hours]

Unit-II: Fourier Transforms

Infinite Fourier transform, Fourier Sine and Cosine transforms, properties, Inverse transforms. [6 hours]

Unit-III: Application of PDE

Various possible solutions of one-dimensional wave and heat equations, two-dimensional Laplace's equation by the method of separation of variables, Solution of all these equations with specified boundary conditions. D'Alembert's solution of one-dimensional wave equation. [6 hours]

Unit-IV: Curve Fitting and Optimisation

Curve fitting by the method of least squares - Fitting of curves of the form $y = ax + b$, $y = ax^2 + bx + c$, $y = ae^{bx}$, $y = ax$

Optimization: Linear programming, mathematical formulation of linear programming problem (LPP), Graphical method and simplex method.

[7 hours]

Note: *In the case of illustrative examples, questions are not to be set.

Part – B

Unit-V: Numerical Methods–1

Numerical Solution of algebraic and transcendental equations: Regula-Falsi method, Newton-Raphson method. Iterative methods of solution of a system of equations: Gauss-Seidel and Relaxation methods. Largest eigenvalue and the corresponding eigenvector by Rayleigh's power method. [6 hours]

Unit-VI: Numerical Methods–2

Finite differences: Forward and backward differences, Newton's forward and backward interpolation formulae. Divided differences – Newton's divided difference formula, Lagrange's interpolation formula and inverse interpolation formula.

Numerical integration: Simpson's one-third, three-eight and Weddle's rules (All formulae/rules without proof) [7 hours]

Unit-VII: Numerical Methods–3

Numerical solutions of PDE – finite difference approximation to derivatives, Numerical solution of two-dimensional Laplace's equation, one-dimensional heat and wave equations. [7 hours]

Unit-VIII: Difference Equations and Z-Transforms

Difference equations: Basic definition; Z-transforms – definition, standard Z-transforms, damping rule, shifting rule, initial value and final value theorems. Inverse Z-transform. Application of Z-transforms to solve difference equations. [6 hours]

CONTENTS

Part – A

UNIT 1

FOURIER SERIES

1.1 INFINITE SERIES

DEFINITION

If a set of real numbers u_1, u_2, u_3,........ u_n.......occurs according to some definite law, then $u_1 + u_2 + u_3 + \ldots\ldots + u_n + \ldots\ldots$ is called a Series.

Finite Series: A series in which, there is a limited number of terms is called a finite series.

Infinite Series: A series in which there is no limit to the number of terms or one in which every term is followed by another is called an infinite series.

The geometric series $1 + r + r^2 + r^3 + \ldots + r^n$ upto n terms is an example of a finite series but when n is made infinitely great the geometric series $1 + r + r^2 + r^3 + \ldots + r^n \ldots\ldots$ becomes an infinite series.

Let us consider an infinite series

$$\Sigma u_n = u_1 + u_2 + u_3 + \ldots\ldots + u_n + \ldots\ldots,$$

and let the sum of the first n term be S_n

i.e., $$S_n = u_1 + u_2 + u_3 + \ldots\ldots + u_n.$$

when n increases indefinitely, there are three possibilities.

a) Convergent Series

If S_n, the sum of the first n terms of an infinite series approaches a certain finite quantity 'S' when n is made infinity, then the infinite series is said to be convergent.

i.e $\lim\limits_{n\to\infty} S_n \to S$ a finite quantity

b) Divergent Series

If S_n, the sum of the first n terms of an infinite series increases with n in such a way that by taking sufficiently great we can make the numerical value of S_n exceed any finite quantity however large, the infinite series is said to be divergent.

i.e., $\lim_{n\to\infty} Sn = \infty$ or $-\infty$

c) Oscillating Series

The series Σu_n is said to oscillate finitely if the sequence $\{S_n\}$ oscillates finitely.

i.e $\lim_{n\to\infty} S_n$ is finite but not unique.

The series Σu_n is said to oscillate infinitely if the sequence $\{S_n\}$ oscillates infinitely

i.e $\lim_{n\to\infty} S_n = +\infty$ or $-\phi . \infty$

Geometric Series

The geometric Series $1 + r + r^2 + r^3 + \ldots\ldots + r^{n-1} + \ldots\ldots\ldots$ converges if $| r | < 1$, diverges if $r \geq 1$ and oscillates if $r \leq -1$.

Properties of an infinite series

1) The convergence or divergence of an infinite series remains unaffected by the addition or removal of a finite number of its terms.
2) If a series in which all the terms are positive and convergent, the series remains convergent even when some or all of its terms are negative.
3) The convergence or divergence of an infinite series remains unaffected by multiplying each term by a finite number.
4) If a positive term series Σu_n is convergent then $\lim_{n\to\infty} u_n = 0$, is the necessary condition for convergence.
5) If $\lim_{n\to\infty} u_n \neq 0$, then the series Σu_n must be divergent.
6) To find n^{th} term we use the formula $l = a + (n - 1)d$
7) $\lim_{n\to\infty} \left(1+\frac{1}{n}\right)^n = e$ and $\lim_{n\to\infty} n^{\frac{1}{n}} = 1$

Comparison Test

a) If two positive term series Σu_n and Σv_n be such that

i) Σv_n is convergent and

ii) $u_n \leq v_n$ for all values of n

Then Σu_n is also convergent.

b) If two positive term series Σu_n and Σv_n be such that

i) Σv_n is divergent and

ii) $u_n \geq v_n$ for all values of n

Then Σu_n is also divergent.

Limit form

If two positive term series Σu_n and Σv_n be such that $\lim\limits_{n\to\infty} \frac{u_n}{v_n}$ = Finite quantity ($\neq 0$) then Σu_n and Σv_n converge or diverge together.

Integral Test

A positive term series

$$f(1) + f(2) + f(3) + \ldots\ldots\ldots + f(n) + \ldots\ldots\ldots$$

Where $f(n)$ decreases as n increases, converges or diverges as the integral $\int_1^{\infty} f(x)\, d_x$ is finite or infinite.

P – Series

The harmonic series of order p

$$\sum_{n=1}^{\infty} \frac{1}{n^p} = \frac{1}{1^p} + \frac{1}{2^p} + \frac{1}{3^p} + \ldots\ldots\ldots\ldots + \frac{1}{n^p} + \ldots\ldots\ldots\ldots\infty$$

Converges for $P > 1$ and diverges for $P \leq 1$

Solution: We use the integral test, to test the P-series. By this test, the series will converge or diverge according as the integral $\int_1^{\infty} \frac{d_x}{x^p}$ is finite or infinite.

$$\text{If } P \neq 1, \int_1^\infty \frac{d_x}{x^p} = \lim_{m\to\infty} \int_1^m \frac{d_x}{x^p}$$

$$= \lim_{m\to\infty} \left(\frac{m^{1-p}-1}{1-p} \right)$$

$$= \frac{1}{p-1} \text{ i.e., finite for } p > 1.$$

$$= \infty \text{ for } p < 1$$

$$\text{If } p = 1, \int_1^\infty \frac{dx}{x} = [\log x]_1^\infty \to \infty$$

Hence the harmonic series $\sum_{n=1}^{\infty} \frac{1}{n^p}$ converges for $p > 1$ and diverges for $p \leq 1$.

Note: If $p = 1$ the series reduces to

$1 + \frac{1}{2} + \frac{1}{3} + \frac{1}{4} + \ldots\ldots\ldots + \frac{1}{n} + \ldots\ldots.$ which is a harmonic series and is always divergent.

Example 1: Test the convergence of the series

$$1 + \frac{1}{4^{2/3}} + \frac{1}{9^{2/3}} + \frac{1}{16^{2/3}} + \ldots\ldots\ldots\ldots\ldots$$

Solution: The given series can be written as

$$1 + \frac{1}{2^{4/3}} + \frac{1}{3^{4/3}} + \frac{1}{4^{4/3}} + \ldots\ldots\ldots\ldots\ldots + \frac{1}{n^{4/3}} + \ldots\ldots\ldots$$

which is a p-series $p = \frac{4}{3} > 1$ known to be convergent.

Example 2: Test for convergence the series

$$\frac{1}{1.2} + \frac{2}{3.4} + \frac{3}{5.6} + \ldots\ldots\ldots\ldots\ldots\infty$$

Solution: Here

$$u_n = \frac{n}{(2n-1)\,2n}$$

$$= \frac{n}{2n^2\left(2-\frac{1}{n}\right)} = \frac{1}{2n\,.\left(2-\frac{1}{n}\right)}$$

Take $v_n = \frac{1}{n}$ as the auxiliary series.

$$\therefore \lim_{n\to\infty} \frac{u_n}{v_n} = \lim_{n\to\infty} \frac{1}{n\,.\,2\left(2-\frac{1}{n}\right)} \times \frac{n}{1} = \frac{1}{4} \text{ which is finite and } \neq 0$$

$\therefore$ Both the series Σu_n and Σu_n converge or diverge together.

But $\Sigma v_n = \Sigma\ \frac{1}{n}$ is a harmonic series and is always divergent.

Hence by comparison test Σu_n is also divergent.

Example 3: Test the series $\sum_{n=1}^{\infty} \frac{1}{\sqrt{n}+\sqrt{n+1}}$ for convergence.

Solution: We have

$$u_n = \frac{1}{\sqrt{n+1}+\sqrt{n}}$$

$$= \frac{\sqrt{n+1}-\sqrt{n}}{(\sqrt{n+1}-\sqrt{n})\,(\sqrt{n+1}+\sqrt{n})}$$

$$= \frac{\sqrt{n+1}-\sqrt{n}}{n+1-n} = \sqrt{n+1}-\sqrt{n}$$

$$= \sqrt{n}\left[\left(1+\frac{1}{n}\right)^{\frac{1}{2}}-1\right]$$

$$= \sqrt{n}\left[1+\frac{1}{2n}-\frac{1}{8n^2}+ \ldots\ldots\ldots\ldots -1\right]$$

$$= \frac{\sqrt{n}}{n}\left[\frac{1}{2}-\frac{1}{8n}+ \ldots\ldots\ldots\ldots\right]$$

$$= \frac{1}{\sqrt{n}}\left[\frac{1}{2}-\frac{1}{8n}+ \ldots\ldots\ldots\ldots\right]$$

Take $v_n = \dfrac{1}{\sqrt{n}}$

$\therefore \quad \lim\limits_{n\to\infty} \dfrac{u_n}{v_n} = \lim\limits_{n\to\infty} \dfrac{1}{\sqrt{n}}\left(\dfrac{1}{2}-\dfrac{1}{8n}\ldots\ldots\ldots\right)\dfrac{\sqrt{n}}{1} = \dfrac{1}{2}$ which is finite.

$\therefore$ Both Σu_n and Σv_n converge or diverge together.

But $\Sigma v_n = \Sigma \dfrac{1}{\sqrt{n}}$ is a p-series, $p = \dfrac{1}{2} < 1$ known to be divergent.

Hence by comparison test Σu_n is also divergent.

Example 4: Test for convergence the series

$$\frac{1}{1.2.3}+\frac{3}{2.3.4}+\frac{5}{3.4.5}+\ldots\ldots\ldots\ldots$$

Solution: We have

$$u_n = \frac{2n-1}{n(n+1)\,(n+2)}$$

$$= \frac{1}{n^2}\,\frac{\left(2-\frac{1}{n}\right)}{\left(1+\frac{1}{n}\right)\left(1+\frac{2}{n}\right)}$$

Take, $v_n = \dfrac{1}{n^2}$ then

$\lim\limits_{n\to\infty} \dfrac{u_n}{v_n} = \lim\limits_{n\to\infty} \dfrac{2-\frac{1}{n}}{n^2\left(1+\frac{1}{n}\right)\left(1+\frac{2}{n}\right)} \times \dfrac{n^2}{1} = 2$ which is finite and non-zero.

$\therefore$ Both Σu_n and Σv_n converge or diverge together.

But $\Sigma v_n = \Sigma \frac{1}{n^2}$ is a p-series $p = 2 > 1$ known to be convergent.

$\therefore$ By comparison test Σu_n is also convergent.

Example 5: Test for convergence the series $\Sigma \frac{1}{\sqrt{n}} \tan\left(\frac{1}{n}\right)$.

Solution: Here $u_n = \frac{1}{\sqrt{n}} \tan \frac{1}{n}$

$$= \frac{1}{\sqrt{n}}\left(\frac{1}{n} + \frac{1}{3n^3} + \frac{2}{15} \cdot \frac{1}{n^5} + \dots\dots\dots\right)$$

Take $v_n = \frac{1}{n^{3/2}}$

$$\therefore \lim_{n\to\infty} \frac{u_n}{v_n} = \lim_{n\to\infty} \frac{1}{\sqrt{n}} \tan\frac{1}{n} \times \frac{n^{3/2}}{1}$$

$$= \lim_{n\to\infty} \frac{\tan\frac{1}{n}}{1/n} = 1 \text{ which is finite and non zero.}$$

$\therefore$ Σu_n and Σv_n converge or diverge together.

But $\Sigma v_n = \Sigma \frac{1}{n^{3/2}}$ is a p-series $p = \frac{3}{2} > 1$ known to be convergent.

$\therefore$ By comparison test Σu_n is also convergent.

Example 6: Test for convergence the series $\sum_{n=0}^{\infty} \frac{2n^2+5}{4n^5+1}$

Solution: Here $u_n = \frac{2n^3+5}{4n^5+1}$

$$= \frac{n^3\left(2+\frac{5}{n^3}\right)}{n^5\left(4+\frac{1}{n^5}\right)} = \frac{1}{n^2} \frac{\left(2+\frac{5}{n^3}\right)}{\left(4+\frac{1}{n^5}\right)}$$

Take $\quad v_n = \frac{1}{n^2}$

$$\therefore \quad \lim_{n\to\infty} \frac{u_n}{v_n} = \lim_{n\to\infty} \frac{\left(2+\frac{5}{n^3}\right)}{\left(4+\frac{1}{n^5}\right)} \times \frac{n^2}{1} = \frac{2}{4} = \frac{1}{2}$$

which is finite and non-zero.

$\therefore \quad \Sigma v_n = \Sigma \frac{1}{n^2}$ is a p-series $p = 2 > 1$ known to be convergent.

Hence by comparison test Σu_n is also convergent.

Example 7: Test the convergence of the following series.

i) $\sum_{n=1}^{\infty} \left(\sqrt{n^3+1} - \sqrt{n^3}\right)$

ii) $1 + \frac{2^2}{2!} + \frac{3^2}{3!} + \frac{4^2}{4!} + \ldots\ldots\ldots\infty$

iii) $\sum \sin\left(\frac{1}{n}\right)$

iv) $\sum_{n=1}^{\infty} \left(\sqrt{n^4+1} - \sqrt{n^4-1}\right)$

v) $\sum_{n=2}^{\infty} \frac{1}{n\log n}$

vi) $\sum_{n=2}^{\infty} \frac{1}{n(\log n)p}, \; p > 0$

D'ALEMBERT'S RATIO TEST

In a positive term series Σu_n if $\lim_{n\to\infty} \frac{u_{n+1}}{u_n} = \lambda$, then the series converges for $\lambda < 1$ and diverges for $\lambda > 1$, but the test fails when $\lambda = 1$.

Example 1: Discuss the convergence of the series $\sum_{n=1}^{\infty} \frac{n^2}{3^n}$.

Solution: Here $u_n = \frac{n^2}{3^n}$ and $u_{n+1} = \frac{(n+1)^2}{3^{n+1}}$

$$\therefore \quad \frac{u_{n+1}}{u_n} = \frac{(n+1)^2}{3^{n+1}} \times \frac{3^n}{n^2}$$

$$= \frac{1}{3}\left(1+\frac{1}{n}\right)^2$$

$$\therefore \quad \lim_{n\to\infty} \frac{u_{n+1}}{u_n} = \lim_{n\to\infty} \frac{1}{3}\left(1+\frac{1}{n}\right)^2 = \frac{1}{3} < 1$$

$\therefore$ By D'Alembert's Ratio test the series Σu_n is convergent.

Example 2: Discuss the convergence of the series $\sum_{n=1}^{\infty} \frac{n!}{n^n}$.

Solution: We have $u_n = \frac{n!}{n^n}$ and $u_{n+1} = \frac{(n+1)!}{(n+1)^{n+1}}$

$$\therefore \quad \frac{u_{n+1}}{u_n} = \frac{(n+1)!}{(n+1)^{n+1}} \times \frac{n^n}{n!}$$

$$= \frac{(n+1)n!}{(n+1)(n+1)^n} \times \frac{n^n}{n!} = \frac{n^n}{(n+1)^n}$$

$$\lim_{n\to\infty} \frac{u_{n+1}}{u_n} = \lim_{n\to\infty} \frac{1}{\left(1+\frac{1}{n}\right)^n} = \frac{1}{e} < 1 \qquad (\because \ e = 2.718)$$

$\therefore$ By DART the series Σu_n is convergent.

Example 3: Test for convergence the series

$$\frac{1}{2\sqrt{1}} + \frac{x^2}{3\sqrt{2}} + \frac{x^4}{4\sqrt{3}} + \ldots\ldots\ldots\ldots \infty$$

Solution: Here $u_n = \dfrac{x^{2n-2}}{(n+1)\sqrt{n}}$ and $u_{n+1} = \dfrac{x^{2n}}{(n+2)\sqrt{n+1}}$

$$\therefore \quad \frac{u_{n+1}}{u_n} = \frac{x^{2n}}{(n+2)\sqrt{n+1}} \times \frac{(n+1)\sqrt{n}}{x^{2n-2}}$$

$$= \frac{n+1}{n+2}\,\frac{\sqrt{n}}{\sqrt{n+1}}\,.\,x^2 = \frac{\left(1+\frac{1}{n}\right)}{\left(1+\frac{2}{n}\right)}\,.\,\frac{1}{\sqrt{1+\frac{1}{n}}}\,x^2$$

$$\therefore \lim_{n\to\infty} \frac{u_{n+1}}{u_n} = \frac{(1+0)}{(1+0)\,\sqrt{1+0}}\,x^2 = x^2$$

$\therefore$ By DART the series Σu_n converges if $x^2 < 1$ and diverges if $x^2 > 1$ but the test fails when $x^2 = 1$.

When $x^2 = 1$, $u_n = \dfrac{1}{(n+1)\,\sqrt{n}} = \dfrac{1}{n^{3/2}\left(1+\frac{1}{n}\right)}$

Take $\quad v_n = \dfrac{1}{n^{3/2}}$

$$\therefore \quad \lim_{n\to\infty} \frac{u_n}{v_n} = \lim_{n\to\infty} \frac{1}{n^{3/2}\left(1+\frac{1}{n}\right)} \times \frac{n^{3/2}}{1} = 1$$

which is finite and non-zero.

But $\Sigma v_n = \Sigma \dfrac{1}{n^{3/2}}$ is p-series $p = \dfrac{3}{2} > 1$ is convergent.

Hence Σu_n is also convergent when $x^2 = 1$.

Thus the series Σu_n converges for $x^2 \le 1$ and diverges for $x^2 > 1$.

Example 4: Discuss the convergence of the series

$$1+\frac{2}{5}x+\frac{6}{9}x^2+\frac{14}{17}x^3+ \,..........+ \frac{2^n-2}{2^n+1}x^{n-1} + \,...........(x>1)$$

Solution: We have

$$u_n = \frac{2^n - 2}{2^n + 1} x^{n-1} \quad \text{and} \quad u_{n+1} = \frac{2^{n+1} - 2}{2^{n+1} + 1} x^n$$

$$\therefore \quad \frac{u_{n+1}}{u_n} = \frac{2^{n+1} - 2}{2^{n+1} + 1} x^n \times \frac{2^n + 1}{2^n - 2} \frac{1}{x^{n-1}}$$

$$= \frac{\left(1 - \frac{2}{2^{n+1}}\right)}{\left(1 + \frac{1}{2^{n+1}}\right)} \cdot \frac{\left(1 + \frac{1}{2^n}\right)}{\left(1 - \frac{2}{2^n}\right)} . x$$

$$\therefore \quad \lim_{n \to \infty} \frac{u_{n+1}}{u_n} = \frac{(1-0)}{(1+0)} \cdot \frac{(1+0)}{(1-0)} . x = x$$

Thus by DART, Σu_n converges for $x < 1$ and diverges for $x > 1$, but the test fails when $x = 1$.

$$\text{When } x{=}1, \ u_n = \frac{2^n - 2}{2^n + 1}$$

$$\therefore \quad \lim_{n \to \infty} u_n = \lim_{n \to \infty} \frac{\left(1 - \frac{2}{2^n}\right)}{\left(1 + \frac{1}{2^n}\right)} = \frac{(1-0)}{(1+0)} = 1 \neq 0$$

Since $\lim\limits_{n \to \infty} u_n = 1 \neq 0$ the series must be divergent.

$\therefore$ The series Σu_n converges if $x < 1$ and diverges if $x \geq 1$.

Example 5: Show that the series

$$1 + \frac{\alpha + 1}{\beta + 1} + \frac{(\alpha+1)(2\alpha+1)}{(\beta+1)(2\beta+1)} + \frac{(\alpha+1)(2\alpha+1)(3\alpha+1)}{(\beta+1)(2\beta+1)(3\beta+1)} + \ldots\ldots\ldots\ldots\ldots$$

converges if $\beta > \alpha > 0$ and diverges if $\alpha \geq \beta > 0$.

Solution: We have

$$u_n = \frac{(\alpha+1)\ (2\alpha+1)\ (3\alpha+1) \ldots\ldots\ldots\ldots\ldots\ldots ((n-1)\,\alpha+1)}{(\beta+1)\ (2\beta+1)\ (3\beta+1) \ldots\ldots\ldots\ldots\ldots\ldots ((n-1)\,\beta+1)}$$

and $$u_{n+1} = \frac{(\alpha+1)\ (2\alpha+1)\ (3\alpha+1)\ ....((n-1)\ \alpha+1)\ (n\alpha+1)}{(\beta+1)\ (2\beta+1)\ (3\beta+1)\ ....\ ((n-1)\ \beta+1)\ (n\beta+1)}$$

$$\therefore \quad \frac{u_{n+1}}{u_n} = \frac{n\alpha+1}{n\beta+1} = \frac{\alpha+\frac{1}{n}}{\beta+\frac{1}{n}}$$

$$\therefore \quad \lim_{n\to\infty} \frac{u_{n+1}}{u_n} = \lim_{n\to\infty} \left(\frac{\alpha+\frac{1}{n}}{\beta+\frac{1}{n}}\right) = \frac{\alpha}{\beta}$$

Thus by DART Σu_n converges if $\frac{\alpha}{\beta} < 1$ i.e., $\alpha < \beta$, $\beta > \alpha > 0$ and diverges if $\frac{\alpha}{\beta} > 1$ or $\alpha > \beta > 0$ and test fails when $\frac{\alpha}{\beta} = 1$

when $\frac{\alpha}{\beta} = 1$ i.e., $\alpha = \beta$ then we have $u_n = 1$.

$$\therefore \quad \lim_{n\to\infty} u_n = 1 \neq 0$$

$\therefore$ Σu_n diverges when $\frac{\alpha}{\beta} = 1$

$\therefore$ the series Σu_n converges if $\beta > \alpha > 0$ and diverges if $\alpha \geq \beta > 0$.

Example 6: Discuss the convergence of the series

$$\frac{1}{3} + \frac{1.2}{3.5} + \frac{1.2.3}{3.5.7} + \frac{1.2.3.4}{3.5.7.9} + ...$$

Solution: Here $$u_n = \frac{1.2.3.\ ....n}{3.5.7............2n+1}$$

and $$u_{n+1} = \frac{1.2.3\ ....n(n+1)}{3.5.7........(2n+1)\ (2n+3)}$$

$$\therefore \quad \lim_{n\to\infty} \frac{u_{n+1}}{u_n} = \lim_{n\to\infty} \frac{n+1}{2n+3} = \lim_{n\to\infty} \frac{\left(1+\frac{1}{n}\right)}{\left(2+\frac{3}{n}\right)} = \frac{1}{2} < 1$$

$\therefore$ By DART the series Σu_n is convergent.

Example 7: Test for the convergence of the following series.

a) $\sum\limits_{n=1}^{\infty} \dfrac{x^n}{(2n)!}$

b) $\sum\limits_{n=1}^{\infty} \dfrac{\sqrt{n}}{\sqrt{n^2+1}}\, x^n$

c) $\sum\limits_{n=1}^{\infty} \dfrac{2^{n-1}}{3^n+1}$

d) $\sum\limits_{n=1}^{\infty} (n+1)^2\, x^n$ omitting the first term.

Raabe's Test

[This test is applied when Ratio test fails]

In the positive term series Σu_n if

$\lim\limits_{n\to\infty} n\left(\dfrac{u_n}{u_{n+1}}-1\right)=k$ then the series converges for $k > 1$ and diverges for $k < 1$ but the test fails for $k = 1$.

Logarithmic Test

In the positive term series Σu_n if

$\lim\limits_{n\to\infty}\left(n\log\dfrac{u_n}{u_{n+1}}\right)=k$, then the series converges for $k > 1$ and diverges for $k < 1$, but the test fails for $k = 1$.

Example 1: Test for convergence the series

$$1+\frac{1}{2}+\frac{1.3}{2.4}+\frac{1.3.5}{2.4.6}+..........$$

Solution: Here

$$u_n = \frac{1.3.5\,.........(2n-1)}{2.4.6\,.........\,2n} \qquad (\because \text{ Omitting the first term})$$

and $$u_{n+1} = \frac{1.3.5 ....(2n-1)(2n+1)}{2.4.6 ....(2n)(2n+2)}$$

$$\therefore \quad \lim_{n\to\infty} \frac{u_{n+1}}{u_n} = \lim_{n\to\infty} \frac{2n+1}{2n+2} = \lim_{n\to\infty} \frac{\left(2+\frac{1}{n}\right)}{\left(2+\frac{2}{n}\right)} = 1$$

$\therefore$ D'Alembert's Ratio Test fails.

Let us try by Raabe's test

For, consider

$$\frac{u_n}{u_{n+1}} - 1 = \frac{2n+2}{2n+1} - 1$$

$$= \frac{2n+2-2n-1}{2n+1} = \frac{1}{2n+1}$$

$$\therefore \quad \lim_{n\to\infty} n\left(\frac{u_n}{u_{n+1}} - 1\right) = \lim_{n\to\infty}\left(\frac{1}{2+\frac{1}{n}}\right) = \frac{1}{2} < 1$$

Hence by Raabe's test the given series Σu_n diverges.

Example 2: Discuss the convergence of the series

$$\frac{x}{1} + \frac{1}{2}\,\frac{x^3}{3} + \frac{1.3}{2.4}\,\frac{x^5}{5} + \frac{1.3.5}{2.4.6}\,\frac{x^7}{7} + \text{............}$$

Solution: Omitting the first term, we have

$$u_n = \frac{1.3.5 ....(2n-1)}{2.4.6............2n}\;\frac{x^{2n+1}}{(2n+1)}$$

and $$u_{n+1} = \frac{1.3.5 ....(2n-1)(2n+1)}{2.4.6 ....2n(2n+2)} \cdot \frac{x^{2n+3}}{(2n+3)}$$

$$\therefore \quad \frac{u_{n+1}}{u_n} = \frac{2n+1}{2n+2} \cdot \frac{2n+1}{2n+3} \cdot x^2$$

$$\therefore \quad \lim_{n\to\infty} \frac{u_{n+1}}{u_n} = \frac{(1+0)(1+0)}{(1+0)(1+0)} \cdot x^2 = x^2$$

By DART the series Σu_n converges if $x^2 < 1$ and diverges if $x^2 > 1$.

But $x^2 = 1$, the test fails

When $x^2 = 1$, let us try by Raabe's test.

Consider,

$$\frac{u_n}{u_{n+1}} = \frac{(2n+2)(2n+3)}{(2n+1)(2n+1)}$$

$$\therefore \quad \frac{u_n}{u_{n+1}} - 1 = \frac{(2n+2)(2n+3)-(2n+1)(2n+1)}{(2n+1)(2n+1)} = \frac{6n+5}{4n^2+4n+1}$$

$$\therefore n\left(\frac{u_n}{u_{n+1}} - 1\right) = \frac{6n^2+5n}{4n^2+4n+1} = \frac{6+\frac{5}{n}}{4+\frac{4}{n}+\frac{1}{n^2}}$$

$$\therefore \quad \lim_{n\to\infty} n\left(\frac{u_n}{u_{n+1}} - 1\right) = \frac{6}{4} = \frac{3}{2} > 1$$

$\therefore$ by Raabe's test the series Σu_n is convergent.

Hence the series Σu_n is convergent if $x^2 \le 1$ and divergent if $x^2 > 1$.

Example 3: Discuss the convergence of $\Sigma \dfrac{4 \,.\, 7 \,.\, 10 \ldots\ldots (3n+1)}{1 \,.\, 2 \,.\, 3 \ldots\ldots n} x^n$

Solution: Here

$$u_n = \frac{4 \,.\, 7 \,.\, 10 \ldots\ldots (3n+1)}{1 \,.\, 2 \,.\, 3 \ldots\ldots n} x^n$$

And

$$u_{n+1} = \frac{4 \,.\, 7 \,.\, 10 \ldots\ldots (3n+1)(3n+4)}{1 \,.\, 2 \,.\, 3 \ldots\ldots n(n+1)} x^{n+1}$$

$$\therefore \quad \frac{u_{n+1}}{u_n} = \frac{3n+4}{n+1} x$$

$$\lim_{n\to\infty} \frac{u_{n+1}}{u_n} = \lim_{n\to\infty} \frac{\left(3+\frac{4}{n}\right)}{\left(1+\frac{1}{n}\right)} x = 3x$$

$\therefore$ By DART the series Σu_n converges if $3x < 1$ or $x < \frac{1}{3}$ and diverges if $3x > 1$ or $x > \frac{1}{3}$ but the test fails when $x = \frac{1}{3}$.

When $x = \frac{1}{3}$. Let us try by Raabe's test

$$\frac{u_n}{u_{n+1}} - 1 = \frac{3n+3}{3n+4} - 1 = \frac{-1}{3n+4}$$

$$\therefore \quad \lim_{n\to\infty} n\left(\frac{u_n}{u_{n+1}} - 1\right) = \lim_{n\to\infty}\left(\frac{-1}{3+\frac{4}{n}}\right) = \frac{-1}{3} < 1$$

$\therefore$ by Raabe's test Σu_n is divergent.

Hence the series Σu_n converges if $x < \frac{1}{3}$ and diverges if $x \geq \frac{1}{3}$.

Example 4: Discuss the convergence of the series

$$x + \frac{2^2 x^2}{2!} + \frac{3^3 x^3}{3!} + \frac{4^4 x^4}{4!} + \frac{5^5 x^5}{5!} + \ldots\ldots\ldots\ldots$$

Solution: Here, $u_n = \dfrac{n^n x^n}{n!}$ and $u_{n+1} = \dfrac{(n+1)^{n+1} x^{n+1}}{(n+1)!}$

$$\therefore \quad \frac{u_{n+1}}{u_n} = \frac{(n+1)^{n+1}}{(n+1)!} \cdot \frac{n!}{n^n} \cdot x$$

$$= \frac{(n+1)\,(n+1)^n}{(n+1)\,n!} \cdot \frac{n!}{n^n} \cdot x$$

$$= \left(1+\frac{1}{n}\right)^n \cdot x$$

$$\therefore \quad \lim_{n\to\infty} \frac{u_{n+1}}{u_n} = \lim_{n\to\infty}\left(1+\frac{1}{n}\right)^n x = ex$$

$\therefore$ By DART, the series Σu_n converges if $ex < 1$ or $x < \frac{1}{e}$ and diverges if $x > \frac{1}{e}$ but it fails if $x = \frac{1}{e}$.

When $x = \frac{1}{e}$, let us try by Logarithmic test.

Consider,

$$\frac{u_n}{u_{n+1}} = \frac{e}{\left(1+\frac{1}{n}\right)^n} \qquad \because \quad x = \frac{1}{e}$$

$$\therefore \quad \log \frac{u_n}{u_{n+1}} = \log e - n \log \left(1+\frac{1}{n}\right)$$

$$= 1 - n\left(\frac{1}{n} - \frac{1}{2n^2} + \frac{1}{3n^3} - \ldots\ldots\ldots\ldots\right)$$

$$= \frac{1}{2n} - \frac{1}{3n^2} + \ldots\ldots\ldots\ldots$$

$$\therefore \quad \lim_{n\to\infty} \left(n \log \frac{u_n}{u_{n+1}}\right) = \frac{1}{2} < 1$$

$\therefore$ By Logarithmic test the series Σu_n diverges. Thus the series Σu_n converges if $x < \frac{1}{e}$ and diverges if $x \geq \frac{1}{e}$.

Example 6: Test for convergence the following series.

i) $1 + \frac{(1!)^2}{2!}x^2 + \frac{(2!)^2}{4!}x^4 + \frac{(3!)^2}{6!}x^6 \ldots\ldots\ldots\ldots \infty,\ x > 0$

ii) $\sum_{n=1}^{\infty} \frac{n+1}{n(n+2)} x^n$

iii) $1 + \frac{a}{b}x + \frac{a(a+1)}{b(b+1)}x^2 + \frac{a(a+1)\,(a+2)}{b(b+1)\,(b+2)}x^3 + \ldots\ldots\ a,\ b > 0,\ x > 0.$

iv) $\Sigma \dfrac{1\,.\,3\,.\,5\,.............(2n-1)}{2\,.\,4\,.\,6\,.........2n}x^n$

v) $\Sigma \dfrac{1\,.\,5\,.\,9\,.............(4n-3)}{3\,.\,7\,.\,11\,.........(4n-1)}$

vi) $1+\dfrac{3}{7}x+\dfrac{3\,.6}{7\,.\,10}x^2+\dfrac{3\,.\,6\,.\,9}{7\,.\,10\,.\,13}x^3+.........\infty$

vii) $x^2+\dfrac{2^2}{3\,.\,4}x^4+\dfrac{2^2\,.\,4^2}{3\,.\,4\,.\,5\,.\,6}x^6+\dfrac{2^2\,.\,4^2\,.\,6^2}{3\,.\,4\,.\,5\,.6\,.7\,.\,8}x^8...........$

viii) $\dfrac{1}{12}+\dfrac{1+2}{1^2+2^2}+\dfrac{1+2+3}{1^2+2^2+3^2}+...........$

Cauchy's Root Test

In a positive term series Σu_n if

$\lim\limits_{n\to\infty}(u_n)^{1/n}=\lambda$. Then the series converges for $\lambda<1$ and diverges for $\lambda>1$ and the test fails for $\lambda=1$.

Example 1: Test for convergence the series $\Sigma\left(\dfrac{nx}{n+1}\right)^n$

Solution: Here $u_n=\left(\dfrac{nx}{n+1}\right)^n$

$$\therefore \quad \lim_{n\to\infty}(u_n)^{1/n} = \lim_{n\to\infty}\left(\frac{n}{n+1}\right)x=x$$

$\therefore$ By Cauchy's root test the series Σu_n converges if $x<1$ and diverges if $x>1$.

If $x=1$ $\quad u_n=\left(\dfrac{n}{n+1}\right)^n=\dfrac{1}{\left(1+\dfrac{1}{n}\right)^n}$

$$\therefore \quad \lim_{n\to\infty} u_n = \lim_{n\to\infty} \frac{1}{\left(1+\frac{1}{n}\right)^n} = \frac{1}{e} \neq 0$$

$\therefore$ the series Σu_n is divergent.

Thus the series Σu_n converges if $x < 1$ and diverges if $x \geq 1$.

Example 2: Test for convergence the series $\dfrac{\Sigma\,[(n+1)\,x]^n}{n^{n+1}}$.

Solution: Here,

$$u_n = \frac{[(n+1)x]^n}{n^{n+1}}$$

$$\therefore \quad (u_n)^{1/4} = \frac{n+1}{n\,n^{1/n}}x = \lim_{n\to\infty}\left(1+\frac{1}{n}\right)\frac{x}{n^{1/n}} = x \qquad \left(\because \lim_{n\to\infty} n^{1/n} = 1\right)$$

$\therefore$ By Cauchy's root test the series Σu_n converges if $x < 1$ and diverges if $x > 1$ but test fails if $x = 1$.

When $x = 1$,

$$u_n = \frac{(n+1)^n}{n^{n+1}} = \left(\frac{n+1}{n}\right)^n \cdot \frac{1}{n} = \left(1+\frac{1}{n}\right)^n \frac{1}{n}$$

Take $v_n = \dfrac{1}{n}$

$$\therefore \quad \lim_{n\to\infty} \frac{u_n}{v_n} = \lim_{n\to\infty}\left(1+\frac{1}{n}\right)^n \frac{1}{n} \cdot \frac{n}{1} = e \neq 0$$

But $\Sigma v_n = \dfrac{1}{n}$ is a harmonic series which is divergent.

$\therefore$ By comparison test Σu_n is also convergent.

Thus the series Σu_n converges if $x \leq 1$ and diverges if $x > 1$.

Example 3: Test for converges the series

$$\frac{1}{2}+\frac{2}{3}x+\left(\frac{3}{4}\right)^2 x^2+\left(\frac{4}{5}\right)^3 x^3+\ldots\ldots\ldots$$

Solution: We have $u_n=\left(\frac{n+1}{n+2}\right)^n x^n$, omitting 1st term.

$$\therefore \quad \lim_{n\to\infty}(u_n)^{1/n}=\lim_{n\to\infty}\frac{n+1}{n+2}\,.\,x=x$$

$\therefore$ By Cauchy's root test, Σu_n converges if $x<1$ and diverges if $x>1$, but test fails if $x=1$.

When $x=1$,

$$u_n=\left(\frac{n+1}{n+2}\right)^n=\frac{\left(1+\frac{1}{n}\right)^n}{\left(1+\frac{2}{n}\right)^n}$$

$$=\frac{\left(1+\frac{1}{n}\right)^n}{\left\{\left(1+\frac{2}{n}\right)^{\frac{n}{2}}\right\}^2}$$

$$\therefore \quad \lim_{n\to\infty}u_n=\lim_{n\to\infty}\frac{\left(1+\frac{1}{n}\right)^n}{\left\{\left(1+\frac{2}{n}\right)^{\frac{n}{2}}\right\}^2}=\frac{e}{e^2}=\frac{1}{e}\neq 0$$

$\therefore$ The series Σu_n diverges.

Thus the series Σu_n converges if $x<1$ and diverges if $x\geq 1$.

Example 4: Test for convergence the series $\Sigma\left(1+\frac{1}{\sqrt{n}}\right)^{-n^{3/2}}$

Solution: Here $u_n=\left(1+\frac{1}{\sqrt{n}}\right)^{-n^{3/2}}$

$$(u_n)^{1/n} = \frac{1}{\left(1+\frac{1}{\sqrt{n}}\right)^{\sqrt{n}}}$$

$$\therefore \quad \lim_{n\to\infty} (u_n)^{1/n} = \lim_{n\to\infty} \frac{1}{\left(1+\frac{1}{\sqrt{n}}\right)^{\sqrt{n}}} = \frac{1}{e} < 1$$

$\therefore$ By Cauchy's root test the series Σu_n is convergent.

Example 5: Test for convergence the series $\Sigma\left(\frac{n}{n+1}\right)^{n^2}$.

Alternating Series

A series Σu_n in which the terms are alternatively positive and negative is said to be alternating series.

i.e. The series of the form

$$\sum_{n=1}^{\infty} (-1)^{n-1} u_n = u_1 - u_2 + u_3 - u_4 + \text{............}$$

is called an alternating series.

Example 1: The series

$1-\frac{1}{\sqrt{2}}+\frac{1}{\sqrt{3}}-\frac{1}{\sqrt{4}}+\text{.........}$ is an alternating series.

Leibnitz's Test

Let us consider an alternating series $\Sigma(-1)^{n-1}u_n$, where $u_n > 0$. Then the series is convergent if

i) $u_{n+1} < u_n$ for all n and

ii) $\lim_{n\to\infty} u_n = 0$

Example 2: Discuss the convergence of the series

$$1-\frac{1}{\sqrt{2}}+\frac{1}{\sqrt{3}}-\frac{1}{\sqrt{4}}\text{ ....}$$

Solution: Here

i) Terms are alternatively positive and negative

ii) $u_{n+1} < u_n$ for all n, $\frac{1}{\sqrt{3}} < \frac{1}{\sqrt{2}}$ etc. and

iii) $\lim_{n\to\infty} u_n = \lim_{n\to\infty} \frac{1}{\sqrt{n}} = 0$

Hence by Leibnitz's test, the series Σu_n is convergent.

Example 3: Test for convergence the series

$$\frac{1}{2} - \frac{2}{3} + \frac{3}{4} - \frac{4}{5} + \ldots\ldots\ldots\ldots$$

Solution: Here

i) The terms are alternatively positive and negative.

ii) $\frac{4}{5} - \frac{3}{4} = \frac{1}{20} > 0$ i.e., $u_{n+1} > u_n$ for all n and

iii) $u_n = \frac{n}{n+1}$ i.e., $\lim_{n\to\infty} u_n = \lim_{n\to\infty} \frac{n}{n+1} = 1$

Hence S_n as $n \to \infty$ is 0 or 1 according as n is even or odd.

Thus the given series is oscillating.

Absolute and Conditional Convergence

Let $\sum_{n=1}^{\infty} u_n$ be a series containing both positive and negative terms. Then Σu_n is said to be absolutely convergent if the series of positive terms $\Sigma \mid u_n \mid$ is convergent.

The series Σu_n is said to be conditionally convergent if Σu_n is convergent and $\Sigma \mid u_n \mid$ is divergent.

Example 1: The series

$1 - \frac{1}{2^2} + \frac{1}{3^2} - \frac{1}{4^2} + \frac{1}{5^2} - \ldots\ldots\ldots\ldots$ is absolutely convergent.

Example 2: The series

$1 - \frac{1}{2} + \frac{1}{3} - \frac{1}{4} + \frac{1}{5}$ .... is conditionally convergent.

Example 3: Examine the absolute convergence of the series

$$\frac{1}{1+1^2}-\frac{x}{1+2^2}+\frac{x^2}{1+3^2}-\frac{x^3}{1+4^2}+\ldots\ldots\ldots\ldots$$

Solution: Here

$$\therefore \qquad u_n = (-1)^{n-1}\frac{x^{n-1}}{1+n^2} \quad \text{and} \quad u_{n+1}=(-1)^n\frac{x^n}{1+(n+1)^2}$$

$$\therefore \qquad \frac{|u_{n+1}|}{|u_n|} = \frac{|x|^n}{1+(n+1)^2}\cdot\frac{1+n^2}{|x|^{n-1}}$$

$$\therefore \qquad \lim_{n\to\infty}\frac{|u_{n+1}|}{|u_n|} = \lim_{n\to\infty}\frac{|1+n^2|}{|1+(n+1)^2|}|x|=|x|$$

$\therefore$ By DART Σu_n is absolutely convergent if $|x|<1$ and is divergent if $|x|>1$ but test fails if $|x|=1$.

When $x = 1$, $|u_n|=\dfrac{1}{1+n^2}$ and $v_n=\dfrac{1}{n^2}$

$\therefore$ By p-series $\Sigma\,|u_n|$ is convergent if $|x|=1$.

Power Series

A series of the form

$a_0+a_1x+a_2x^2+a_3x^3+\ldots\ldots\ldots\ldots+a_n x^n+\ldots\ldots\ldots\ldots$

Where a_0, a_1, a_2 .... a_n are constants, is called a power series.

Interval of Convergence

Let the power series be

$$\sum_{n=0}^{\infty} a_n x^n = a_0+a_1x+a_2x^2+\ldots\ldots\ldots$$

Here $u_n = an\, x^n$ and $u_{n+1}=a_{n+1}\,x^{n+1}$

$$\therefore \quad \lim_{n\to\infty}\frac{u_{n+1}}{u_n} = \left(\lim_{n\to\infty}\frac{a_{n+1}}{a_n}\right)x = lx$$

where $\displaystyle\lim_{n\to\infty}\frac{a_{n+1}}{a_n}=l$

Then by DART the power series converges when $| lx | < 1$ or $| x | < \frac{1}{l}$

i.e., $-\frac{1}{l} < x < \frac{1}{l}$ and diverges when $| x | > \frac{1}{l}$.

In the interval $\left(-\frac{1}{l}, \frac{1}{l}\right)$ the power series converges and diverges for all values of x outside the interval. Hence $\left(-\frac{1}{l}, \frac{1}{l}\right)$ is called the interval of convergence.

1. Convergence of Logarithmic Series

The series

$x - \frac{x^2}{2} + \frac{x^3}{3} - \frac{x^4}{4} + \ldots\ldots\ldots\ldots$ is called Logarithmic series. Here

$$u_n = (-1)^{n-1} \frac{x^n}{n} \text{ and } u_{n+1} = (-1)^n \frac{x^{n+1}}{n+1}$$

$$\therefore \lim_{n\to\infty} \frac{|u_{n+1}|}{|u_n|} = \lim_{n\to\infty} \frac{|x|^{n+1}}{n+1} \cdot \frac{n}{|x|^n} = |x|$$

By generalised DART the series is absolutely convergent for $| x | < 1$ and is divergent for $| x | > 1$ when $x = 1$, the series becomes

$1 - \frac{1}{2} + \frac{1}{3} - \frac{1}{4} + \ldots\ldots\ldots\ldots$ which is convergent by Leibnitz's test.

When $x = -1$ the series becomes

$-\left\{1 + \frac{1}{2} + \frac{1}{3} + \frac{1}{4} + \frac{1}{5} \ldots\ldots\ldots\ldots\right\}$ which is divergent.

Thus the logarithmic series is convergent for $-1 < x \le 1$.

2. Convergence of exponential series

The series

$$1 + \frac{x}{11} + \frac{x^2}{2!} + \frac{x^3}{3!} + \ldots\ldots\ldots\ldots + \frac{x^n}{n!} + \ldots\ldots\ldots\ldots$$

is called exponential series.

Here, $u_n = \dfrac{x^n}{n!}$ and $u_{n+1} = \dfrac{x^{n+1}}{(n+1)!}$ omitting 1st term.

$$\therefore \quad \lim_{n\to\infty} \frac{|u_{n+1}|}{|u_n|} = \lim_{n\to\infty} \frac{|x|}{n+1} = 0 < 1$$

$\therefore$ By DART the series is absolutely convergent for all x.

3. Convergence of Binomial Series

The series

$$1 + nx + \frac{n(n-1)}{2!}x^2 + \frac{n(n-1)(n-2)}{3!}x^3 + \ldots\ldots\ldots$$

where n is a real number is called binomial series.

$$\text{Here} \qquad u_n = \frac{n(n-1)(n-2)\ldots\ldots\ldots(n-4+1)}{r!}x^r$$

$$\text{and} \qquad u_{r+1} = \frac{n(n-1)(n-2)\ldots\ldots\ldots(n-r+1)(n-r)}{(r+1)!}x^{r+1}$$

$$\therefore \qquad \frac{|u_{r+1}|}{|u_r|} = \frac{|n-r|}{|r+1|}|x|$$

$$\therefore \lim_{r\to\infty} \frac{|u_{r+1}|}{|u_r|} = |x|$$

Hence the binomial series converges for $|x| < 1$ i.e., in the interval $(-1, 1)$

Example 1: Test for convergence the series $\sum_{n=1}^{\infty} \left(1 - \frac{3}{n}\right)^{n^2}$

Example 2: Test for convergence the series

$$\sum_{n=1}^{\infty} \left[\sqrt[3]{n^3+1} - n\right]$$

Example 3: Test for convergence the series $\Sigma(-1)^{n-1}\dfrac{\sin nx}{n^3}$.

1.2 FOURIER SERIES

Periodic functions are of common occurrence in many engineering problems, for example, in conduction of heat and mechanical vibrations. It is useful to express these functions in a series of sine and cosine. Most of the single valued functions which occur in applied mathematics can be expressed in the form of sine and cosine series.

Periodic function: If at equal interval of abscissa x, the value of each ordinate $f(x)$ repeat itself i.e., $f(x) = f(x + T)$ for all x, then $y = f(x)$ is called a periodic function having period T.

$$f(x) = f(x+T) = f(x+2T) = \dots\dots\dots = f(x+nT)$$

or $$f(x) = f(x+nT) \quad n = 1, 2, 3, 4 \dots\dots\dots\dots$$

Definition: The series

$$\frac{a_0}{2} + a_1 \cos 1x + a_2 \cos 2x + a_3 \cos 3x + \dots\dots\dots$$
$$+ b_1 \sin 1x + b_2 \sin 2x + b_3 \sin 3x + \dots\dots\dots$$

Within a required range of values of the variables, such a series is known as Fourier Series.

Thus any function $f(x)$ defined in the interval $c_1 \leq x \leq c_2$ can be expressed in the fourier series.

$$f(x) = \frac{a_0}{2} + \sum_{n=1}^{\infty} (a_n \cos nx + b_n \sin nx)$$

(where a_0, a_n, b_n are constants) provided in the interval

i) $f(x)$ and its integrals are finite and single valued.

ii) $f(x)$ has discontinuities finite in number.

iii) $f(x)$ has maxima and minima finite in number.

These conditions are known as Dirichlet conditions.

Euler's Formulae

The Fourier series for the function $f(x)$ in the interval $\alpha < x < \alpha + 2\pi$ is given by

$$f(x) = \frac{a_0}{2} + \sum_{n=1}^{\infty} a_n \cos nx + \sum_{n=1}^{\infty} b_n \sin nx$$

Where,

$$a_0 = \frac{1}{\pi} \int_{\alpha}^{\alpha+2\pi} f(x)\,dx$$

$$a_n = \frac{1}{\pi}\int_{\alpha}^{\alpha+2\pi} f(x)\cos nx\, dx$$

and $$b_n = \frac{1}{\pi}\int_{\alpha}^{\alpha+2\pi} f(x)\sin nx\, dx$$

These values of a_0, a_n, b_n are known as Euler's formulae.

Corollary 1: When $\alpha = 0$, the interval becomes $0 < x < 2\pi$ and Fourier series is

$$f(x) = \frac{a_0}{2} + \sum_{n=1}^{\infty} a_n \cos nx + \sum_{n=1}^{\infty} b_n \sin nx$$

Where $$a_0 = \frac{1}{\pi}\int_{0}^{2\pi} f(x)\, dx$$

$$a_n = \frac{1}{\pi}\int_{0}^{2\pi} f(x)\cos nx\, dx$$

and $$b_n = \frac{1}{\pi}\int_{0}^{2\pi} f(x)\sin nx\, dx$$

Corollary 2: When $\alpha = -\pi$, the interval becomes $-\pi < x < \pi$ and Fourier series is

$$f(x) = \frac{a_0}{2} + \sum_{n=1}^{\infty} a_n \cos nx + \sum_{n=1}^{\infty} b_n \sin nx$$

Where, $$a_0 = \frac{1}{\pi}\int_{-\pi}^{\pi} f(x)\, dx$$

$$a_n = \frac{1}{\pi}\int_{-\pi}^{\pi} f(x)\cos nx\, dx$$

and $$b_n = \frac{1}{\pi}\int_{-\pi}^{\pi} f(x)\sin nx\, dx$$

Corollary 3: If $f(x)$ is discontinuous at $x = x_0$ where $\alpha < x_0 < \alpha + 2\pi$ then $f(x)$ can be expressed as

$$f(x) = \frac{a_0}{2} + \sum_{n=1}^{\infty} a_n \cos nx + \sum_{n=1}^{\infty} b_n \sin nx$$

Where

$$a_0 = \frac{1}{\pi}\left[\int_{\alpha}^{x_0} f(x)dx + \int_{x_0}^{\alpha+2\pi} f(x)dx\right]$$

$$a_n = \frac{1}{\pi}\left[\int_{\alpha}^{x_0} f(x)\cos nx\, dx + \int_{x_0}^{\alpha+2\pi} f(x)\cos nx\, dx\right]$$

and
$$b_n = \frac{1}{\pi}\left[\int_{\alpha}^{x_0} f(x)\sin nx\, dx + \int_{x_0}^{\alpha+2\pi} f(x)\sin nx\, dx\right]$$

Change of interval

The Fourier expansion of $f(x)$ in the interval $(\alpha, \alpha + 2L)$ is given by

$$f(x) = \frac{a_0}{2} + \sum_{n=1}^{\infty} a_n \cos \frac{n\pi x}{L} + \sum_{n=1}^{\infty} b_n \sin \frac{n\pi x}{L}$$

Where,
$$a_0 = \frac{1}{L}\int_{\alpha}^{\alpha+2L} f(x)\, dx$$

$$a_n = \frac{1}{L}\int_{\alpha}^{\alpha+2L} f(x)\cos\left(\frac{n\pi x}{L}\right) dx$$

and
$$b_n = \frac{1}{L}\int_{\alpha}^{\alpha+2L} f(x)\sin\left(\frac{n\pi x}{L}\right) dx$$

Corollary : Follows as above.

1.3 EVEN AND ODD FUNCTIONS

Even Function: $f(x)$ is said to be an even function if $f(x) = f(-x)$ and examples of such types are x^2, x^4, $\cos x$ etc. Graphically an even function is symmetric about y-axis. Since an even function is symmetric about y-axis, origin is the midpoint of the interval 2L and we take $c = -L$ in the general expression for 'a_n' and 'b_n'.

Thus for even function

$$a_n = \frac{1}{L}\int_{-L}^{L} f(x)\cos\frac{n\pi x}{L}dx$$

$$= \frac{1}{L}\int_{-L}^{0} f(x)\cos\frac{n\pi x}{L}dx + \frac{1}{L}\int_{0}^{L} f(x)\cos\frac{n\pi x}{L}dx$$

put $x = -x$ when $x = -L,\ -x = L$

$dx = -dx$ $\quad x = 0,\ -x = 0$

$$= \frac{1}{L}\int_{L}^{0} f(-x)\cos\frac{n\pi(-x)}{L}(-dx) + \frac{1}{L}\int_{0}^{L} f(x)\cos\frac{n\pi x}{L}dx$$

$$= \frac{1}{L}\int_{0}^{L} f(x)\cos\frac{n\pi x}{L}dx + \frac{1}{L}\int_{0}^{L} f(x)\cos\frac{n\pi x}{L}dx$$

$$\therefore \quad a_n = \frac{2}{L}\int_{0}^{L} f(x)\cos\frac{n\pi x}{L}dx$$

And,

$$b_n = \frac{1}{L}\int_{-L}^{L} f(x)\sin\frac{n\pi x}{L}dx$$

$$= \frac{1}{L}\int_{-L}^{0} f(x)\sin\frac{n\pi x}{L}dx + \frac{1}{L}\int_{0}^{L} f(x)\sin\frac{n\pi x}{L}dx$$

Put $x = -x$

$$= -\frac{1}{L}\int_{0}^{L} f(x)\sin\frac{n\pi x}{L}dx + \frac{1}{L}\int_{0}^{L} f(x)\sin\frac{n\pi x}{L}dx$$

$$\therefore \quad b_n = 0$$

Thus if a periodic function $f(x)$ is even, its Fourier series is

$$f(x) = \frac{a_0}{2} + \sum_{n=1}^{\infty} a_n\cos\frac{n\pi x}{L}$$

Where, $a_0 = \frac{2}{L}\int_0^L f(x)dx$, $a_n = \frac{2}{L}\int_0^L f(x)\cos\frac{n\pi x}{L}dx$

And $b_n = 0.$

Odd function

$f(x)$ is said to be an odd function if $f(-x) = -f(x)$ and examples of such types are x, $x^3 \sin x$ etc. Graphically an odd function is symmetric about the origin. To evaluate the coefficients a_0, a_n and b_n put $c = -L$ in general expressions.

$$a_n = \frac{1}{L}\int_{-L}^{L} f(x)\cos\frac{n\pi x}{L}dx$$

$$= \frac{1}{L}\int_{-L}^{0} f(x)\cos\frac{n\pi x}{L}dx + \frac{1}{L}\int_0^L f(x)\cos\frac{n\pi x}{L}dx$$

Put $x = -x \quad f(-x) = -f(x)$

$$= -\frac{1}{L}\int_0^L f(x)\cos\frac{n\pi x}{L}dx + \frac{1}{L}\int_0^L f(x)\cos\frac{n\pi x}{L}dx$$

$\therefore \quad a_n = 0$

And, $b_n = \frac{1}{L}\int_{-L}^{L} f(x)\sin\frac{n\pi x}{L}dx$

$$= \frac{1}{L}\int_{-L}^{0} f(x)\sin\frac{n\pi x}{L}dx + \frac{1}{L}\int_0^L f(x)\sin\frac{n\pi x}{L}dx$$

put $x = -x \quad f(x) = -f(x) \quad \sin\frac{n\pi(-x)}{L} = -\sin\frac{n\pi x}{L}$

$dx = -dx$

$$= \frac{1}{L}\int_0^L f(x)\sin\frac{n\pi x}{L}dx + \frac{1}{L}\int_0^L f(x)\sin\frac{n\pi x}{L}dx$$

$$\therefore \quad b_n = \frac{2}{L}\int_0^L f(x)\sin\frac{n\pi x}{L}dx$$

Hence if a periodic function is odd its Fourier series is

$$f(x) = \sum_{n=1}^{\infty} b_n \sin \frac{n\pi x}{L}$$

Where $\quad b_n = \frac{2}{L}\int_0^L f(x)\sin \frac{n\pi x}{L} dx \quad$ and $\quad a_n = 0, \; a_0 = 0$

1.4 HALF RANGE SERIES

Cosine Series: If it is required to express $f(x)$ as a cosine series in $0 < x < L$. We extend the function reflecting it in the y-axis so that $f(-x) = f(x)$. Then the extended function is even in $(-L, L)$ and its expansion will give the required fourier cosine series.

$$f(x) = \frac{a_0}{2} + \sum_{n=1}^{\infty} a_n \cos \frac{n\pi x}{L}$$

Where $\quad a_0 = \frac{2}{L}\int_0^L f(x)dx$

and, $\quad a_n = \frac{2}{L}\int_0^L f(x) \cos \frac{n\pi x}{L} dx$

Sine Series: It is required to expand $f(x)$ as a sine series in $0 < x < L$, then we extend the function reflecting it in the origin so that $f(-x) = -f(x)$. Then the extended function is odd in $(-L, L)$ and its expansion will give the desired Fourier sine series

$$f(x) = \sum_{n=1}^{\infty} b_n \sin \frac{n\pi x}{L}$$

Where, $\quad b_n = \frac{2}{L}\int_0^L f(x)\sin \frac{n\pi x}{L} dx.$

Note: To determine a_0, a_n and b_n we need the following results.

1) $\int e^{ax} \sin bx \, dx = \frac{e^{ax}(a\sin bx - b\cos bx)}{a^2 + b^2}$

2) $\int e^{ax}\cos bx\,dx = \dfrac{e^{ax}(a\cos bx + b\sin bx)}{a^2+b^2}$

3) $\int_a^b f(x)dx = [F(x)]_a^b = F(b) - F(a)$

4) $\int \sin nx\,dx = -\dfrac{\cos nx}{n}, \int \cos nx\,dx = \dfrac{\sin nx}{n}$

5) $\sin n\pi = 0, \qquad \cos n\pi = (-1)^n$;

6) $\sin 2n\pi = 0, \quad \cos 2n\pi = 1$;

7) $\sin 0 = 0, \qquad \cos 0 = 1$

8) $\sin\left(n+\dfrac{1}{2}\right)\pi = (-1)^n, \qquad \cos\left(n+\dfrac{1}{2}\right)\pi = 0$

9) $\sin(-\theta) = -\sin\theta, \quad \cos(-\theta) = \cos\theta.$

10) To integrate the product of two functions, one of which is a power of x, we apply the generalised rule of integration by parts.

$$\int uv\,dx = uv_1 - u'v_2 + u''v_3 - u'''v_4 + \text{............}$$

Where suffixes denote integrations and dashes denote differentiations. For examples,

i) $\int x^2\cos nx\,dx = x^2\left(\dfrac{\sin nx}{n}\right) - 2x\left(-\dfrac{\cos nx}{n^2}\right) + 2\left(\dfrac{-\sin nx}{n^3}\right)$

ii) $\int x^3 e^{2x}\,dx = x^3\,\dfrac{e^{2x}}{2} - 3x^2\left(\dfrac{e^{2x}}{2^2}\right) + 6x\left(\dfrac{e^{2x}}{2^3}\right) - 6\left(\dfrac{e^{2x}}{2^4}\right)$

11) Integration by parts.

$$\int uv\,dx = u\int v\,dx - \int\left(\int v\,dx\right)\frac{d(u)}{dx}\,dx.$$

Example 1: Find a Fourier series to represent $f(x) = x - x^2$ from $x = -\pi$ to $x = \pi$.

Solution: Let the Fourier series in $(-\pi, \pi)$ be

$$f(x) = \frac{a_0}{2} + \sum_{n=1}^{\infty} a_n\cos nx + \sum_{n=1}^{\infty} b_n\sin nx$$

Where,

$$a_0 = \frac{1}{\pi}\int_{-\pi}^{\pi} f(x)\,dx$$

$$= \frac{1}{\pi}\int_{-\pi}^{\pi} (x-x^2)\,dx$$

$$= \frac{1}{\pi}\left[\frac{x^2}{2}-\frac{x^3}{3}\right]_{-\pi}^{\pi}$$

$$= \frac{1}{\pi}\left[\left(\frac{\pi^2}{2}-\frac{\pi^3}{3}\right)-\left(\frac{\pi^2}{2}+\frac{\pi^3}{3}\right)\right]$$

$$\therefore \quad a_0 = -\frac{2\pi^2}{3}$$

$$a_n = \frac{1}{\pi}\int_{-\pi}^{\pi} (x-x^2)\cos nx\,dx$$

$$= \frac{1}{\pi}\left[(x-x^2)\frac{\sin nx}{n}-(1-2x)\left(\frac{-\cos nx}{n^2}\right)+(-2)\frac{(-\sin nx)}{n^3}\right]_{-\pi}^{\pi}$$

$$\therefore \quad a_n = -\frac{4}{n^2}(-1)^n \qquad \left(\because \cos n\pi = (-1)^n\right)$$

And, $b_n = \frac{1}{\pi}\int_{-\pi}^{\pi} (x-x^2)\sin nx\,dx$

$$= \left[(x-x^2)\left(-\frac{\cos nx}{n}\right)-(1-2x)\left(\frac{-\sin nx}{n^2}\right)+(-2)\frac{(\cos nx)}{n^3}\right]_{-\pi}^{\pi}$$

$$b_n = \frac{-2}{n}(-1)^n$$

$\therefore$ Fourier series is

$$f(x) = -\frac{\pi^2}{3}+4\left(\frac{1}{1^2}\cos 1x-\frac{1}{2^2}\cos 2x+\frac{1}{3^2}\cos 2x - \ldots\ldots\ldots\ldots\right)$$

$$+\,2\left(\sin 1x-\frac{1}{2}\sin 2x+\frac{1}{3}\sin 3x-\text{..............}\right)$$

Now, $\quad f(x) = x - x^2, \qquad$ put $x = 0$

$$0 = -\frac{\pi^2}{3}+4\left(\frac{1}{1^2}-\frac{1}{2^2}+\frac{1}{3^2}-\text{..........}\right)$$

$$\therefore \quad \frac{1}{1^2}-\frac{1}{2^2}+\frac{1}{3^2}-\frac{1}{4}\text{ ....} = \frac{\pi^2}{12}$$

or $\quad \displaystyle\sum_{n=0}^{\infty} (-1)^{n-1}\frac{1}{n^2} = \frac{\pi^2}{12}.$

Example 2: Find a Fourier series for $f(x) = x + x^2$ in $-\pi < x < \pi$.

Solution: Let the Fourier series in $-\pi < x < \pi$ be

$$f(x) = \frac{a_0}{2}+\sum_{n=1}^{\infty} a_n \cos nx+\sum_{n=1}^{\infty} b_n \sin nx$$

Where

$$a_0 = \frac{1}{\pi}\int_{-\pi}^{\pi} (x+x^2)\,dx$$

$$= \frac{1}{\pi}\left[\frac{x^2}{2}+\frac{x^3}{3}\right]_{-\pi}^{\pi}$$

$$= \frac{2\pi^2}{3}$$

$$a_n = \frac{1}{\pi}\int_{-\pi}^{\pi} (x+x^2)\cos nx\,dx$$

$$= \frac{1}{\pi}\left[(x+x^2)\frac{\sin nx}{n}-(1+2x)\left(\frac{-\cos nx}{n^2}\right)+2\,\frac{(-\sin nx)}{n^3}\right]_{-\pi}^{\pi}$$

$$= \frac{1}{\pi n^2}\left[1+2\pi-1+2\pi\right]\cos n\pi$$

$$= \frac{4}{n^2}(-1)^n$$

And,

$$b_n = \frac{1}{\pi}\int_{-\pi}^{\pi} (x+x^2)\sin nx\, dx$$

$$= \frac{1}{\pi}\left[(x+x^2)\frac{(-\cos nx)}{n} - (1+2x)\frac{(-\sin nx)}{n^2} + 2\frac{(\cos nx)}{n^3}\right]_{-\pi}^{\pi}$$

$$= \frac{1}{\pi n}\left[\pi+\pi^2+\pi-\pi^2\right](-\cos n\pi)$$

$$= \frac{-2}{n}(-1)^n$$

$\therefore$ Fourier series is

$$f(x) = \frac{\pi^2}{3} - 4\left(\frac{1}{1^2}\cos 1x - \frac{1}{2^2}\cos 2x + \frac{1}{3^2}\cos 3x - \dots\dots\dots\right)$$

$$+ 2\left(1\sin 1x - \frac{1}{2}\sin 2x + \frac{1}{3}\sin 3x + \dots\dots\dots\right)$$

Put $x = 0$ $\quad\therefore\quad f(0) = 0$

$$0 = \frac{\pi^2}{3} - 4\left(\frac{1}{1^2} - \frac{1}{2^2} + \frac{1}{3^2} - \dots\dots\dots\right)$$

$$\therefore \quad \frac{\pi^2}{12} = \frac{1}{1^2} - \frac{1}{2^2} + \frac{1}{3^2} - \dots\dots\dots$$

$$= \sum_{n=1}^{\infty} (-1)^{n-1}\cdot\frac{1}{n^2}$$

Example 3: Obtain the Fourier expansion of e^x in the interval $-\pi < x < \pi$.

Solution: Let $f(x) = e^x$ and Fourier series in $(-\pi, \pi)$ be

$$f(x) = \frac{a_0}{2} + \sum_{n=1}^{\infty} a_n \cos nx + \sum_{n=1}^{\infty} b_n \sin nx$$

$$\text{Where,}\quad a_n = \frac{1}{\pi}\int_{-\pi}^{\pi} e^x \cos nx\, dx$$

$$= \frac{1}{\pi}\left[e^x \frac{(1\cos nx + n\sin nx)}{1^2+n^2}\right]_{-\pi}^{\pi}$$

$$= \frac{e^{\pi}-e^{-\pi}}{\pi}\,\frac{\cos n\pi}{(n^2+1)} \qquad \cos n\pi = (-1)^n$$

$$\therefore\quad a_0 = \frac{e^{\pi}-e^{-\pi}}{\pi},\ a_1 = -\frac{e^{\pi}-e^{-\pi}}{\pi}\,\frac{1}{2},\ a_2 = \frac{e^{\pi}-e^{-\pi}}{\pi}\cdot\frac{1}{5^2}\ \text{etc.}$$

$$\text{And,}\quad b_n = \frac{1}{\pi}\int_{-\pi}^{\pi} e^x \sin nx\, dx$$

$$= \frac{1}{\pi}\left[e^x \frac{(1\sin nx - n\cos nx)}{1^2+n^2}\right]_{-\pi}^{\pi}$$

$$= \frac{e^{\pi}-e^{-\pi}}{\pi}\left(\frac{-n}{n^2+1}\right)\cos n\pi \quad \text{and} \quad \cos n\pi = (-1)^n$$

$$\therefore\quad b_1 = \frac{e^{\pi}-e^{-\pi}}{\pi}\,\frac{1}{2},\ b_2 = -\frac{e^{\pi}-e^{-\pi}}{\pi}\,\frac{2}{5},\ b_3 = \frac{e^{\pi}-e^{-\pi}}{\pi}\,\frac{3}{10}\ \text{etc.}$$

$\therefore$ The fourier series is

$$f(x) = \frac{e^{\pi}-e^{-\pi}}{\pi}\left[\frac{1}{2}-\left(\frac{1}{2}\cos x - \frac{1}{5}\cos 2x + \frac{1}{10}\cos 3x - \ldots\ldots\ldots\right)\right.$$

$$\left. + \left(\frac{1}{2}\sin x - \frac{2}{5}\sin 2x + \frac{3}{10}\sin 3x - \ldots\ldots\ldots\ldots\right)\right]$$

Example 4: Obtain the Fourier series for $f(x) = e^{-x}$ in the interval $0 < x < 2\pi$.

Solution: Let $f(x) = e^{-x}$ and Fourier series in $0 < x < 2\pi$ be

$$f(x) = \frac{a_0}{2} + \sum_{n=1}^{\infty} a_n \cos nx + \sum_{n=1}^{\infty} b_n \sin nx$$

Where, $$a_n = \frac{1}{\pi}\int_0^{2\pi} e^{-x}\cos nx\, dx$$

$$= \frac{1}{\pi}\left[\frac{e^{-x}(-1\cos nx + n\sin nx)}{1^2+n^2}\right]_0^{2\pi}$$

$$= \frac{1-e^{-2\pi}}{\pi}\cdot\frac{1}{n^2+1} \qquad \because \cos 2n\pi = 1 \text{ and } \cos 0 = 1$$

$$\therefore\quad a_1 = \frac{(1-e^{-2\pi})}{\pi},\quad a_1 = \frac{(1-e^{-2\pi})}{\pi}\cdot\frac{1}{2},\quad a_2 = \frac{(1-e^{-2\pi})}{\pi}\cdot\frac{1}{5} \text{ etc}$$

And,

$$b_n = \frac{1}{\pi}\int_0^{2\pi} e^{-x}\sin nx\, dx$$

$$= \frac{1}{\pi}\left[e^{-x}(-1\sin nx - n\cos nx)\right]_0^{2\pi}$$

$$= \frac{(1-e^{-2\pi})}{\pi}\cdot\frac{n}{n^2+1}$$

$$\therefore\quad b_1 = \frac{1-e^{-2\pi}}{\pi}\cdot\frac{1}{2},\quad b_2 = \frac{1-e^{-2\pi}}{\pi}\cdot\frac{2}{5} \text{ etc.}$$

$\therefore$ The Fourier series is

$$e^{-x} = \frac{1-e^{-2\pi}}{\pi}\left[\frac{1}{2}+\left(\frac{1}{2}\cos x+\frac{1}{5}\cos 2x+\frac{1}{10}\cos 3x+\ldots\ldots.\right)\right.$$

$$\left.+\left(\frac{1}{2}\sin x+\frac{2}{5}\sin 2x+\frac{3}{10}\sin 3x+\ldots\ldots\ldots.\right)\right]$$

Example 5: Obtain the Fourier series for $f(x)$ defined as follows

$$f(x) = x + \frac{\pi}{2} \qquad -\pi < x < 0$$

$$= \frac{\pi}{2} - x \qquad 0 < x < \pi$$

Solution: Let the Fourier series in $(-\pi, \pi)$ be

$$f(x) = \frac{a_0}{2} + \sum_{n=1}^{\infty} a_n \cos nx + \sum_{n=1}^{\infty} b_n \sin nx$$

$$\text{Here } f(-x) = -x + \frac{\pi}{2}$$

$$= \frac{\pi}{2} - x = f(x) \text{ in } 0 < x < \pi$$

$\therefore$ $f(x)$ is an even function and hence $b_n = 0$

$$\therefore \quad a_0 = \frac{2}{\pi}\int_0^{\pi}\left(\frac{\pi}{2} - x\right) dx$$

$$= \frac{2}{\pi}\left[\frac{\pi x}{2} - \frac{x^2}{2}\right]_0^{\pi} = 0$$

$$\text{and,} \quad a_n = \frac{2}{\pi}\int_0^{\pi}\left(\frac{\pi}{2} - x\right) \cos nx \, dx$$

$$= \frac{2}{\pi}\left[\left(\frac{\pi}{2} - x\right)\frac{\sin nx}{n} - (-1)\frac{(-\cos nx)}{n^2}\right]_0^{\pi}$$

$$= \frac{2}{\pi n^2}[1 - \cos n\pi]$$

$$= \frac{4}{\pi n^2} \text{ if } n \text{ is odd}$$

$$= 0 \text{ if } n \text{ is even.}$$

$\therefore$ Fourier series is

$$f(x) = \frac{4}{\pi}\left(\frac{1}{1^2}\cos 1x + \frac{1}{3^2}\cos 3x + \frac{1}{5^2}\cos 5x + \dots\dots\dots\right)$$

When $x = 0$ $f(0) = \dfrac{\pi}{2}$

$$\therefore \quad \frac{\pi}{2} = \frac{4}{\pi}\left(\frac{1}{1^2} + \frac{1}{3^2} + \frac{1}{5^2} + \dots\dots\dots\right)$$

$$\therefore \quad \frac{\pi^2}{8} = \frac{1}{1^2}+\frac{1}{3^2}+\frac{1}{5^2}+\frac{1}{7^2}+................$$

Example 6: Obtain the Fourier series for the function $f(x)$ defined as follows

$$f(x) = -k \qquad -\pi < x < 0$$
$$= k \qquad 0 < x < \pi$$

Solution: Let the Fourier series be

$$f(x) = \frac{a_0}{2}+\sum_{n=1}^{\infty} a_n \cos nx + \sum_{n=1}^{\infty} b_n \sin nx$$

Where,

$$a_0 = \frac{1}{\pi}\left[\int_{-\pi}^{0} -k\, dx + \int_{0}^{\pi} k\, dx\right]$$

$$= \frac{1}{\pi}\left[(-kx)_{-\pi}^{0} + (kx)_0^{\pi}\right]$$

$$= 0$$

$$a_n = \frac{1}{\pi}\left[\int_{-\pi}^{0} -k\cos nx\, dx + \int_{0}^{\pi} k\cos nx\, dx\right]$$

$$= \frac{1}{\pi}\left[\left(-\frac{k\sin nx}{n}\right)_{-\pi}^{0} + \left(\frac{k\sin nx}{n}\right)_0^{\pi}\right]$$

$$= 0$$

And,

$$b_n = \frac{1}{\pi}\left[\int_{-\pi}^{0} -k\sin nx\, dx + \int_{0}^{\pi} k\sin nx\, dx\right]$$

$$= \frac{k}{\pi}\left[\left(\frac{\cos nx}{n}\right)_{-\pi}^{0} + \left(\frac{-\cos nx}{n}\right)_0^{\pi}\right]$$

$$= \frac{2k}{\pi n}\left(1-(-1)^n\right)$$

$$= \frac{4k}{n\pi} \text{ if } n \text{ is odd}$$

$$= 0 \text{ if } n \text{ is even.}$$

∴ Fourier series is

$$f(x) = \frac{4k}{\pi}\left(1\sin x + \frac{1}{3}\sin 3x + \frac{1}{5}\sin 5x + \ldots\ldots\ldots\right)$$

Example 7: Obtain the Fourier expansion of the function $f(x)$ defined by

$$f(x) = -\pi \qquad -\pi < x < 0$$

$$= x \qquad 0 < x < \pi$$

Solution: Let the Fourier series be

$$f(x) = \frac{a_0}{2} + \sum_{n=1}^{\infty} a_n \cos nx + \sum_{n=1}^{\infty} b_n \sin nx$$

Where,

$$a_0 = \frac{1}{\pi}\left[\int_{-\pi}^{0} (-\pi)\, dx + \int_{0}^{\pi} x\, dx\right] = -\frac{\pi}{2}$$

$$a_n = \frac{1}{\pi}\left[\int_{-\pi}^{0} -\pi\cos nx\, dx + \int_{0}^{\pi} x\cos nx\, dx\right]$$

$$= \frac{1}{n^2\pi}[\cos n\pi - 1]$$

$$= -\frac{2}{\pi n^2} \quad \text{if } n \text{ is odd}$$

$$= 0 \quad \text{if } n \text{ is even}$$

And,

$$b_n = \frac{1}{\pi}\left[\int_{-\pi}^{0} -\pi\sin nx\, dx + \int_{0}^{\pi} x\sin nx\, dx\right]$$

$$= \frac{1}{n\pi}[1 - 2\cos n\pi] = \frac{1}{n\pi}\left[1 - 2(-1)^n\right]$$

∴ Fourier series is

$$f(x) = -\frac{\pi}{4} - \frac{2}{\pi}\left(\frac{1}{1^2}\cos 1x + \frac{1}{3^2}\cos 3x + \frac{1}{5^2}\cos 5x + \dots\dots\dots\right)$$

$$+3\left(\sin x - \frac{1}{2}\sin 2x + \frac{3}{3}\sin 3x - \frac{1}{4}\sin 4x \dots\dots\right)$$

put $x = 0 \qquad f(0) = 0$

$$0 = -\frac{\pi}{4} - \frac{2}{\pi}\left(\frac{1}{1^2} + \frac{1}{3^2} + \frac{1}{5^2} + \dots\dots\dots\dots\right)$$

$$\therefore \quad \frac{\pi^2}{8} = \frac{1}{1^2} + \frac{1}{3^2} + \frac{1}{5^2} + \dots\dots\dots\dots\dots + \frac{1}{(2n-1)^2} + \dots\dots\dots$$

Example 8: Obtain the Fourier series for the function $f(x)$ given by

$$f(x) = 1 + \frac{2x}{\pi} \qquad -\pi < x < 0$$

$$= 1 - \frac{2x}{\pi} \qquad 0 < x < \pi$$

Solution: Here $f(-x) = 1 + \frac{2}{\pi}(-x)$

$$= 1 - \frac{2x}{\pi} = f(x) \text{ in } 0 < x < \pi$$

$\therefore$ $f(x)$ is an even function and the Fourier series is

$$f(x) = \frac{a_0}{2} + \sum_{n=1}^{\infty} a_n \cos nx \quad \text{and} \quad b_n = 0$$

$$a_0 = \frac{2}{\pi}\int_0^{\pi}\left(1 - \frac{2x}{\pi}\right)dx = 0$$

and $$a_n = \frac{2}{\pi}\int_0^{\pi}\left(1 - \frac{2x}{\pi}\right)\cos nx\, dx$$

$$= \frac{8}{\pi n^2} \qquad \text{if } n \text{ is odd}$$

$$= 0 \qquad \text{if } n \text{ is even.}$$

∴ Fourier series is

$$f(x) = \frac{8}{\pi^2}\left(\frac{1}{12}\cos 1x + \frac{1}{3^2}\cos 3x + \frac{1}{5^2}\cos 5x + ...\right)$$

When $x = 0$, $f(0) = 1$

$$\therefore \quad \frac{\pi^2}{8} = \frac{1}{1^2} + \frac{1}{3^2} + \frac{1}{5^2} + \frac{1}{7^2} + ...$$

Example 9: Obtain the Fourier series to represent the function $f(x)$ defined as follows

$$f(x) = \pi x \qquad 0 \le x \le 1$$
$$= \pi(2 - x) \qquad 1 \le x \le 2$$

Solution: Here $2L = 2$ ∴ $L = 1$, the Fourier series is

$$f(x) = \frac{a_0}{2} + \sum_{n=1}^{\infty} a_n \cos\frac{n\pi x}{L} + \sum_{n=1}^{\infty} b_n \sin\frac{n\pi x}{L}$$

Where,

$$a_0 = \frac{1}{L}\int_0^{2L} f(x)\,dx$$

$$= \frac{1}{1}\int_0^1 \pi x\,dx + \frac{1}{1}\int_1^2 \pi(2 - x)\,dx$$

$$= \left[\frac{\pi x^2}{2}\right]_0^1 + \left[\pi\left(2x - \frac{x^2}{2}\right)\right]_1^2$$

$$= \frac{\pi}{2} + \pi(4 - 2) - \pi\left(2 - \frac{1}{2}\right)$$

$$a_0 = \pi$$

$$a_n = \frac{1}{L}\int_0^{2L} f(x)\cos\frac{n\pi x}{L}\,dx \qquad L = 1$$

$$= \int_0^1 \pi x \cos n\pi x\,dx + \int_1^2 \pi(2 - x)\cos n\pi x\,dx$$

$$= \left[\pi x \frac{\sin n\pi x}{n\pi} - \pi \frac{(-\cos n\pi x)}{n^2\pi^2}\right]_0^1$$

$$+ \left[\pi(2-x) \frac{(-\sin n\pi x)}{n\pi} - (-\pi) \frac{(-\cos n\pi x)}{n^2\pi^2}\right]_1^2$$

$$a_n = \frac{2}{\pi n^2}\left[(-1)^n - 1\right]$$

$$= 0 \text{ if } n \text{ is even}$$

$$= \frac{-4}{\pi n^2} \text{ if } n \text{ is odd}$$

And,

$$b_n = \frac{1}{L}\int_0^{2L} f(x) \sin \frac{n\pi x}{L}\, dx$$

$$= \int_0^1 \pi x \sin n\pi x\, dx + \int_1^2 \pi(2-x) \sin n\pi x\, dx$$

$$= \left[\pi x \frac{(-\cos n\pi x)}{n\pi} - \pi \frac{(-\sin n\pi x)}{n^2\pi^2}\right]_0^1$$

$$\left[\pi(2-x) \frac{(-\cos n\pi x)}{n\pi} - (-\pi) \frac{(-\sin n\pi x)}{n^2\pi^2}\right]_1^2$$

$$= \frac{-\pi \cos n\pi}{n\pi} + \frac{\pi \cos n\pi}{n\pi}$$

$\therefore$ $\quad b_n = 0$

$\therefore$ Fourier series is

$$f(x) = \frac{\pi}{2} - \frac{4}{\pi}\left(\frac{\cos \pi x}{1^2} + \frac{\cos 3\pi x}{3^2} + \frac{\cos 5\pi x}{5^2} + \ldots\ldots\ldots\ldots\right)$$

put $\quad x = 0, \quad f(0) = 0$

$\therefore$
$$\pi^2 = 8\sum_{n=1}^{\infty} \frac{1}{(2n-1)^2}$$

Example 10: Obtain the Fourier series of $f(x) = |x|$ in $-l < x < l$. Hence deduce that

$$\frac{1}{1^2}+\frac{1}{3^2}+\frac{1}{5^2}+\ldots\ldots\ldots\ldots=\frac{\pi^2}{8}$$

Solution: The period of $f(x)$ is $l-(-l) = 2l$ and the Fourier series of period $2l$ is given by

$$f(x) = \frac{a_0}{2}+\sum_{n=1}^{\infty} a_n \cos\frac{n\pi x}{l}+\sum_{n=1}^{\infty} b_n \sin\frac{n\pi x}{l}$$

We have $f(-x) = |-x| = |x| = f(x)$

Hence $f(x)$ is an even function and $b_n = 0$

$$\therefore \quad a_0 = \frac{2}{l}\int_0^l x\,dx = \frac{2}{l}\left[\frac{x^2}{2}\right]_0^l = l$$

$$a_n = \frac{2}{l}\int_0^l x\cos\left(\frac{n\pi x}{l}\right)dx$$

$$= \frac{2}{l}\left[x\frac{\sin\frac{n\pi x}{l}}{\frac{n\pi}{l}} - 1\frac{\left(-\cos\frac{n\pi x}{l}\right)}{\left(\frac{n\pi}{l}\right)^2}\right]_0^l$$

$$= -\frac{2l}{n^2\pi^2}\left[1-(-1)^n\right]$$

$\therefore$ Fourier series is

$$f(x) = \frac{l}{2}+\sum_{n=1}^{\infty}\frac{-2l}{n^2\pi^2}\left[1-(-1)^n\right]\cos\frac{n\pi x}{l}$$

put $x = 0, \quad f(0) = 0$

$$\therefore \quad \frac{\pi^2}{8} = \frac{1}{1^2}+\frac{1}{3^2}+\frac{1}{5^2}+\frac{1}{7^2}+\ldots\ldots\ldots\ldots$$

Example 11: Find the Fourier series of the periodic function defined by $f(x) = 2x - x^2$ in the interval $0 < x < 3$.

Solution: The period of $f(x)$ is $3 - 0 = 3$

$$\therefore \quad 2l = 3 \quad \therefore \quad l = \frac{3}{2}$$

The Fourier series of period $2l$ is given by

$$f(x) = \frac{a_0}{2} + \sum_{n=1}^{\infty} a_n \frac{\cos n\pi x}{l} + \sum_{n=1}^{\infty} b_n \sin\frac{n\pi x}{l}$$

$$= \frac{a_0}{2} + \sum_{n=1}^{\infty} a_n \cos\frac{2n\pi x}{3} + \sum_{n=1}^{\infty} b_n \sin\frac{2n\pi x}{3}$$

Now, $$a_0 = \frac{2}{3}\int_0^3 (2x - x^2)\, dx$$

$$= \frac{2}{3}\left[\frac{2x^2}{2} - \frac{x^3}{3}\right]_0^3 = 0$$

$$a_n = \frac{2}{3}\int_0^3 (2x - x^2)\cos\frac{2\pi nx}{3}\, dx$$

$$= \frac{2}{3}\left[(2x - x^2)\frac{\sin\frac{2n\pi x}{3}}{\frac{2n\pi}{3}} - (2-2x)\frac{\left(-\cos\frac{2n\pi x}{3}\right)}{\left(\frac{2n\pi}{3}\right)^2} + (-2)\frac{\left(-\sin\frac{2n\pi x}{3}\right)}{\left(\frac{2n\pi}{3}\right)^3}\right]_0^3$$

$$= \frac{2}{3}\cdot\frac{9}{4n^2\pi^2}\left[(2-6)\cos 2n\pi - (2-0)\cos 0\right]$$

$$= \frac{3}{2n^2\pi^2}(-6) = \frac{-9}{n^2\pi^2}$$

and $$b_n = \frac{2}{3}\int_0^3 (2x - x^2)\sin\frac{2n\pi x}{3}dx$$

$$= \frac{2}{3}\left[\frac{-3}{2n\pi}(-3)\right] = \frac{3}{n\pi}$$

$\therefore$ Fourier series is

$$f(x) = \sum_{n=0}^{\infty}\left(\frac{-9}{n^2\pi^2}\right)\cos\frac{2n\pi x}{3} + \sum_{n=1}^{\infty}\left(\frac{3}{n\pi}\right)\sin\frac{2n\pi x}{3}$$

Example 12: Find a series of cosines of multiples of x which will represent $x \sin x$ in $(0, \pi)$

Solution: Let $f(x) = x \sin x$ and Fourier cosine series be

$$f(x) = \frac{a_0}{2} + \sum_{n=1}^{\infty} a_n \cos nx$$

Then $$a_0 = \frac{2}{\pi}\int_0^{\pi} x\sin x\, dx$$

$$= \frac{2}{\pi}\left[x(-\cos x) - 1(-\sin x)\right]_0^{\pi}$$

$$= \frac{2}{\pi}\left[\pi(-\cos\pi) - 0\right]$$

$$a_0 = 2$$

$$a_n = \frac{2}{\pi}\int_0^{\pi} x\sin x\cos nx\, dx$$

$$= \frac{1}{\pi}\int_0^{\pi} x\,[\sin(n+1)x - \sin(n-1)x]\, dx$$

$$= \frac{1}{\pi}\left[x\left(\frac{-\cos(n+1)x}{n+1} + \frac{\cos(n-1)x}{n-1}\right) - \left(\frac{-\sin(n+1)x}{(n+1)^2} - \frac{\sin(n-1)x}{(n-1)^2}\right)\right]_0^{\pi}$$

$$= \frac{1}{\pi}\left[\frac{-\pi\cos(n+1)\pi}{n+1} + \frac{\cos(n-1)\pi}{n-1}\right]$$

$$= \frac{(-1)(-1)^{n+1}}{n+1} + \frac{(-1)^{n-1}}{n-1}$$

$$= (-1)^{n-1}\left(\frac{1}{n-1} - \frac{1}{n+1}\right)$$

$$\therefore \quad a_n = \frac{2(-1)^{n-1}}{(n-1)(n+1)}$$

When $n = 1, \ a_1 = \infty$

$$\therefore \quad a_1 = \frac{2}{\pi}\int_0^{\pi} x \sin x \cos x \, dx = \frac{1}{\pi}\int_0^{\pi} x \sin 2x \, dx$$

$$= \frac{1}{\pi}\left[x\frac{(-\cos 2x)}{2} - \frac{(-\sin 2x)}{4}\right]_0^{\pi}$$

$$= -\frac{1}{2}$$

$\therefore$ Fourier cosine series is

$$x \sin x = 1 - \frac{1}{2}\cos x - 2\left(\frac{\cos 2x}{1.3} - \frac{\cos 3x}{2.4} + \frac{\cos nx}{3.5} \ldots\ldots\ldots\right)$$

Example 13: Find the Fourier series for the function

$$f(x) = |\cos x| \text{ in } (-\pi, \pi)$$

Solution: Here $f(-x) = |\cos(-x)| = \cos x$ in $(-\pi, \pi)$

It is even function and also we know that

$$f(x) = \cos x \quad \text{in } (-\pi, 0)$$
$$= \cos x \quad \text{in } (0, \pi)$$

$$\therefore \quad |\cos x| = \frac{a_0}{2} + \sum_{n=1}^{\infty} a_n \cos nx$$

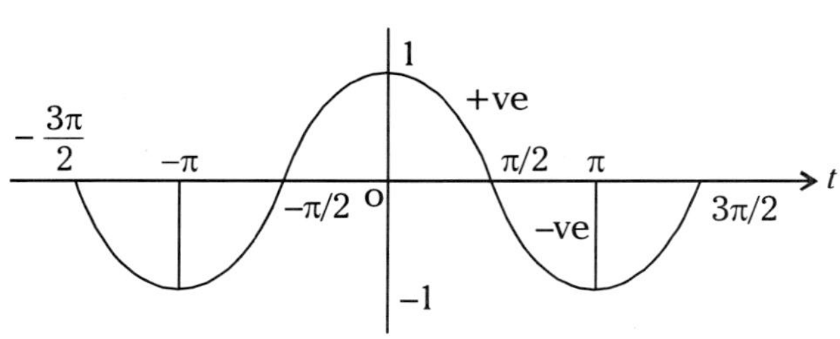

$$a_0 = \frac{2}{\pi}\int_0^{\pi} f(x)\,dx$$

$$= \frac{2}{\pi}\int_0^{\pi/2} \cos x\,dx + \frac{2}{\pi}\int_{\pi/2}^{\pi} -\cos x\,dx$$

$$= \frac{2}{\pi}[\sin x]_0^{\pi/2} + \frac{2}{\pi}[-\sin x]_{\pi/2}^{\pi}$$

$$= \frac{2}{\pi} + \frac{2}{\pi} = \frac{4}{\pi}$$

$$\therefore \quad a_0 = \frac{4}{\pi}$$

Now,

$$a_n = \frac{2}{\pi}\int_0^{\pi} f(x)\cos nx\,dx$$

$$= \frac{2}{\pi}\int_0^{\pi/2} \cos x \cos nx\,dx + \frac{2}{\pi}\int_{\pi/2}^{\pi} -\cos x\,\cos nx\,dx$$

$$= \frac{2}{\pi}\cdot\frac{1}{2}\int_0^{\pi/2} \{\cos(1+n)x + \cos(1-n)x\}\,dx$$

$$- \frac{2}{\pi}\cdot\frac{1}{2}\int_{\pi/2}^{\pi} \{\cos(1+n)\,x + \cos\,(1-n)x\}\,dx$$

$$= \frac{1}{\pi}\left[\frac{\sin(1+n)x}{1+n} + \frac{\sin(1-n)x}{1-n}\right]_0^{\pi/2}$$

$$- \frac{1}{\pi}\left[\frac{\sin(1+n)x}{1+n} + \frac{\sin(1-n)x}{1-n}\right]_{\pi/2}^{\pi}$$

$$= \frac{2}{\pi}\left[\frac{\sin(1+n)}{1+n}\frac{\pi}{2} + \frac{\sin(n-1)}{n-1}\frac{\pi}{2}\right]$$

For $n = 1$, 2nd term becomes $\frac{\infty}{\infty}$

Putting $n = 2, 3, 4, 5, \ldots\ldots\ldots$

$$a_2 = \frac{2}{\pi}\left(-\frac{1}{3}+1\right) = \frac{4}{3\pi} \qquad a_4 = \frac{-4}{\pi\,.\,15} \qquad a_6 = \frac{4}{\pi\,.\,35} \text{ etc}$$

and $\quad a_3 = a_5 = a_7 = \ldots\ldots\ldots\ldots\ldots\ldots = 0$ odd.

Also $$a_1 = \frac{2}{\pi}\int_0^{\pi/2} \cos x \cos x\, dx - \frac{2}{\pi}\int_{\pi/2}^{\pi} \cos x \cos x\, dx$$

$$a_1 = 0$$

$\therefore$ Fourier series is

$$|\cos x| = \frac{2}{\pi} + \frac{4}{\pi}\left(\frac{\cos 2x}{1\,.\,3} - \frac{\cos 4x}{3\,.\,5} + \frac{\cos 6x}{5\,.\,7} - \frac{\cos 8x}{7\,.\,9} + \ldots\ldots\ldots\right)$$

Example 14: Obtain a half range cosine series for

$$f(x) = kx \qquad 0 < x < l/2$$

$$= k\,(l - x) \qquad l/2 < x < l$$

and hence deduce that

$$\frac{1}{1^2} + \frac{1}{3^2} + \frac{1}{5^2} + \ldots\ldots\ldots\ldots = \frac{\pi^2}{8}$$

Solution: Here $L = l$ and Fourier cosine half range series is

$$f(x) = \frac{a_0}{2} + \sum_{n=1}^{\infty} a_n \cos\frac{n\pi x}{l}$$

where, $$a_0 = \frac{2}{l}\int_0^{l} f(x)\, dx$$

$$= \frac{2}{l}\int_0^{l/2} kx\, dx + \frac{2}{l}\int_{l/2}^{l} k\,(l - x)\, dx$$

$$= \frac{2}{l}\left[k\,\frac{x^2}{2}\right]_0^{l/2} + \frac{2}{l}\left[k\left(lx - \frac{x^2}{2}\right)\right]_{l/2}^{l}$$

$$= \frac{2}{l}\left[\frac{kl^2}{8}\right] + \frac{2}{l}\left[k\left(l^2 - \frac{l^2}{2}\right) - k\left(\frac{l^2}{2} - \frac{l^2}{8}\right)\right]$$

$$a_0 = \frac{kl}{2}$$

And,

$$a_n = \frac{2}{l}\int_0^l f(x)\cos\frac{n\pi x}{l}\,dx$$

$$= \frac{2}{l}\int_0^{l/2} kx\cos\frac{n\pi x}{l}\,dx + \frac{2}{l}\int_{l/2}^{l} k(l-x)\cos\frac{n\pi x}{l}\,dx$$

$$= \frac{2}{l}\left[kx\,\frac{\sin\frac{n\pi x}{l}}{\frac{n\pi}{l}} - k\,\frac{-\cos\frac{n\pi x}{l}}{\left(\frac{n\pi}{l}\right)^2}\right]_0^{l/2}$$

$$+ \frac{2}{l}\left[k(l-x)\frac{\sin\frac{n\pi x}{l}}{\frac{n\pi}{l}} - k(-1)\,\frac{\left(-\cos\frac{n\pi x}{l}\right)}{\left(\frac{n\pi}{l}\right)^2}\right]_{l/2}^{l}$$

$$= \frac{2kl}{n^2\pi^2}\left[2\cos\frac{n\pi}{2} - 1 - \cos n\pi\right]$$

When n is odd $\cos\frac{n\pi}{2} = 0$, and $\cos n\pi = (-1)^n$

$\therefore$ $a_1 = a_3 = a_5 = \ldots\ldots\ldots\ldots = 0$

$$a_2 = \frac{8kl}{2^2\pi^2},\quad a_4 = 0,\quad a_6 = \frac{-8kl}{6^2\pi^2}\quad \text{etc.}$$

$$\therefore \quad f(x) = \frac{kl}{4} - \frac{8kl}{\pi^2}\left(\frac{1}{2^2}\cos\frac{2\pi x}{l} + \frac{1}{6^2}\cos\frac{6\pi x}{l} + \ldots\ldots\ldots\ldots\right)$$

Put $x = 0$

$$0 = \frac{kl}{4} - \frac{8kl}{\pi^2}\left(\frac{1}{2^2} + \frac{1}{6^2} + \ldots\ldots\ldots\ldots\right)$$

$$\therefore \quad \frac{1}{2^2} + \frac{1}{6^2} + \ldots\ldots\ldots = \frac{\pi^2}{32}$$

$$\frac{1}{2^2}\left(\frac{1}{1^2} + \frac{1}{3^2} + \frac{1}{5^2} + \ldots\ldots\ldots\ldots\right) = \frac{\pi^2}{32}$$

$$\therefore \quad \frac{1}{1^2} + \frac{1}{1^2} + \frac{1}{5^2} + \ldots\ldots\ldots\ldots = \frac{\pi^2}{8}$$

Example 15: Expand $f(x) = \frac{1}{4} - x \quad \text{if } 0 < x < \frac{1}{2}$

$= x - \frac{3}{4} \quad \text{if } \frac{1}{2} < x < 1$ in the Fourier series of sine terms.

Solution: Let $f(x)$ represent an odd function in $(-1, 1)$ so that

$$f(x) = \frac{2}{1}\int_0^1 f(x)\sin\frac{n\pi x}{1}\,dx$$

$$= \left[\int_0^{1/2}\left(\frac{1}{4} - x\right)\sin n\pi x\,dx + \int_{1/2}^{1}\left(x - \frac{3}{4}\right)\sin n\pi x\,dx\right]$$

$$= 2\left[\left(\frac{1}{4} - x\right)\left(\frac{-\cos n\pi x}{n\pi}\right) - (-1)\,\frac{(-\sin n\pi x)}{(n\pi)^2}\right]_0^{1/2}$$

$$+ 2\left[\left(x - \frac{3}{4}\right)\left(-\frac{\cos n\pi x}{n\pi}\right) - (1)\,\frac{(-\sin n\pi x)}{(n\pi)^2}\right]_{1/2}^{1}$$

$$= 2\left[\frac{1}{4n\pi}\cos\frac{n\pi}{2} + \frac{1}{4n\pi} - \frac{\sin\frac{n\pi}{2}}{n^2\pi^2}\right]$$

$$+ 2\left[-\frac{1}{4n\pi}\cos n\pi - \frac{1}{4n\pi}\cos\frac{n\pi}{2} - \frac{\sin\frac{n\pi}{2}}{n^2\pi^2}\right]$$

$$\therefore \quad b_n = \frac{1}{2n\pi}\left[1-(-1)^n\right] - \frac{4\sin\frac{n\pi}{2}}{n^2\pi^2}$$

Thus

$$b_1 = \frac{1}{\pi} - \frac{4}{\pi^2} \qquad b_2 = 0$$

$$b_3 = \frac{1}{3\pi} + \frac{4}{3^2\pi^2} \qquad b_4 = 0$$

$$b_5 = \frac{1}{5\pi} - \frac{4}{5^2\pi^2} \qquad b_6 = 0 \text{ etc.}$$

$$\therefore \quad f(x) = \left(\frac{1}{\pi} - \frac{4}{\pi^2}\right)\sin \pi x + \left(\frac{1}{3\pi} + \frac{4}{3^2\pi^2}\right)\sin 3\pi x + \left(\frac{1}{5\pi} - \frac{4}{5^2\pi^2}\right)\sin 5\pi x + \ldots\ldots\ldots\ldots$$

Example 16: Express $f(x) = x$ as a half range sine series in $0 < x < 2$.

Solution: The Fourier series for $f(x)$ over the full period $(-2, 2)$ will contain only sine terms is given by

$$f(x) = \sum_{n=1}^{\infty} b_n \sin\frac{n\pi x}{2} \qquad L = 2$$

Where,

$$b_n = \frac{2}{2}\int_0^2 f(x)\sin\frac{n\pi x}{2}\,dx$$

$$= \int_0^2 x\sin\frac{n\pi x}{2}\,dx$$

$$= \left[\frac{x\left(-\cos\frac{n\pi x}{2}\right)}{\frac{n\pi}{2}} - \frac{\left(-\sin\frac{n\pi x}{2}\right)}{\left(\frac{n\pi}{2}\right)^2} \right]_0^2$$

$$\therefore \qquad b_n = -\frac{4}{n\pi}(-1)^n$$

$\therefore$ Fourier sine series for $f(x)$ over the half range (0, 2) is

$$f(x) = \frac{4}{\pi}\left(\sin\frac{\pi x}{2} - \frac{1}{2}\sin\frac{2\pi x}{2} + \frac{1}{3}\sin\frac{3\pi x}{2} - \ldots\ldots\ldots\ldots \right)$$

Example 17: Express $f(x) = x$ as a half range cosine series in $0 < x < 2$

Solution: The Fourier series for $f(x)$ over the full period (–2, 2) will contain only cosine terms is given by

$$f(x) = \frac{a_0}{2} + \sum_{n=1}^{\infty} a_n \cos\frac{n\pi x}{2}, \quad L = 2$$

Where

$$a_0 = \frac{2}{2}\int_0^2 x\,dx = \left[\frac{x^2}{2}\right]_0^2 = 2$$

$$a_n = \frac{2}{2}\int_0^2 x\cos\frac{n\pi x}{2}\,dx$$

$$= \left[\frac{x\sin\frac{n\pi x}{2}}{\frac{n\pi}{2}} - 1\frac{\left(-\cos\frac{n\pi x}{2}\right)}{\left(\frac{n\pi}{2}\right)^2} \right]_0^2$$

$$= \frac{4}{\pi^2 n^2}\left[(-1)^n - 1\right]$$

$$= \frac{-8}{\pi^2 n^2} \qquad \text{if } n \text{ is odd}$$

$$= 0 \qquad \text{if } n \text{ is even}$$

$\therefore$ Fourier cosine series for $f(x)$ over the half range is

$$f(x) = 1 - \frac{8}{\pi^2}\left(\frac{1}{12}\cos\frac{\pi x}{2} + \frac{1}{3^2}\cos\frac{3\pi x}{2} + \frac{1}{5^2}\cos\frac{5\pi x}{2} + \ldots\ldots\ldots\right)$$

EXERCISE

1. Obtain a Fourier series to represent e^{-ax} in $(-\pi, \pi)$

Ans.: $\dfrac{2\sinh a\pi}{\pi}\left[\left(\dfrac{1}{2a} - \dfrac{a\cos x}{1^2 + a^2} + \dfrac{a\cos 2x}{2^2 + 4^2}\ldots\right) + \left(\dfrac{\sin x}{1^2 + 4^2} - \dfrac{2\sin 2x}{2^2 + 4^2} + \ldots\right)\right]$

2. If $f(x) = \left(\dfrac{\pi - x}{2}\right)^2$ in the range 0 to 2π show that

$$f(x) = \frac{\pi^2}{12} + \sum_{n=1}^{\infty} \frac{\cos nx}{n^2}$$

3. Prove that $x^2 = \dfrac{\pi^2}{3} + 4\sum_{n=1}^{\infty} (-1)^n \dfrac{\cos nx}{n^2}$ in $-\pi < x < \pi$.

4. Obtain the Fourier series for $f(x) = \pi x$ in $0 \le x \le 2$.

Ans.: $f(x) = -\dfrac{2}{\pi}\sum_{n=1}^{\infty} \sin\dfrac{n\pi x}{2}$

5. Find a Fourier series for $f(x) = 1 - x^2$ in $-1 \le x \le 1$.

Ans.: $f(x) = \dfrac{2}{3} + \dfrac{4}{\pi^2}\left(\cos \pi x - \dfrac{\cos 2\pi x}{2^2} + \dfrac{\cos 3\pi x}{3^2} - \ldots\ldots\ldots\right)$

6. If $f(x) = \pi x \qquad 0 \le x \le 1$

$\qquad = \pi(2 - x) \qquad 1 \le x \le 2$

Show that

$$f(x) = \frac{\pi}{2} - \frac{4}{\pi}\left(\frac{\cos 1\pi x}{1^2} + \frac{\cos 3\pi x}{3^2} + \frac{\cos 5\pi x}{5^2} + \ldots\ldots\ldots\right)$$

7. Find the Fourier series of the function

$f(x) = 0$ when $-2 < x < -1$ Here $L = 2$

$\qquad = k$ when $-1 < x < 1$

$\qquad = 0$ when $1 < x < 2$

Ans.: $f(x) = \frac{k}{2} + \frac{2k}{\pi}\left(\frac{\cos\pi x}{2} - \frac{1}{3}\frac{\cos 3\pi x}{2} + \frac{1}{5}\frac{\cos 5\pi x}{2}\text{.........}\right)$

8. Prove that $x\cos x = -\frac{1}{2}\sin x + 2\sum_{n=2}^{\infty}\frac{n(-1)^n}{n^2-1}\sin nx \quad -\pi < x < \pi$

9. Given $f(x) = -x + 1$ for $-\pi < x < 0$
   $= x + 1$ for $0 \le x \le \pi$

   Is the function even or odd? Find the Fourier series for $f(x)$ and deduce

   $\frac{1}{1^2} + \frac{1}{3^2} + \frac{1}{5^2} + \text{..........} = \frac{\pi^2}{8}$.

Ans.: $\frac{\pi}{2} + 1 - \frac{\pi}{4}\left(\cos x + \frac{\cos 3x}{3^2} + \frac{\cos 5x}{5^2} + \text{.............}\right)$

10. A function is defined as follows

$$f(x) = -x \quad \text{when} \quad -\pi < x \le 0$$
$$= x \quad \text{when} \quad 0 < x < \pi$$

Show that

$$f(x) = \frac{\pi}{2} - \frac{4}{\pi}\left[\frac{1}{1^2}\cos x + \frac{1}{3^2}\cos 3x + \frac{1}{5^2}\cos 5x + \text{..........}\right]$$

11. Find the half range cosine series for the function $f(x) = x^2$ in the range $0 \le x \le \pi$.

Ans.: $f(x) = \frac{\pi^2}{3} - 4\left[\cos x - \frac{1}{2^2}\cos 2x + \frac{1}{3^2}\cos 3x - \frac{1}{4^2}\cos 4x + \text{..........}\infty\right]$

12. Find the half range cosine series for the function $f(x) = (x - 1)^2$ in the interval $0 < x < 1$.

Ans: $f(x) = \frac{1}{3} + \frac{4}{\pi^2}\sum_{1}^{\infty}\cos n\pi x$

13. Find the half range sine series for the function $f(x) = x - x^2$ in $0 < x < 1$.

Ans.: $f(x) = \frac{8}{\pi^3}\left(\frac{\sin\pi x}{13} + \frac{\sin 3\pi x}{3^3} + \frac{\sin 5\pi x}{5^3} + \text{..........}\right)$

14. Obtain the half range sine series for $f(x) = e^x$ in $0 < x < 1$.

Ans.: $$f(x) = \sum_{n=1}^{\infty} \frac{2n\pi}{1+n^2\pi^2}(1 - e\cos n\pi)\sin n\pi x$$

15. An alternating current after passing through a rectifier has the form

$$i = I_0 \sin x \quad \text{for} \quad 0 \le x \le \pi$$
$$= 0 \quad \text{for} \quad \pi \le x \le 2\pi$$

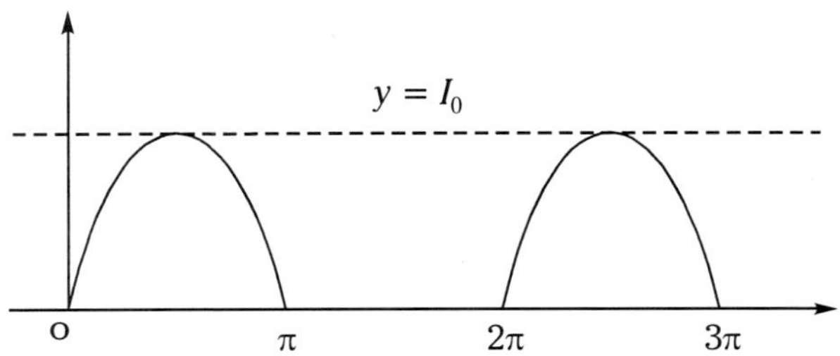

Where I_0 is the maximum and the period is 2π Express i as a Fourier series.

Ans.: $$i = \frac{I_0}{\pi} + \frac{1}{2} I_0 \sin x - \frac{2I_0}{\pi} \sum_{n=1}^{\infty} \frac{\cos 2n\pi}{4n^2 - 1}$$

16. For the function defined by the graph OAB in the following figure find the half range fourier sine series.

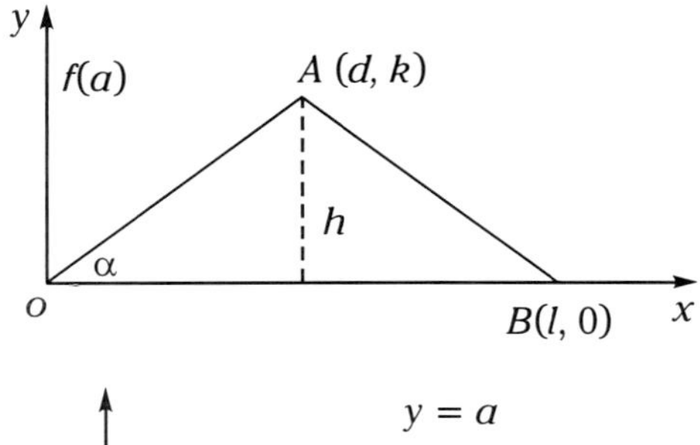

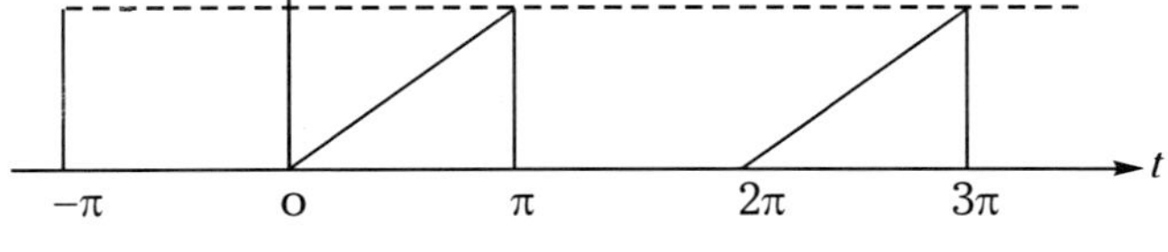

Ans.:

$$\frac{21^2 h}{\alpha(l-\alpha)\pi^2}\left[\frac{\sin \pi\alpha}{l}\sin\frac{\pi x}{l} + \frac{1}{2^2}\sin\frac{2\pi\alpha}{l}\sin\frac{2\pi x}{l} + \frac{1}{3^2}\sin\frac{3\pi\alpha}{l}\sin\frac{3\pi x}{l} + \ldots\right]$$

1.5 WAVEFORMS AND THEIR FOURIER EXPANSIONS

In communication engineering usually we meet the following typical wave forms

1. Square waveform

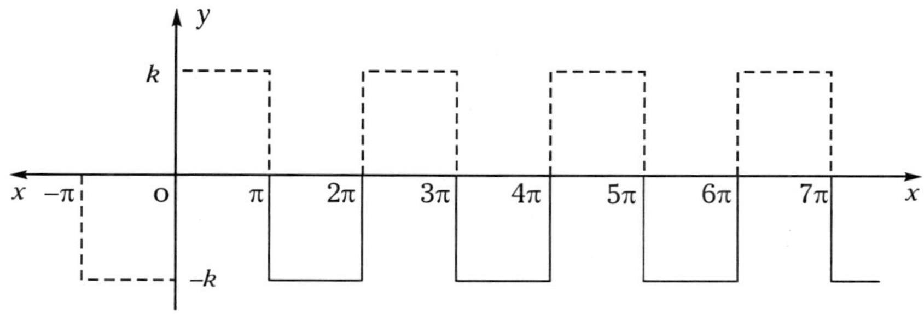

The Fourier series of the periodic function $f(x)$

$$f(x) = -k \quad \text{when} \quad -\pi < x < 0$$
$$= k \quad \text{when} \quad 0 < x < \pi$$

and $f(x) = f(x + 2\pi)$ is a square wave form and its Fourier series expansion is

$$f(x) = \frac{4k}{\pi}\left(\frac{\sin 1x}{1} + \frac{\sin 3x}{3} + \frac{\sin 5x}{5} + \ldots\ldots\ldots\ldots\right)$$

$$\therefore \quad \sum_{n=1}^{\infty} \frac{(-1)^{n+1}}{2n-1} = \frac{\pi}{4}$$

2. Saw-toothed waveform

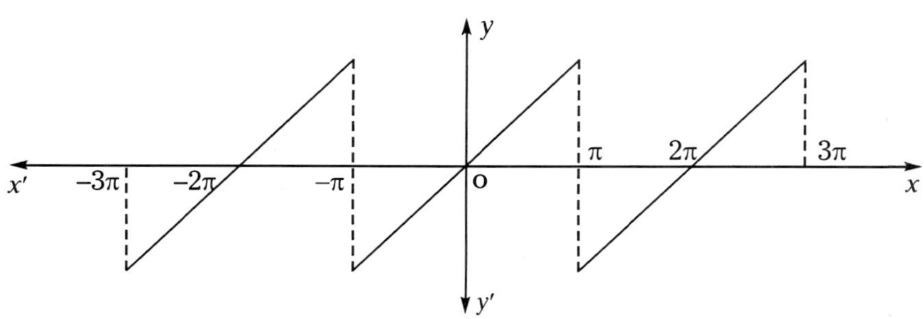

The Fourier series for $f(x) = x$ in the interval $-\pi < x < \pi$ is a saw-toothed wave form. The series is

$$f(x) = x = 2\left(\sin x - \frac{1}{2}\sin 2x + \frac{1}{3}\sin 3x - \frac{1}{4}\sin 4x..........\right) \quad ...(1)$$

The graph of $y = 2 \sin x$, $\quad y = 2\left(\sin x - \frac{1}{2}\sin 2x\right)$

$$y = 2\left(\sin x - \frac{1}{2}\sin 2x + \frac{1}{3}\sin 3x\right)$$ are shown in the figure.

The curves in the figure illustrate the manner in which the successive approximations to the series (i) approach more and more closely to $y = x$ for all values of x in $-\pi < x < \pi$ but not for $x = \pm\,\pi$.

As the series has a period 2π, it represents the discontinuous function, called saw-toothed wave form. It is important to note that the given function $y = x$ is continuous and each term of the series (1) is continuous, but the function represented by the series (1) has finite discontinuities at $x = \pm\,\pi, \pm\,3\pi, \pm\,5\pi$ etc.

3. Modified saw-toothed waveform

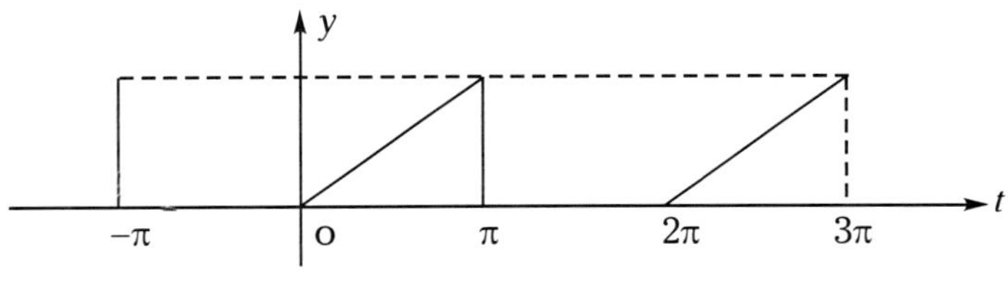

The function defined by

$$f(x) = 0 \qquad -\pi \le x \le 0$$
$$= x \qquad 0 \le x \le \pi$$

is a modified saw-toothed wave form and its Fourier expansion is

$$f(x) = \frac{a}{4} - \frac{2a}{\pi^2}\left(\cos x + \frac{\cos 3x}{3^2} + \frac{\cos 5x}{5^2} + ...\right)$$
$$+ \frac{a}{\pi}\left(\sin x - \frac{\sin 2x}{2} + \frac{\sin 3x}{3} - ....\right)$$

4. Triangular waveform

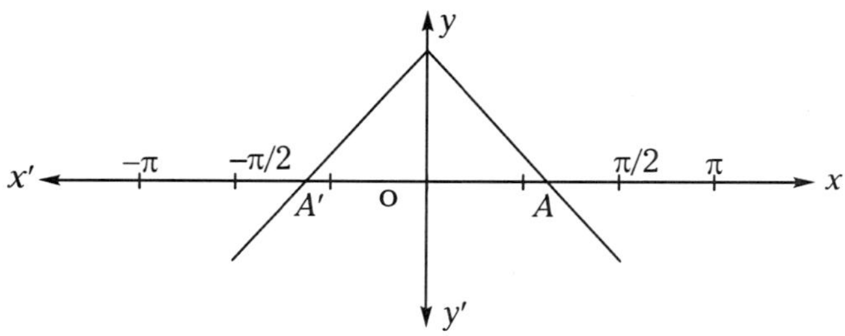

The function defined by

$$f(x) = 1+\frac{2x}{\pi} \qquad -\pi \le x \le 0$$

$$= 1-\frac{2x}{\pi} \qquad 0 \le x \le \pi$$ is a triangular

wave form and its Fourier expansion is

$$f(x) = \frac{8}{\pi^2}\left[\frac{\cos x}{1^2}+\frac{\cos 3x}{3^2}+\frac{\cos 5x}{5^2}+\ldots\ldots\ldots\ldots\right]$$

5. Halfwave rectifier

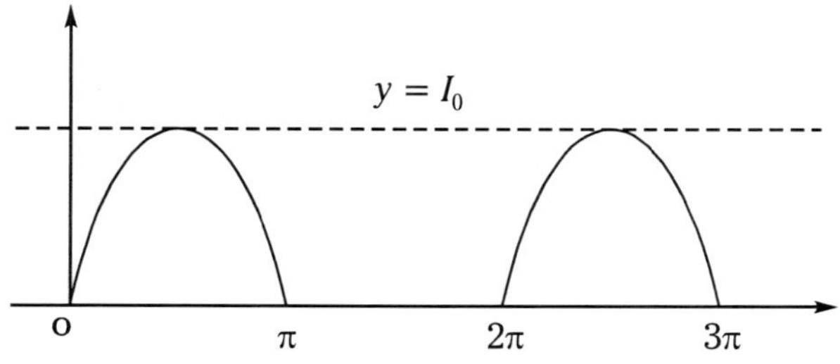

An alternating current after passing through a rectifier has the form

$$i = I_0 \sin x \qquad \text{for} \qquad 0 \le x \le \pi$$

$$= 0 \qquad \text{for} \qquad \pi \le x \le 2\pi$$

where I_0 is the maximum current and the period is 2π and its Fourier series is

$$i = \frac{I_0}{\pi}+\frac{1}{2}I_0 \sin x-\frac{2I_0}{\pi}\sum_{n=1}^{\infty}\frac{\cos 2n\pi}{4n^2-1}$$

6. Full-wave rectifier

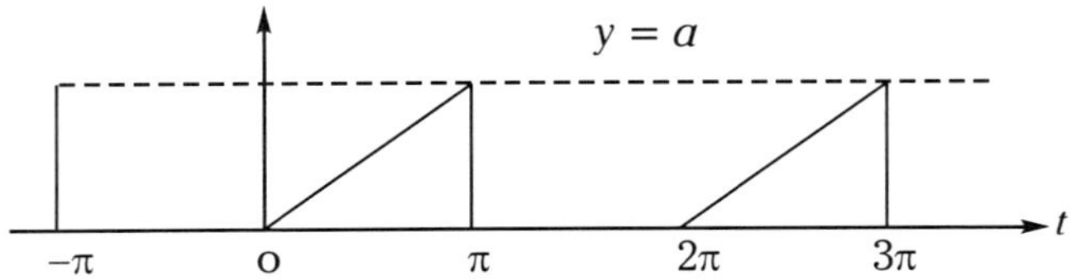

The function defined by

$$f(x) = a \sin x \qquad 0 \le x \le \pi$$

is a Full-wave rectifier and its Fourier expansion is

$$f(x) = \frac{4a}{\pi}\left[\frac{1}{2} - \frac{1}{1.3}\cos 2x - \frac{1}{3.5}\cos 4x - \frac{1}{5.7}\cos 6x - ....\right]$$

1.6 COMPLEX FORM OF FOURIER SERIES

The complex form of Fourier series is given by

$$f(x) = \sum_{n=-\infty}^{\infty} c_n \, e^{\frac{in\pi x}{l}}$$

Where $$c_n = \frac{1}{2l}\int_{-l}^{l} f(x)\, e^{-\frac{in\pi x}{l}}\, dx$$

and $$n = 0 \ \pm 1, \ \pm 2, \ \pm 3 \ ....$$

Note that

i) $\cos\theta = \frac{1}{2}\left(e^{i\theta} + e^{-i\theta}\right)$ ii) $\sin\theta = \frac{1}{2}\left(e^{i\theta} - e^{-i\theta}\right)$

$\therefore$ $e^{i\theta} = \cos\theta + i\sin\theta$ and $e^{-i\theta} = \cos\theta - i\sin\theta$.

The complex form of Fourier series is useful in problems on electrical circuits having impressed periodic voltage.

Example 1: Find the complex form of Fourier series of $f(x) = e^{-x}$ in $-1 \le x \le 1$.

Solution: We have the complex form of Fourier series is

$$f(x) = \sum_{n=-\infty}^{\infty} c_n \, e^{in\pi x} \qquad \because \ l = 1$$

Where $$c_n = \frac{1}{2}\int_{-1}^{1} e^{-x}\, e^{-in\pi x}\,dx$$

$$= \frac{1}{2}\int_{-1}^{1} e^{-(1+\text{ in }\pi)x}\,dx$$

$$= \frac{1}{2}\left[\frac{e^{-(1+in\pi)x}}{-(1+in\pi)}\right]_{-1}^{1}$$

$$= \frac{1}{2}\left[\frac{e^{-(1+in\pi)} - e^{(1+in\pi)}}{-(1+in\pi)}\right]$$

$$= \frac{e^{1+in\pi} - e^{-(1+in\pi)}}{2(1+in\,\pi)}$$

$$= \frac{e^{1}(\cos n\pi + i\sin n\pi) - e^{-1}(\cos n\pi - i\sin n\pi)}{2(1+in\pi)}$$

$$= \frac{e^{1}-e^{-1}}{2}(-1)^n \cdot \frac{1-in\pi}{1+n^2\pi^2} \qquad \because\ i^2 = -1,$$

$\cos n\pi = (-1)^n$ and $\sin n\,\pi = 0$

$$= (-1)^n\, \frac{(1-in\pi)}{1+n^2\pi^2}\sinh 1$$

$\therefore$ $$f(x) = \sum_{n=-\infty}^{\infty} (-1)^n\, \frac{(1-in\pi)}{1+n^2\pi^2}\sinh 1\, e^{in\pi x}$$ is the complex form of e^{-x} in $-1 < x < 1$.

Example 2: Find the complex form of Fourier series of

$f(x) = e^{ax}\ -l < x < l.$

Solution: Let $f(x) = e^{ax}$ in $-l < x < l$ and complex form of Fourier series is

$$f(x) = \sum_{n=-\infty}^{\infty} c_n\, e^{\frac{in\pi x}{l}}$$

where $$c_n = \frac{1}{2l}\int_{-l}^{l} e^{ax}\, e^{-\frac{in\pi x}{l}}\, dx$$

$$= \frac{1}{2l}\int_{-l}^{l} e^{\left(a - \frac{in\pi}{l}\right)x}\, dx$$

$$= \frac{1}{2l}\,\frac{\left[e^{\left(a - \frac{in\pi}{l}\right)x}\right]_{-l}^{l}}{\left(a - \frac{in\pi}{l}\right)}$$

$$= \frac{1}{2l\,\frac{(al - in\pi)}{l}}\Big[e^{al}(\cos n\pi - i\sin n\pi)$$

$$- e^{-al}(\cos n\pi + i\sin n\pi)\Big]$$

$$= \frac{1}{2\,(al - in\pi)}(e^{al} - e^{-al})\cos n\pi$$

$$= (-1)^n \frac{\sinh al}{(al - in\pi)}$$

$$\therefore \quad f(x) = \sum_{n=-\infty}^{\infty} (-1)^n \frac{\sinh al}{(al - in\pi)} e^{\frac{in\pi x}{l}}$$

Example 3: Find the complex form of Fourier series of $f(x) = \cos ax$ in $-\pi < x < \pi$.

Solution: We have $f(x) = \cos ax$ in $-\pi < x < \pi$ and complex form of Fourier series is

$$f(x) = \sum_{n=-\infty}^{\infty} c_n\, e^{inx} \qquad \because l = \pi$$

Where

$$c_n = \frac{1}{2\pi}\int_{-\pi}^{\pi} \cos ax\, e^{-inx}\, dx$$

$$= \frac{1}{2\pi}\left[\frac{e^{-inx}(-in\cos ax + a\sin ax)}{(-in)^2 + a^2}\right]_{-\pi}^{\pi}$$

$$= \frac{1}{2\pi}\left[\frac{e^{-in\pi}(-in)\cos a\pi + e^{in\pi}(in)\cos a\pi}{a^2 - n^2}\right]$$

$$= \frac{in}{2\pi\,(a^2 - n^2)}(e^{in\pi} - e^{-in\pi})\cos a\pi$$

$$= \frac{in}{\pi\,(a^2 - n^2)}\sin n\pi\,\cos a\pi$$

$$\therefore \quad f(x) = \sum_{n=-\infty}^{\infty} \frac{in}{\pi\,(a^2 - n^2)}\sin n\pi\,\cos a\pi\, e^{inx}$$

Example 4: Find the complex form of Fourier series of

$$f(x) = \sin t \quad 0 < t < \pi$$

Ans.: $\frac{2}{\pi}\left\{1 - \frac{e^{it} + e^{-it}}{1.3} - \frac{e^{4it} + e^{-4it}}{3.5} - \frac{e^{5it} + e^{-6it}}{5.7} - \ldots\ldots\ldots\right\}$

Example 5: Find the complex form of Fourier series of

$$f(x) = \begin{cases} 0 & \text{when} \quad 0 < x < l \\ a & \text{when} \quad l < x < 2l \end{cases}$$

Ans.: $\frac{a}{2} - \frac{a}{\pi}\left[(e^{u} - e^{-u}) + \frac{1}{3}(e^{3u} - e^{-3u}) + \frac{1}{5}(e^{5u} - e^{-5u}) + \ldots\ldots\right]$

where $u = \frac{i\pi x}{l}$

1.7 HARMONIC ANALYSIS

We know the problem of expanding $y = f(x)$ as Fourier series

$$f(x) = \frac{a_0}{2} + \sum_{n=1}^{\infty} (a_n \cos nx + b_n \sin nx) \qquad \ldots.(1)$$

where

$$\left.\begin{aligned} a_0 &= \frac{1}{\pi}\int_0^{2\pi} f(x)\,dx \\ a_n &= \frac{1}{\pi}\int_0^{2\pi} f(x)\cos nx\,dx \\ \text{and} \qquad b_n &= \frac{1}{\pi}\int_0^{2\pi} f(x)\sin nx\,dx \end{aligned}\right\} \quad ....(2)$$

In practice, however, the function is often given not by a formula but by a graph or by a table of corresponding values. In such cases the integrals in (2) cannot be evaluated and instead, the following alternative forms (2) are employed.

Let $f(x)$ defined in the interval $(\alpha, \alpha + 2c)$ be given in the tabular form as follows:

x :	α	$\alpha + h$	$\alpha + 2h$	....	$\alpha + nh = \alpha + 2c$
y :	y_0	y_1	y_2	....	y_n

Where $\quad nh = 2c \qquad \therefore \; h = \dfrac{2c}{n}$

Let the Fourier series be

$$f(x) = \frac{a_0}{2} + \sum_{n=1}^{\infty} a_n \cos\frac{n\pi x}{c} + \sum_{n=1}^{\infty} b_n \sin\frac{n\pi x}{c}$$

$$= \frac{a_0}{2} + \left(a_1\cos\frac{1\pi x}{c} + b_1\sin\frac{1\pi x}{c}\right) + \left(a_2\cos\frac{2\pi x}{c} + b_2\sin\frac{2\pi x}{c}\right)$$
$$+ \left(a_3\cos\frac{3\pi x}{c} + b_3\sin\frac{3\pi x}{c}\right) + \ldots\ldots\ldots\ldots$$

Here

$\left(a_1\cos\dfrac{\pi x}{c} + b_1\sin\dfrac{\pi x}{c}\right)$ is known as first harmonic

$\left(a_2\cos\dfrac{2\pi x}{c} + b_2\sin\dfrac{2\pi x}{c}\right)$ is known as second harmonic

$\left(a_3\cos\dfrac{3\pi x}{c} + b_3\sin\dfrac{3\pi x}{c}\right)$ is known as third harmonic etc.

Now,

$$a_0 = \frac{1}{c} \int_{\alpha}^{\alpha+2c} f(x)\, dx$$

$$= \frac{1}{c} \cdot \frac{h}{2} \left[(y_0 + y_n) + 2(y_1 + y_2 + y_3 + ... + y_{n-1})\right]$$

by Trapoezoidal rule.

$$= \frac{2}{n} \left\{ \frac{1}{2}(y_0 + y_n) + (y_1 + y_2 + ... + y_{n-1}) \right\}$$

$$\because h = \frac{2c}{n}$$

Since $f(x)$ is periodic function, we have

$$\therefore \frac{1}{2}(y_0 + y_n) = \frac{1}{2}(y_0 + y_0) = \frac{2y_0}{2} = y_0$$

$$\therefore \quad a_0 = \frac{2}{n}(y_0 + y_1 + y_2 + ... + y_{n-1})$$

$$= 2\,\frac{(y_0 + y_1 + y_2 + ... + y_{n-1})}{n}$$

$$= 2 \text{ (mean of } f(x))$$

$$a_1 = 2\left(\text{mean of } y \times \cos \frac{\pi x}{c}\right)$$

$$a_2 = 2\left(\text{mean of } y \times \cos \frac{2\pi x}{c}\right)$$

$$a_3 = 2\left(\text{mean of } y \times \cos \frac{3\pi x}{c}\right) \quad \text{etc.}$$

And

$$b_1 = 2\left(\text{mean of } y \times \sin \frac{1\pi x}{c}\right)$$

$$b_2 = 2\left(\text{mean of } y \times \sin \frac{2\pi x}{c}\right) \quad \text{etc.}$$

Example 1: Analyse harmonically the data given below and express y in terms of a Fourier series upto third harmonic.

x:	0	$\frac{\pi}{3}$	$\frac{2\pi}{3}$	π	$\frac{4\pi}{3}$	$\frac{5\pi}{3}$	2π
y:	1.0	1.4	1.9	1.7	1.5	1.2	1.0

Solution: We form the table as follows

x	y	$y \cos x$	$y \sin x$	$y \cos x$	$y \sin 2x$	$y \cos 3x$	$y \sin 3x$
0	1.0	1.0	0	1	0	1	0
$\frac{\pi}{3}$	1.4	0.7	1.21	–.7	1.212	–1.4	0
$\frac{2\pi}{3}$	1.9	–0.95	1.645	–.95	–1.64	1.9	0
π	1.7	–1.7	0	1.7	0	–1.7	0
$\frac{4\pi}{3}$	1.5	0.75	–1.29	–.75	1.3	1.5	0
$\frac{5\pi}{3}$	1.2	0.6	–1.039	–0.6	–1	–1.2	0
2π	–	–	–	–	–	–	–
Total	**8.7**	**–1.1**	**0.526**	**–0.3**	**–0.18**	**.1**	**0**

Here $2\pi = 2c \quad \therefore \quad c = \pi$ and the Fourier series is

$$f(x) = \frac{a_0}{2} + \sum_{n=1}^{\infty} (a_n \cos nx + b_n \sin nx)$$

$$= \frac{a_0}{2} + (a_1 \cos x + b_1 \sin x) + (a_2 \cos 2x + b_2 \sin 2x) + \ldots$$

Where

$$a_0 = \frac{2\Sigma y}{n} = \frac{2}{6} \times 8.7 = 2.9$$

$$a_1 = \frac{2}{n} \Sigma y \cos x = \frac{2}{6} \times -1.1 = -0.37$$

$$b_1 = \frac{2}{n} \Sigma y \sin x = \frac{2}{6} \times .526 = 0.17$$

$$a_2 = \frac{2}{n}\Sigma y \cos 2x = \frac{2}{6} \times -0.3 = -0.1$$

$$b_2 = \frac{2}{n}\Sigma y \sin 2x = \frac{2}{6}(-0.18) = -0.06$$

$$a_3 = \frac{2}{6}\Sigma y \cos 3x = \frac{2}{6} \times 0.1 = 0.03$$

and $b_3 = 0$

∴ The Fourier series is

$$f(x) = 1.45 + (-0.37 \cos x + 0.17 \sin x)$$
$$+ (-0.1 \cos 2x - 0.06 \sin 2x) + 0.03 \cos 3x + 0) + \ldots\ldots\ldots\ldots$$

Example 2: The following values of y give the displacement in inches of a certain machine part for the rotation x of the flywheel. Express y in terms of Fourier series.

x:	0	30	60	90	120	150	180
y:	0	9.2	14.4	17.8	17.3	11.7	0

Solution: The Fourier series for $f(x)$ is given by

$$f(x) = \frac{a_0}{2} + \sum_{n=1}^{\infty} (a_n \cos nx + b_n \sin nx)$$

$$= \frac{a_0}{2} + (a_1 \cos 1x + a_2 \cos 2x)$$
$$+ (a_2 \cos 2x + b_2 \sin 2x) + \ldots\ldots\ldots\ldots\ldots$$

To find the constants, we form the table as follows:

x	y	$y \cos x$	$y \sin x$	$y \cos 2x$	$y \sin 2x$
0	0	0	0	0	0
30	9.2	6.96	4.6	4.6	7.9674
60	14.4	7.2	12.47	–7.2	12.47
90	17.8	0	17.8	–17.8	0
120	17.3	–8.65	14.98	–8.65	–14.98
150	11.7	–10.17	5.88	5.85	–10.13
180	–	–	–	–	–
Total	70.4	–3.62	55.70	–23.2	–4.673

Here, $n = 6$

$$\therefore \quad a_0 = \frac{2}{n}\Sigma y = \frac{2}{6}(70.4) = 23.46$$

$$a_1 = \frac{2}{n}\Sigma y \cos x = \frac{2}{6} \times -3.62 = -1.2$$

$$b_1 = \frac{2}{n}\Sigma y \sin x = \frac{2}{6} \times 55.70 = 18.46$$

$$a_2 = \frac{2}{n}\Sigma y \cos 2x = \frac{2}{6} \times -23.2 = -7.73$$

$$b_2 = \frac{2}{n}\Sigma y \sin 2x = \frac{2}{6} \times -4.673 = -1.557 \quad \text{etc.}$$

$\therefore$ The Fourier series is

$$f(x) = 11.73 - 1.2 \cos 1x + 18.46 \sin x - 7.73 \cos 2x - 1.55 \sin 2x$$

Example 3: Obtain the constant term and the coefficients of the first sine and cosine terms in the fourier expansion of y as given in the following table.

x:	0	1	2	3	4	5
y:	9	18	24	28	26	20

Solution: Here $n = 6$ and $h = 1$

$\therefore \quad nh = 2c \qquad \therefore \quad c = 3$

Let the Fourier series to represent y in (0, 5) be

$$y = \frac{a_0}{2} + \left(a_1 \cos\frac{\pi x}{3} + b_1 \sin\frac{\pi x}{3}\right) + \left(a_2 \cos\frac{2\pi x}{3} + b_2 \sin\frac{2\pi x}{3}\right) + \ldots\ldots$$

We form the table to calculate the constants as follows

x	y	$y\cos\frac{\pi x}{3}$	$y\sin\frac{\pi x}{3}$
0	9	9	0
1	18	9	15.588
2	24	–12	20.78
2	24	–12	20.78
3	28	–28	0
4	26	–13	–22.51
5	20	10	–17.32
Total	125	–25	–3.464

$$\therefore \quad a_0 = \frac{2}{n}\Sigma y = 2\times\frac{125}{6} = 41.66$$

$$a_1 = \frac{2}{n}\Sigma y\times\cos\frac{1\pi x}{3} = \frac{2}{6}\times -25 = -8.33$$

$$b_1 = \frac{2}{n}\Sigma y\times\sin\frac{1\pi x}{3} = \frac{2}{6}\times -3.464 = -1.15$$

$\therefore$ Fourier series is

$$y = \frac{41.66}{2} - 8.33\cos\frac{\pi x}{3} - 1.15\sin\frac{\pi x}{3} + \ldots\ldots\ldots\ldots$$

Example 4: The following table gives the variations of periodic current over a period

t sec:	0	$T/6$	$T/3$	$T/2$	$2T/3$	$5T/6$	T
A amp:	1.98	1.30	1.05	1.30	–0.88	–0.25	1.98

Show that there is a direct current part of 0.75 Amp. in the variable current and obtain the amplitude of the first harmonic.

Solution: Here, length of the interval is *T*.

$$\therefore \quad 2c = T \qquad \therefore \quad c = T/2, \quad \text{Then}$$

$$A = \frac{a_0}{2} + a_1\cos\frac{2\pi t}{T} + b_1\sin\frac{2\pi t}{T} + a_2\cos\frac{4\pi t}{T} + b_2\sin\frac{4\pi t}{T} + \ldots\ldots\ldots$$

To calculate the constants, we form the table as follows

t	A	$A\cos\frac{2\pi t}{T}$	$A\sin\frac{2\pi t}{T}$
0	1.98	1.980	0.000
$T/6$	1.30	0.650	1.126
$T/3$	1.05	–0.525	0.909
$T/2$	1.30	–1.300	0.000
$2T/3$	–0.88	0.440	0.762
$5T/6$	–0.25	–0.125	0.717
Total	4.5	1.12	3.014

Now,
$$a_0 = \frac{2}{n}\Sigma A = \frac{2}{6}\times 4.5 = 1.5$$
$$a_1 = \frac{2}{n}\Sigma A\cos\frac{2\pi t}{T} = \frac{2}{6}\times 1.12 = 0.373$$
$$b_1 = \frac{2}{n}\Sigma A\sin\frac{2\pi t}{T} = \frac{2}{6}\times 3.014 = 1.005$$

$$\therefore \quad A = 0.75 + 0.373\cos\frac{2\pi t}{T} + 1.005\sin\frac{2\pi t}{T} + \ldots\ldots\ldots.$$

Thus the direct current part in the variable current

$= \frac{a_0}{2} = 0.75$ Amp. and the amplitude of the first harmonic

$$= \sqrt{a_1^2 + b_1^2}$$
$$= \sqrt{(0.373)^2 + (1.005)^2}$$
$$= 1.072.$$

Example 5: Obtain the first three coefficients in the Fourier cosine series for y where y is given in the following table:

x:	0	1	2	3	4	5
y:	4	8	15	7	6	2

Solution: Here $n = 6$, $h = 1$

$\therefore \quad 2c = nh \qquad \therefore \quad c = 3$

Let the Fourier cosine series in (0, 5) be

$$f(x) = \frac{a_0}{2} + a_1\cos\frac{\pi x}{3} + a_2\cos\frac{2\pi x}{3} + a_3\cos\frac{3\pi x}{3} + \ldots\ldots\ldots\ldots$$

To find the constants, we form the table as follows

x	y	$y\cos\frac{\pi x}{3}$	$y\cos\frac{2\pi x}{3}$	$y\cos\frac{3\pi x}{3}$
0	4	4	4	4
1	8	4	–4	–8
2	15	–7.5	–7.5	15
3	7	–7	7	–7
4	6	–3	–3	6
5	2	1	–1	–2
Total	42	–8.5	–4.5	8

$$\therefore \quad a_0 = \frac{2}{6}\Sigma y = \frac{2}{6} \times 42 = 14$$

$$a_1 = \frac{2}{6}\Sigma y \cos\frac{\pi x}{3} = \frac{2}{6} \times -8.5 = -2.8$$

$$a_2 = \frac{2}{6}\Sigma y \cos\frac{2\pi x}{3} = \frac{2}{6} \times -4.5 = -1.5$$

$$a_3 = \frac{2}{6}\Sigma y \cos\frac{3\pi x}{3} = \frac{2}{6} \times 8 = 2.7$$

$\therefore$ Fourier cosine series is

$$y = 7 - 2.8\cos\frac{\pi x}{3} - 1.5\cos\frac{2\pi x}{3} + 2.7\cos\frac{3\pi x}{3} + \ldots\ldots\ldots\ldots$$

Example 6: The turning moment T on the crank shaft of a steam engine for the crank angle θ degrees is given as follows:

θ:	0	30	60	90	120	150	180
T:	0	5224	8097	7850	5499	2626	0

Expand T in a series of sines upto the second harmonic and calculate T for $\theta = 75$.

Solution: Let the Fourier sine series to represent T in (0,180) be

$$T = b_1 \sin\theta + b_2 \sin 2\theta + b_3 \sin 3\theta + b_4 \sin 4\theta + \ldots\ldots$$

To find the coefficients, we form the table as follows:

$\theta°$	T	sin θ	sin 2θ	sin 3θ	sin 4θ
0	0	0	0	0	0
30	5224	0.500	0.866	1	0.866
60	8097	0.866	0.866	0	–0.866
90	7850	1.000	0	1	0
120	5499	0.866	–0.866	0	0.866
150	2626	0.500	–0.866	1	–0.866
180	–	–	–	–	–
Total					

$$\therefore \quad b_1 = \frac{2}{6}\Sigma T \times \sin\theta = \frac{1}{3}[(5224 + 2626)\, 0.5$$

$$+ (8097 + 5499)\, 0.866 + 7850] = 785$$

$$b_2 = \frac{2}{6}\Sigma T \times \sin 2\theta = \frac{1}{3}\ [(5224 + 8097)\ 0.866 + (5499 + 2626)\ (-0.866)$$

$$= 150$$

$$b_3 = \frac{2}{6}\Sigma T \times \sin 3\theta = \frac{1}{3}[5224 - 7850 + 2626] = 0$$

$$b_4 = \frac{2}{6}\Sigma T \times \sin 4\theta = \frac{1}{3}\ [(5224 + 5499)\ (0.866)$$

$$+ (8097 + 2626)\ (-0.866)] = 0$$

Hence

$$T = 785 \sin\theta + 150 \sin 2\theta$$

For $\theta = 75°$ $T = 785 \sin 75° + 150 \sin 150$

$$= 785\ (0.9659) + 150\ (0.5)$$

$$= 8332.$$

1.8 QUESTION BANK

1. Find the Fourier series for the function $f(x) = x + x^2$ from $x = -\pi$ to $x = \pi$ and deduce that

$$\frac{\pi^2}{8} = \frac{1}{1^2} + \frac{1}{3^2} + \frac{1}{5^2} + \ldots\ldots\ldots\ldots$$

2. Obtain the Fourier series for the function

$$f(x) = \begin{cases} \pi x & \text{for } 0 \le x \le 1 \\ \pi(2 - x) & \text{for } 1 \le x \le 2 \end{cases}$$ and hence deduce

that $\dfrac{\pi^2}{8} = \displaystyle\sum_{n=1}^{\infty} \frac{1}{(2n-1)^2}$

3. Expand the function $f(x) = x \sin x$ as a Fourier series in the interval $-\pi \le x \le \pi$

4. Obtain the Fourier series of $f(x) = \dfrac{\pi - x}{2}$ in $0 < x < 2\pi$ and deduce

$$1 - \frac{1}{3} + \frac{1}{5} - \frac{1}{7} \ldots\ldots\ldots\ldots = \frac{\pi}{4}$$

5. Find the Fourier series of $f(x) = \begin{cases} 1 + \dfrac{2x}{\pi} & \text{in } -\pi < x \le 0 \\ 1 - \dfrac{2x}{\pi} & \text{in } 0 \le x < \pi \end{cases}$

Hence deduce that $\dfrac{\pi^2}{8} = \dfrac{1}{1^2} + \dfrac{1}{3^2} + \dfrac{1}{5^2} + \ldots\ldots\ldots\ldots$

6. Expand in Fourier series

$$\begin{aligned} f(x) &= 0 & -2 \le x \le -1 \\ &= 1 + x & -1 \le x \le 0 \\ &= 1 - x & 0 \le x \le 1 \\ &= 0 & 1 \le x \le 2 \end{aligned}$$

7. An alternating current after passing through a half wave rectifier has the form

$$i = \begin{cases} I_0 \sin\theta & \text{for } 0 \le \theta \le \pi \\ 0 & \text{for } \pi \le \theta \le 2\pi \end{cases}$$

where I_0 is the maximum current.
Express i in a Fourier series.

8. Obtain the Fourier series for $f(x)$ defined as

$$f(x) = \begin{cases} -\pi & -\pi < x < 0 \\ x & 0 < x < \pi \end{cases}$$

and deduce that $\dfrac{1}{1^2} + \dfrac{1}{3^2} + \dfrac{1}{5^2} + \ldots\ldots\ldots\ldots = \dfrac{\pi^2}{8}$.

9. Expand the function $f(x) = x(2\pi - x)$ in Fourier series over the interval $(0, 2\pi)$. Hence deduce that $\displaystyle\sum_{n=1}^{\infty} \frac{1}{n^2} = \frac{\pi^2}{6}$.

10. Expand $f(x) = x - x^2$ in the interval $-\pi < x < \pi$. Deduce that

$$\frac{\pi^2}{12} = \frac{1}{1^2} - \frac{1}{2^2} + \frac{1}{3^2} - \frac{1}{4^2} + \ldots\ldots\ldots\ldots$$

11. Obtain the cosine half-range Fourier series for

$$\begin{aligned} f(x) &= kx & \text{in } 0 \le x \le l/2 \\ &= k(l - x) & \text{in } l/2 \le x \le l \end{aligned}$$

12. Obtain the half-range cosine series for the function $f(x) = \sin x$ in $0 \le x \le \pi$.

13. Obtain the cosine half-range series of $f(x) = \begin{cases} \frac{1}{4} - x & \text{in } 0 < x < \frac{1}{2} \\ x - \frac{3}{4} & \text{in } \frac{1}{2} < x < 1 \end{cases}$

14. Obtain half range cosine series expression of $f(x) = x \sin x$ in $(0, \pi)$.

15. Expand $f(x) = (x-1)^2$ as half-range cosine series over $0 \le x \le 1$. Hence show that $\frac{\pi^2}{8} = \left(\frac{1}{1^2} + \frac{1}{3^2} + \frac{1}{5^2} + \dots\dots\dots\right)$

16. Find the complex form of Fourier series of $f(x) = e^{-x}$ in $-1 \le x \le 1$.

17. Find the complex form of Fourier series of $f(x) = \cos ax$ in $-\pi < x < \pi$.

18. Obtain the constant term and the coefficients of the first sine and cosine terms in the fourier expansion of y as given in the following table.

x:	0	1	2	3	4	5
y:	9	18	24	28	26	20

Ans.: $a_0 = 41.66$ $a_1 = -8.33$ $b_1 = -1.15$.

19. In a machine the displacement y of a given point is given for a certain angle θ as follows:

$\theta°$	0	30	60	90	120	150	180	210	240	270	300	330
y:	7.9	8.0	7.2	5.6	3.6	1.7	0.5	0.2	0.9	2.5	4.7	6.8

Find the coefficient of $\sin 2\theta$ in the fourier series representing the above variation.

Ans.: $b_2 = -0.0731$

20. Determine the first two harmonics of the Fourier series for the following values.

x:	30	60	90	120	150	180	210	240	270	300	330	360
y:	2.34	3.01	3.68	4.15	3.69	2.20	0.83	0.51	0.88	1.09	1.19	1.64

Ans.: $y = 2.102 + 0.558 \cos x + 1.531 \sin x + 0.354 \cos 2x + 0.145 \sin 2x\dots$

UNIT 2

FOURIER TRANSFORMS

2.1 DEFINITION

The integral transform of a function $f(x)$ denoted by $I[f(x)]$, is defined by

$$\bar{f}(s) = \int_{x_1}^{x_2} f(x)\, k(s, x)\, dx$$

where $k(s, x)$ is called the KERNEL of the transform and is a known function of s and x. The function $f(x)$ is called the inverse transform of $\bar{f}(s)$.

The examples of a kernel are

i) When $k(s, x) = e^{-sx}$, it leads to the Laplace transform of $f(x)$ i.e.,

$$\bar{f}(s) = \int_{0}^{\infty} e^{-sx} f(x)\, dx$$

ii) When $k(s, x) = e^{isx}$ we have the Fourier transform of $f(x)$ i.e.,

$$F(s) = \int_{-\infty}^{\infty} f(x)\, e^{isx} dx$$

iii) When $k(s, x) = x^{s-1}$, it gives the Mellin transform of $f(x)$ i.e.,

$$M(s) = \int_{0}^{\infty} f(x)\, x^{s-1} dx$$

iv) Hankel transform $K(s, x) = x\, J_n(sx)$

$$H_n\{f(x)\} = \bar{f}(s) = \int_0^\infty f(x) x\, J_n(sx)\, dx$$

v) Fourier sine transform

$$F_s\{f(x)\} = \bar{f}(s) = \sqrt{\frac{2}{\pi}} \int_0^\infty f(x) \sin sx\, dx$$

vi) Fourier cosine transform

$$F_c\{f(x)\} = F(s) = \sqrt{\frac{2}{\pi}} \int_0^\infty f(x) \cos sx\, dx$$

2.2 FOURIER INTEGRAL

If $f(x)$ is piecewise continuous in every interval of the form $(-l, l)$ and has right hand and left hand derivatives at every point and if $\int_{-\infty}^{\infty} |f(x)|\, dx$ exists.

Then the integral

$$f(x) = \frac{1}{\pi} \int_0^\infty \int_{t=-\infty}^{\infty} f(t) \cos u(t-x)\, dt\, du \qquad \text{....(1)}$$

is known as the Fourier integral of $f(x)$.

Further,

$$f(x) = \frac{1}{2\pi} \int_{-\infty}^{\infty} \int_{-\infty}^{\infty} f(t) e^{iu(t-x)}\, dt\, du \qquad \text{....(2)}$$

is known as the complex form of Fourier integral.

The Fourier cosine integral is defined as

$$f(x) = \frac{2}{\pi} \int_0^\infty \int_0^\infty f(t) \cos ut \cos xu\, dt\, du \qquad \text{...(3)}$$

And, the Fourier sine integral is defined as

$$f(x) = \frac{2}{\pi} \int_0^\infty \int_0^\infty f(t) \sin ut \sin xu\, dt\, du \qquad \text{....(4)}$$

2.3 INFINITE FOURIER TRANSFORMS

Consider,

$$f(x) = \frac{1}{2\pi} \int_{u=-\infty}^{\infty} \int_{t=-\infty}^{\infty} f(t)\, e^{iu(t-x)} dt\, du$$

$$= \frac{1}{2\pi} \int_{u=-\infty}^{\infty} \left(\int_{t=-\infty}^{\infty} f(t) e^{iut} dt \right) e^{-ixu} du$$

We define

$$F(u) = \int_{t=-\infty}^{\infty} f(t)\, e^{iut} dt \quad \text{as the complex.}$$

Fourier transform or simply Fourier transform of $f(t)$ or $f(x)$.
Further,

$$f(x) = \frac{1}{2\pi} \int_{u=-\infty}^{\infty} F(u) e^{-ixu} du \quad \text{is called the inverse Fourier}$$

transform of $F(u)$.

Note: We may define the Fourier integral as follows also

$$f(x) = \frac{1}{\sqrt{2\pi}} \int_{-\infty}^{\infty} \left[\frac{1}{\sqrt{2\pi}} \int_{-\infty}^{\infty} f(t) e^{iut} dt \right] e^{-ixu}\, du$$

and the transformation pair will be

$$F(u) = \frac{1}{\sqrt{2\pi}} \int_{t=-\infty}^{\infty} f(t)\, e^{iut} dt$$

and
$$f(x) = \frac{1}{\sqrt{2\pi}} \int_{u-\infty}^{\infty} F(u)\, e^{-ixu} du$$

2.4 FOURIER COSINE AND SINE TRANSFORMS

1) We define

$$F_c(u) = \int_{t=0}^{\infty} f(t) \cos ut\, dt \quad \text{as a Fourier cosine transform of } f(t)$$

and

$$f(x) = \frac{2}{\pi} \int_{u=0}^{\infty} F_c(u) \cos xu \, du, \text{ as the inverse.}$$

Fourier cosine transform

ii) We define

$$F_s(u) = \int_{t=0}^{\infty} f(t) \sin ut \, dt \text{ as the Fourier sine transform of } f(t)$$

and

$$f(x) = \frac{2}{\pi} \int_{u=0}^{\infty} F_s(u) \sin xu \, du, \text{ as the inverse}$$

Fourier sine transform.

Note: Also we may define the above formulae as follows

$$F_c(u) = \sqrt{\frac{2}{\pi}} \int_{0}^{\infty} f(t) \cos ut \, dt$$

$$F_s(u) = \sqrt{\frac{2}{\pi}} \int_{0}^{\infty} f(t) \sin ut \, dt$$

and,
$$f(x) = \sqrt{\frac{2}{\pi}} \int_{u=0}^{\infty} F_c(u) \cos xu \, du$$

$$f(x) = \sqrt{\frac{2}{\pi}} \int_{u=0}^{\infty} F_s(u) \sin xu \, du.$$

Example 1: Express the function $f(x) = \begin{cases} 1 & \text{for } |x| \leq 1 \\ 0 & \text{for } |x| > 1 \end{cases}$ as a Fourier integral. Hence evaluate $\int_{0}^{\infty} \frac{\sin \lambda \cos \lambda x}{\lambda} d\lambda$

Solution: The Fourier integral is given by

$$f(x) = \frac{1}{2\pi} \int_{-\infty}^{\infty} \int_{-\infty}^{\infty} f(t) \cos u \, (x - t) \, dt \, du$$

$$= \frac{1}{2\pi} \int_{-\infty}^{\infty} \int_{-\infty}^{\infty} f(t)\left[\cos ut \, \cos xu + \sin ut \, \sin xu\right] dt \, du$$

$$= \frac{1}{2\pi} \int_{-\infty}^{\infty} \cos xu \left(\int_{-\infty}^{\infty} f(t) \, \cos ut \, dt \right) du$$

$$+ \frac{1}{2\pi} \int_{-\infty}^{\infty} \sin xu \left(\int_{-\infty}^{\infty} f(t) \sin ut \, dt \right) du$$

The limits $-\infty$ to ∞ w.r.t. 't', the integration variable 't' has to be changed to -1 to $+1$, since the given range is $|x| \leq 1$.

$$\therefore \qquad f(x) = \frac{1}{2\pi} \int_{-\infty}^{\infty} \cos xu \left(\int_{-1}^{1} f(t)\cos ut \, dt \right) du$$

$$+ \frac{1}{2\pi} \int_{-\infty}^{\infty} \sin xu \left(\int_{-1}^{1} f(t)\sin ut \, dt \right) du$$

$$= \frac{2}{2\pi} \int_{-\infty}^{\infty} \cos xu \, (I_1) du + \frac{1}{2\pi} \int_{-\infty}^{\infty} \sin(xu) \, (I_2) \, du \quad ...(1)$$

Consider

$$I_1 = \int_{-1}^{1} f(t)\cos ut \, dt = \int_{-1}^{1} \cos ut \, dt \qquad (\because \; f(t) = 1)$$

$$= \left[\frac{\sin ut}{u} \right]_{-1}^{1} = \frac{\sin u + \sin u}{u} = \frac{2\sin u}{u} = \varphi(u)$$

and

$$I_2 = \int_{-1}^{1} f(t)\sin ut \, dt = \int_{-1}^{1} \sin ut \, dt$$

$$= \left[-\frac{\cos ut}{u} \right]_{-1}^{1} = \frac{-\cos u + \cos u}{u} = 0$$

$$(\because \; \cos(-u) = \cos u)$$

$\therefore$ From (1)

$$f(x) = \frac{1}{2\pi}\int_{-\infty}^{\infty} \cos(xu)\frac{2\sin u}{u}du + 0 \qquad (\because\ I_2 = 0)$$

$$= \frac{2}{\pi}\int_{0}^{\infty} \frac{\cos ux \sin u}{u}du$$

Changing u to λ

$$f(x) = \frac{2}{\pi}\int_{0}^{\infty} \frac{\cos\lambda x \sin\lambda}{\lambda}d\lambda$$

$$\therefore \quad \int_{0}^{\infty} \frac{\cos\lambda x\sin\lambda}{\lambda}d\lambda = \frac{\pi}{2}f(x)$$

$$= \frac{\pi}{2}\begin{cases} 1 & \text{for } |x| \le 1 \\ 0 & \text{for } |x| > 1 \end{cases}$$

put $x = 0$

$$\int_{0}^{\infty} \frac{\sin\lambda}{\lambda}d\lambda = \frac{\pi}{2}, \text{ or } \int_{0}^{\infty} \frac{\sin x}{x}dx = \frac{\pi}{2}$$

Example 2: Find the Fourier transform of

$$f(x) = 1 - x^2 \qquad \text{for } |x| \le 1$$
$$= 0 \qquad \text{for } |x| > 1$$

Solution:

We have, Fourier transform of $f(x)$ is

$$F\{f(x)\} = \int_{-\infty}^{\infty} f(x)e^{isx}dx = F(s)$$

$$= \int_{-\infty}^{-1} (0)\, e^{isx}dx + \int_{-1}^{1} (1-x^2)e^{isx}dx + \int_{1}^{\infty} (0)e^{isx}dx$$

$$= \int_{-1}^{1} (1-x^2)\, e^{isx}dx$$

$$= \left[(1-x^2)\,\frac{e^{isx}}{is} - (-2x)\,\frac{e^{isx}}{(is)^2} + 2\,\frac{e^{isx}}{(is)^3}\right]_{-1}^{1}$$

$$= 2\,\frac{(e^{is}+e^{-is})}{-s^2} - 2\,\frac{(e^{is}-e^{-is})}{-is^3}$$

$$= -\frac{4}{s^3}(s\cos s - \sin s)$$

Now, by inversion formula, we have

$$f(x) = \frac{1}{2\pi}\int_{-\infty}^{\infty} F(s)\,e^{-isx}ds$$

$$\therefore \quad -\frac{1}{2\pi}\int_{-\infty}^{\infty} \frac{4}{s^3}(s\cos s - \sin s)\,e^{-isx}ds = \begin{cases} 1-x^2 & |x| \le 1 \\ 0 & |x| > 0 \end{cases}$$

Putting $x = \frac{1}{2}$ we get

$$-\frac{1}{2\pi}\int_{-\infty}^{\infty} \frac{4}{5^3}(s\cos s - \sin s)e^{-\frac{is}{2}}ds = 1 - \frac{1}{4} = \frac{3}{4}$$

$$\therefore \quad \int_{-\infty}^{\infty} \frac{s\cos s - \sin s}{s^3}\left(\cos\frac{s}{2} - i\sin\frac{s}{2}\right)ds = \frac{-3\pi}{8}$$

$$\therefore \quad \int_{-\infty}^{\infty} \frac{s\cos s - \sin s}{s^3}\cos\frac{s}{2}\,ds = -\frac{3\pi}{8} \qquad \text{(Since the integral is even)}$$

Changing the dummy variable s to x we get

$$\int_{0}^{\infty} \frac{x\cos x - \sin x}{x^3}\cos\frac{x}{2}\,dx = -\frac{3\pi}{16}$$

Example 3: Find the Fourier sine transform of $\dfrac{e^{-ax}}{x}$.

Solution: Let $f(x) = \dfrac{e^{-ax}}{x}$ then its Fourier sine transform is

$$F_s\{f(x)\} = \int_0^{\infty} \frac{e^{-ax}}{x} \sin sx\, dx = F(s) \quad \text{(say).}$$

Differentiating both sides w.r.t. s underintegral sign, we get

$$\frac{d}{ds} F(s) = \int_0^{\infty} \frac{xe^{-ax}\cos sx}{x} dx$$

$$= \int_0^{\infty} e^{-ax} \cos sx\, dx$$

$$= \left[e^{-ax} \frac{(-a\cos sx + s\sin sx)}{a^2 + s^2} \right]_0^{\infty}$$

$$= \frac{a}{s^2 + a^2}$$

Integrating w.r.t. s we get

$$F(s) = \int \frac{a}{s^2 + a^2} ds = \tan^{-1}\frac{s}{a} + c$$

But $F(s) = 0$ when $s = 0$ $\therefore$ $c = 0$

$$\therefore \quad F(s) = \tan^{-1}\frac{s}{a}$$

Example 4: Find the Fourier transform of $f(t) = e^{-|t|}$

Solution: Fourier transform of $f(t)$ is given by

$$F(u) = \int_{-\infty}^{\infty} f(t) e^{iut}\, dt$$

Here,
$$f(t) = \begin{cases} e^{-t} & \text{for } t \geq 0 \\ e^{t} & \text{for } t \leq 0 \end{cases}$$

$$\therefore \quad F(u) = \int_{-\infty}^{0} e^{t} e^{iut} dt + \int_0^{\infty} e^{-t} e^{iut}\, dt$$

$$= \int_{-\infty}^{0} e^{(1+iu)t} dt + \int_0^{\infty} e^{-(1-iu)t} dt$$

$$= \left[\frac{e^{(1+iu)t}}{1+iu}\right]_{-\infty}^{0} + \left[\frac{e^{-(1-iu)t}}{-(1-iu)}\right]_{0}^{\infty}$$

$$= \frac{1}{1+iu} - 0 + 0 \frac{-1}{-(1-iu)} = \frac{2}{1-i^2u^2} = \frac{2}{1+u^2} \quad (\because i^2 = -1)$$

Example 5: Find the Fourier sine and cosine transform of $f(x) = e^{-ax}$, $a > 0$

Solution: Fourier sine and cosine transforms are

$$F_s(u) = \int_0^\infty f(t) \sin ut \, dt$$

and $$F_c(u) = \int_0^\infty f(t) \cos ut \, dt$$

Here $$f(t) = e^{-at}$$

$\therefore$ $$F_s(u) = \int_0^\infty e^{-at} \sin ut \, dt$$

$$= \left[\frac{e^{-at}}{a^2+u^2}(-a \sin ut - u \cos ut\right]_0^\infty$$

$$= \frac{u}{a^2+u^2}$$

Also,

$$F_c(u) = \int_0^\infty e^{-at} \cos ut \, dt$$

$$= \left[\frac{e^{-at}}{a^2+u^2}(-a \cos ut + u \sin ut)\right]_0^\infty$$

$$= \frac{a}{a^2+u^2}$$

Example 6: Find the Fourier sine transform of $f(x) = e^{-|x|}$ and hence evaluate $\int_0^\infty \frac{x \sin mx}{1+x^2} dx, \; m > 0.$

Solution:

Fourier sine transform is given by

$$F_s(u) = \int_0^\infty f(t) \sin ut \; dt$$

$$= \int_0^\infty e^{-|t|} \sin ut \; dt \qquad \because \; e^{-|t|} = e^{-t}$$

$$= \int_0^\infty e^{-t} \sin ut \; dt$$

$$= e^{-t} \left[\frac{(-1 \sin ut - u \cos ut)}{1^2 + u^2} \right]_0^\infty$$

$$F_s(u) = \frac{u}{1+u^2}$$

By inverse Fourier sine transform we have

$$f(x) = \frac{2}{\pi} \int_0^\infty F_s(u) \sin xu \; du$$

$$= \frac{2}{\pi} \int_0^\infty \frac{u}{1+u^2} \sin xu \; du$$

Putting $x = m, \quad m > 0$ we have

$$f(x) = \frac{2}{\pi} \int_0^\infty \frac{u \sin mu}{1+u^2} du$$

$$\therefore \quad \int_0^\infty \frac{u \sin mu \; du}{1+u^2} = \frac{\pi}{2} e^{-m} \qquad \because \; f(x) = e^{-x}, \quad f(m) = e^{-m}$$

Changing the dummy variable u to x we get

$$\int_0^\infty \frac{x \sin mx}{1+x^2} dx = \frac{\pi}{2} e^{-m}$$

Example 7: Find the Fourier cosine transform of

$$f(x) = \begin{cases} 4x & 0 < x < 1 \\ 4 - x & 1 < x < 4 \\ 0 & x > 4 \end{cases}$$

Solution: The Fourier cosine transform is

$$F_c\{f(x)\} = \int_0^\infty f(x) \cos ux \, dx$$

$$= \int_0^1 4x \cos ux \, dx + \int_1^4 (4-x) \cos ux \, dx + 0$$

$$= 4\left[\frac{x \sin ux}{u} + \frac{1 \cos ux}{u^2}\right]_0^1 + \left[(4-x)\frac{\sin ux}{u} - \frac{(-1)(-\cos ux)}{u^2}\right]_1^4$$

$$= \frac{4 \sin u}{u} + \frac{4 \cos u}{u^2} - \frac{4}{u^2} - \frac{\cos 4u}{u^2} - \frac{3 \sin u}{u} + \frac{\cos u}{u^2}$$

$$= \frac{\sin u}{u} + \frac{5 \cos u}{u^2} - \frac{\cos 4u}{u^2} - \frac{4}{u^2}$$

Example 8: Find the Fourier cosine transform of

$$f(x) = \begin{cases} 1 & \text{for } 0 \le x \le 1 \\ 0 & \text{for } x > 1 \end{cases}$$

and hence evaluate $\displaystyle\int_0^\infty \frac{\sin x}{x} dx$

Solution: The Fourier cosine transform is

$$F_c\{f(x)\} = \int_0^\infty f(x) \cos ux \, dx$$

$$= \int_0^1 1 \cos ux \, dx + 0$$

$$= \left[\frac{\sin ux}{u}\right]_0^1 = \frac{\sin u}{u} = \varphi(u)$$

Hence from inverse Fourier cosine transform

We have, $f(x) = \frac{2}{\pi}\int_0^\infty \varphi(u)\cos xu \, du$

$$= \frac{2}{\pi}\int_0^\infty \frac{\sin u}{u}\cos xu \, du$$

Put $x = 0$

$$f(0) = \frac{2}{\pi}\int_0^\infty \frac{\sin u}{u} du$$

$$\therefore \int_0^\infty \frac{\sin u}{u} du = \frac{\pi}{2} f(0)$$

$$\therefore \int_0^\infty \frac{\sin u}{u} du = \frac{\pi}{2} \qquad \because f(0) = 1 \qquad 0 \le x \le 1$$

Changing the dummy variable u to x we get

$$\int_0^\infty \frac{\sin x}{x} dx = \frac{\pi}{2}.$$

Example 9: Find the Fourier cosine transform of $f(x) = \frac{1}{1+x^2}$. Hence derive Fourier cosine transform of $\varphi(x) = \frac{x}{1+x^2}$

Solution: Let the Fourier cosine transform be

$$F_c\{f(x)\} = \int_0^\infty \frac{\cos sx}{1+x^2} dx = I \quad \text{(say)} \qquad ....(1)$$

$$\therefore \quad \frac{dI}{ds} = -\int_0^\infty \frac{x \sin sx}{1+x^2} dx$$

$$= -\int_0^\infty \frac{x^2 \sin sx}{x(1+x^2)} dx \qquad ....(2)$$

$$= -\int_0^\infty \frac{(1+x^2)-1}{x\,(1+x^2)} \sin sx\, dx$$

$$= -\int_0^\infty \frac{\sin sx}{x} dx + \int_0^\infty \frac{\sin sx}{x(1+x^2)} dx$$

$$\therefore \quad \frac{dI}{ds} = -\frac{\pi}{2} + \int_0^\infty \frac{\sin sx}{x(1+x^2)} dx \qquad ....(3)$$

$$\therefore \quad \frac{d^2 I}{ds^2} = 0 + \int_0^\infty \frac{x \cos sx}{x(1+x^2)} dx = \int_0^\infty \frac{\cos sx}{1+x^2} dx = I$$

$$\therefore \quad \frac{d^2 I}{ds^2} - I = 0 \quad \text{or} \quad (D^2 - 1)\, I = 0 \quad \text{where } D = \frac{d}{ds}$$

It is solution is

$$I = c_1\, e^s + c_2\, e^{-s} \qquad ....(4)$$

$$\frac{dI}{ds} = c_1\, e^s - c_2\, e^{-s} \qquad ....(5)$$

When $s = 0$ (1) and (4) gives

$$c_1 + c_2 = \int_0^\infty \frac{dx}{1+x^2} = \frac{\pi}{2}$$

Also when $s = 0$ (3) and (5) gives

$$c_1 - c_2 = -\frac{\pi}{2}$$

Solving $c_1 = 0$ and $c_2 = \frac{\pi}{2}$

Thus from (1) and (4) we have

$$F_c\{f(x)\} = I = \frac{\pi}{2}e^{-s}$$

$$\text{Now } F_s\{\varphi(x)\} = \int_0^\infty \frac{x\sin sx}{1+x^2}dx = -\frac{dI}{ds} \quad \text{from (2)}$$

$$= \frac{\pi}{2}e^{-s} \quad \text{from (5)}$$

Where $c_1 = 0$ and $c_2 = \frac{\pi}{2}$.

Example 10: Solve the integral equation

$$\int_0^\infty f(\theta)\cos\alpha\theta\, d\theta = \begin{cases} 1-\alpha & 0 \le \alpha < 1 \\ 0 & \alpha > 1 \end{cases}$$

and hence evaluate $\int_0^\infty \frac{\sin^2 t}{t^2}dt$

Solution:

Consider the inverse Fourier cosine transform with

$$F(\alpha) = \begin{cases} 1-\alpha & 0 \le \alpha < 1 \\ 0 & \alpha > 0 \end{cases}$$

$$\therefore \quad f(\theta) = \frac{2}{\pi}\int_0^\infty F(\alpha)\cos\alpha\theta\, d\alpha$$

$$= \frac{2}{\pi}\int_0^1 (1-\alpha)\cos\alpha\theta\, d\alpha \qquad \because F(\alpha)=0 \text{ for } \alpha>0$$

$$= \frac{2}{\pi}\left[(1-\alpha)\frac{\sin\alpha\theta}{\theta} - (-1)\frac{(-\cos\alpha\theta)}{\theta^2}\right]_0^1$$

$$= \frac{2}{\pi}\frac{(1-\cos\theta)}{\theta^2} = \frac{2\,.\,2\sin^2\theta/2}{\theta^2} = \frac{4\sin^2\theta/2}{\theta^2}$$

Thus we have

$$\int_0^\infty \frac{4\sin^2\theta/2}{\pi\,\theta^2}\cos\alpha\theta\, d\theta = F(\alpha)$$

$$\int_0^\infty \frac{\sin^2\left(\frac{\theta}{2}\right)}{\left(\frac{\theta}{2}\right)^2}\cos\alpha\theta\, d\theta = \pi\, F(\alpha)$$

Put $\frac{\theta}{2} = t \qquad \therefore \quad d\theta = 2dt \quad t \to 0 \text{ to } \infty$

$$\therefore \quad \int_0^\infty \frac{\sin^2 t}{t^2}\cos(2\alpha t)\, 2dt = \pi F(\alpha)$$

Put $\qquad \alpha = 0 \qquad F(\alpha) = 1 - 0 = 1$

$$\therefore \quad \int_0^\infty \frac{\sin^2 t}{t^2} dt = \frac{\pi}{2}$$

Example 11: Find the Fourier cosine transform of e^{-x^2}

Solution:

Fourier cosine transform of e^{-x^2} is given by

$$F_c\left\{e^{-x^2}\right\} = \sqrt{\frac{2}{\pi}}\int_0^\infty e^{-x^2}\cos sx\, dx = I \quad \text{(say)} \qquad ....(1)$$

Differentiating w.r.t. s under integral sign

$$\frac{dI}{ds} = -\sqrt{\frac{2}{\pi}} \int_0^\infty e^{-x^2} x \sin sx \, dx$$

$$= \frac{1}{2}\sqrt{\frac{2}{\pi}} \int_0^\infty (\sin sx)(-2xe^{-x^2})\, dx$$

Integrating by parts

$$= \frac{1}{2}\sqrt{\frac{2}{\pi}} \left[\left(\sin sx \, e^{-x^2} \right)_0^\infty - s \int_0^\infty \cos sx \, e^{-x^2} dx \right]$$

$$= 0 - \frac{s}{2}\sqrt{\frac{2}{\pi}} \int_0^\infty e^{-x^2} \cos x \, dx = -\frac{s}{2} I$$

$$\therefore \quad \frac{dI}{ds} = -\frac{sI}{2} \quad \text{or} \quad \frac{dI}{I} = -\frac{s}{2} ds$$

Integrating we get

$$\therefore \quad \log I = -\frac{s^2}{4} + \log k$$

$$\therefore \quad I = K e^{-\frac{s^2}{4}} \qquad \text{....(2)}$$

Now, when $s = 0$, from (1)

$$I = \sqrt{\frac{2}{\pi}} \int_0^\infty e^{-x^2} dx = \sqrt{\frac{2}{\pi}} \cdot \frac{\sqrt{\pi}}{2} = \frac{1}{\sqrt{2}}$$

From (2) $\frac{1}{\sqrt{2}} = k$

$$\therefore \quad I = F_c\{f(x)\} = \frac{1}{\sqrt{2}} e^{-\frac{s^2}{4}}$$

2.5 QUESTION BANK

1. Find the Fourier sine and cosine transform of

$$f(x) = \begin{cases} x & \text{for} \quad 0 < x < 1 \\ 2-x & \text{for} \quad 1 < x < 2 \\ 0 & \text{for} \quad x > 2 \end{cases}$$

Ans.: $\frac{2}{s^2}\sqrt{\frac{2}{\pi}} \sin s\,(1-\cos s)$; $\frac{2}{s^2}\sqrt{\frac{2}{\pi}} \cos s(1-\cos s)$

2. Find the Fourier cosine transform of $f(x)$ if

$$f(x) = \begin{cases} \cos x & 0 < x < a \\ 0 & x > a \end{cases}$$

Ans.: $\frac{1}{\sqrt{2\pi}}\left[\frac{\sin(1+s)a}{1+s} + \frac{\sin(1-s)a}{1-s}\right]$

3. Using Fourier integrals show that

$$\int_0^\infty \frac{\sin \pi\lambda \, \sin x\lambda}{1-\lambda^2} d\lambda = \begin{vmatrix} \frac{1}{2}\pi \sin x & 0 \le x \le \pi \\ 0 & x > \pi \end{vmatrix}$$

4. Solve the integral equation

$$\int_0^\infty f(x) \cos \lambda x \, dx = e^{-\lambda}$$

Ans.: $f(x) = \frac{2}{\pi\,(1+x^2)}$

5. Find the Fourier transform of

$$f(x) = \begin{cases} x & |x| < a \\ 0 & |x| < a \end{cases}$$

Ans.: $\sqrt{\frac{2}{\pi}}\left(\frac{as \cos as - \sin a}{s^2}\right)$

6. Find the Fourier sine and cosine transform of

$$f(x) = \begin{cases} x & 0 < x < 2 \\ 0, & \text{other wise} \end{cases}$$

7. Find the finite Fourier sine transform of the function $f(x) = \cos kx$, where k is a non-integer over $(0, \pi)$.

8. Find the inverse Fourier sine transform of $\dfrac{s}{1+s^2}$

9. Find the Fourier transform of the function

$$f(x) = \begin{cases} a^2 - x^2 & \text{for } |x| \le a \\ 0 & \text{for } |x| > a \end{cases}$$

UNIT 3

APPLICATION OF PARTIAL DIFFERENTIAL EQUATIONS

THE WAVE EQUATION

Vibrations in a membrane or drumhead, or oscillations induced in a guitar or violin string, are governed by a partial differential equation called the **wave equation.**

3.1 DERIVATION OF ONE-DIMENSIONAL WAVE EQUATION

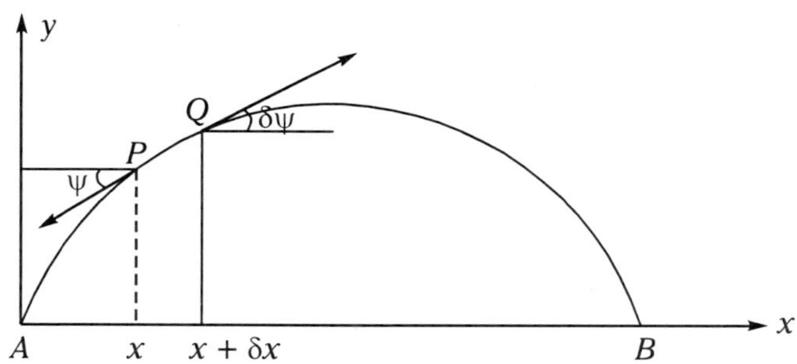

Consider an elastic string stretched between two fixed ends A and B of length l and subjected to constant tension T. To begin with neglect damping forces such as air resistance and weight of the string and assume that the tension $T(x, t)$ in the string always acts tangentially to the string and that individual particles of the string move only vertically.

Let the string be released from rest and allowed to vibrate. We shall study the motion of the string with no external forces acting on it. Assuming that each point of the string makes small vibrations at right angles to the equilibrium position AB of the string entirely in one plane.

Taking the end A, as the origin, AB as the x-axis and Ay perpendicular to it as the y-axis, so that the motion takes place entirely in the xy-plane. The above figure shows string in the position APB at time t. Consider the motion of the element PQ of the string between the points $P(x, y)$ and $Q(x + \delta x, y + \delta y)$, where the tangents make angles ψ and $\psi + \delta\psi$ with x-axis. Clearly this element is moving upwards with acceleration $\dfrac{\partial^2 y}{\partial t^2}$. Also the vertical component of the force acting on this element

$$= T\sin(\psi + \delta\psi) - T\sin\psi$$

$$= T[\sin(\psi + \delta\psi) - \sin\psi]$$

$$= T[\tan(\psi + \delta\psi) - \tan\psi] \qquad \text{(since } \psi \text{ is small)}$$

$$= T\left[\left(\frac{\partial y}{\partial x}\right)_{x+\delta x} - \left(\frac{\partial y}{\partial x}\right)_x\right]$$

If m is the mass per unit length of the string then by Newton's second law of motion, we have

$$m\delta x \frac{\partial^2 y}{\partial t^2} = T\left[\left(\frac{\partial y}{\partial x}\right)_{x+\delta x} - \left(\frac{\partial y}{\partial x}\right)_x\right]$$

$$\therefore \quad \frac{\partial^2 y}{\partial t^2} = \frac{T}{m}\,\frac{\left[\left(\frac{\partial y}{\partial x}\right)_{x+\delta x} - \left(\frac{\partial y}{\partial x}\right)_x\right]}{\delta x}$$

Taking the limit as $Q \to P$, $\delta x \to 0$, we have

$$\frac{\partial^2 y}{\partial t^2} = c^2 \frac{\partial^2 y}{\partial x^2}, \quad \text{where} \quad c^2 = \frac{T}{m}$$

This is the partial differential equation giving the transverse vibrations of the string. It is also called the one-dimensional wave equation.

3.2 SOLUTION OF THE WAVE EQUATION

Let $\dfrac{\partial^2 y}{\partial t^2} = c^2 \dfrac{\partial^2 y}{\partial x^2}$ .....(1) be the one-dimensional wave equation.

Let us assume that

$$Y(t, x) = T(t)\,X(x)$$

Where T is the function of t only and X is the function of x alone, be the solution of equation (1).

$$\therefore \quad \frac{\partial^2 y}{\partial t^2} = XT'' \text{ and } \frac{\partial^2 y}{\partial x^2} = X''T$$

Substituting in (1)

$$XT'' = C^2 X''T$$

$$\therefore \quad \frac{T''}{c^2 T} = \frac{X''}{X} = k \text{ (say)}$$

$$\therefore \quad T'' - c^2 kT = 0 \text{ and } X'' - kX = 0$$

Thus the wave equation has separated into two ordinary differential equations.

$$\therefore \quad (D^2 - c^2 k)\, T = 0 \text{ and } \left(D_1^2 - k\right) X = 0$$

Where $\frac{d}{dt} = D$, and $\frac{d}{dx} = D_1$

$\therefore$ Auxiliary equations are

$$D^2 - c^2 k = 0 \text{ and } D_1^2 - k = 0$$

$$\therefore \quad D = \pm c\sqrt{k} \text{ and } D_1 = \pm\sqrt{k}$$

Case 1: When $k > 0$ (say $= p^2$) the roots are real and distinct.

$$\therefore \quad D = \pm\, cp \quad \text{and} \quad D_1 = \pm\, p$$

$$\therefore \quad \left.\begin{aligned} T &= c_1\, e^{cpt} + c_2 e^{-cpt} \\ X &= c_3\, e^{px} + c_4 e^{-px} \end{aligned}\right\} \qquad \text{....(ii)}$$

Case II: When $k < 0$ (say $= -p^2$), the roots are imaginary

$$\therefore \quad D = \pm\, cpi \quad \text{and} \quad D_1 = \pm\, pi$$

$$\therefore \quad \left.\begin{aligned} T &= c_5 \cos cpt + c_6 \sin cpt \\ X &= c_7 \cos px + c_8 \sin px \end{aligned}\right\} \qquad \text{.....(iii)}$$

Case III: When $k = 0$ roots are equal

$$\left.\begin{aligned} T &= c_9 t + c_{10} \\ X &= c_{11} x + c_{12} \end{aligned}\right\} \qquad \text{....(iv)}$$

These are the three cases depending upon the particular problem. Here we are dealing with the wave motion ($k < 0$) always. Hence the suitable solution is

$$\begin{aligned} Y(t, x) &= T(t)\, X(x) \\ &= (c_5 \cos cpt + c_6 \sin cpt)(c_7 \cos px + c_8 \sin px) \end{aligned}$$

is the required solution.

The boundary conditions and initial conditions which the equation $\frac{\partial^2 y}{\partial t^2} = c^2 \frac{\partial^2 y}{\partial x^2}$ has to satisfy are:

i) If the ends of the string are fixed then

$$y(0, t) = y(l, t) = 0 \qquad \text{for} \qquad t \geq 0$$

These are the boundary conditions.

ii) The initial conditions specify the initial (at time zero) position.

$$y(x, 0) = f(x) \qquad \text{for} \qquad 0 \leq x \leq l$$

and the initial velocity

$$\frac{\partial y}{\partial t}(x,0) = g(x) \qquad \text{for} \qquad 0 < x < l$$

in which $f(x)$ and $g(x)$ are given functions satisfying certain compatibility conditions.

For example, if the string is fixed at its ends, then the initial position function must reflect this by satisfying.

$$f(0) = f(l) = 0$$

If the initial velocity is zero (the string is released from rest) then $g(x) = 0$.

The wave equation, together with the boundary and initial conditions, constitute a **boundary value problem** for the position function $y(x, t)$

Example 1: A string is stretched and fastened to two points l apart, motion is started by displacing the string in the form $y = a \sin\left(\frac{\pi x}{l}\right)$ from which it is released at time $t = 0$. Show that the displacement of any point at a distance x from one end at time t is given by $y(x,t) = a \sin\left(\frac{\pi x}{l}\right) \cos\left(\frac{\pi ct}{l}\right)$.

Solution: The vibration of the string is given by

$$\frac{\partial^2 y}{\partial t^2} = c^2 \frac{\partial^2 y}{\partial x^2} \quad ....(i)$$

As the end points of the string are fixed for all time

$$y(0, t) = 0 \quad ....(ii)$$

and $$y(l, t) = 0 \quad ....(iii)$$

Since the initial transverse velocity of any point of the string is zero

$$\left(\frac{\partial y}{\partial t}\right)_{t=0} = 0 \quad ....(iv)$$

and $$y(x, 0) = a \sin\left(\frac{\pi x}{l}\right) \quad ....(v)$$

We have to solve the equation (1) subject to the boundary conditions (ii) and (iii) and initial conditions (iv) and (v).

Since the vibration of the string is periodic therefore the solution of (1) is of the form

$$y(x, t) = (c_1 \cos px + c_2 \sin px)(c_3 \cos cpt + c_4 \sin cpt) \quad ....(vi)$$

By (ii), $y(0, t) = c_1 (c_3 \cos cpt + c_4 \sin cpt) = 0$

$\therefore \quad c_1 = 0$ for all the time.

$$\therefore \quad y(x, t) = c_2 \sin px (c_3 \cos cpt + c_4 \sin cpt) \quad ...(vii)$$

Now,

$$\frac{\partial y}{\partial t} = c_2 \sin px [c_3 (-cp \sin cpt + c_4 (cp \cos cpt)]$$

By (iv),

$$\left(\frac{\partial y}{\partial t}\right)_{t=0} = c_2 \sin px (c_4\, cp) = 0$$

$$\therefore \quad c_2\, c_4\, cp = 0$$

If $c_2 = 0$, (vii) will lead to the trivial solution $y(x, t) = 0$

$\therefore$ The only possibility is that $c_4 = 0$

Thus (vii) reduces to

$$y(x, t) = c_2\, c_3 \sin px \cos cpt.$$

By (iii), $y(l, t) = c_2\, c_3 \sin pl \cos cpt = 0$, for all t since $c_2 \neq 0$ and $c_3 \neq 0$, we have $\sin pl = 0$

$$\therefore \qquad pl = n\pi \quad \text{i.e., } p = \frac{n\pi}{l} \text{ where } n \text{ is an integer.}$$

$$\therefore \qquad y(x,\ t) = c_2 c_3 \sin\frac{n\pi x}{l}\cos\frac{n\pi ct}{l}$$

$$\therefore \qquad y\,(x,\ t) = a\sin\left(\frac{\pi x}{l}\right)\cos\left(\frac{\pi ct}{l}\right) \text{ where } c_2c_3 = a,\ n = 1 \text{ which is}$$

the desired solution.

Note: We have

$y(x,\ t) = c_2 c_3 \sin\frac{n\pi x}{l}\cos\frac{n\pi ct}{l}$ are the solutions of equation (1) satisfying the boundary conditions. These functions are called the eigen-functions corresponding to the eigenvalues $\lambda_n = \frac{cn\pi}{l}$ of the vibrating string. The set of values λ_1, λ_2 ....λ_n ....is called its spectrum.

Example 2: A tightly stretched string of length l with fixed ends is initially in equilibrium position. It is set vibrating by giving each point a velocity $v_0 \sin^3\frac{\pi x}{l}$. Find the dispalcement $y\,(x,\ t)$.

Solution: The equation of the vibrating string is

$$\frac{\partial^2 y}{\partial t^2} = c^2\frac{\partial^2 y}{\partial x^2} \qquad ....(i)$$

The boundary conditions are

$$y\,(0,\ t) = 0, \qquad y\,(l,\ t) = 0 \qquad ....(ii)$$

Also the initial conditions are $y(x,\ 0) = 0$ ....(iii)

$$\text{and} \quad \left(\frac{\partial y}{\partial t}\right)_{t=0} = v_0 \sin^3\frac{\pi x}{l}$$

Since the vibration of the string is periodic, therefore the solution of (1) is of the form

$$y\,(0,\ t) = c_1\,(c_3 \cos cpt + c_4 \sin cpt) = 0$$

$$\therefore \qquad c_1 = 0 \quad \text{for all values of } t.$$

$$\therefore \qquad y\,(x,\ t) = c_2 \sin px\,(c_3 \cos cpt + c_4 \sin cpt)$$

Also, $y\,(l,\ t) = c_2 \sin pl\,(c_3 \cos cpt + c_4 \sin cpt) = 0$ for all t.

This gives $pl = n\pi$ or $p = \frac{n\pi}{l}$, n being an integer.

Thus

$$y(x, t) = c_2 \sin\frac{n\pi x}{l}\left(c_3 \cos\frac{cn\pi t}{l} + c_4 \sin\frac{cn\pi t}{l}\right)$$

By (iii), $0 = c_2 c_3 \sin\frac{n\pi x}{l}$ for all x i.e., $c_2\, c_3 = 0$

$$\therefore \quad y(x, t) = b_n \sin\frac{n\pi x}{l}\sin\frac{cn\pi t}{l} \quad \text{where } b_n = c_2\, c_4.$$

Adding all such solutions, the general solution of (1) is

$$y(x, t) = \Sigma b_n \sin\frac{n\pi x}{l}\sin\frac{cn\pi t}{l} \qquad \text{....(v)}$$

Now, $$\frac{\partial y}{\partial t} = \Sigma\, b_n \sin\frac{n\pi x}{l}\cdot\left(\frac{cn\pi}{l}\right)\cos\frac{cn\pi t}{l}$$

$$\therefore \quad \frac{v_0}{4}\left(3\sin\frac{\pi x}{l} - \sin\frac{3\pi x}{l}\right) = \Sigma\frac{cn\pi}{l} b_n \sin\frac{n\pi x}{l}$$

$$(\because \sin 3\theta = 3\sin\theta - 4\sin^3\theta)$$

$$= \frac{c\pi}{l} b_1 \sin\frac{\pi x}{l} + \frac{2c\pi}{l} b_2 \sin\frac{2\pi x}{l} + \frac{3c\pi}{l} b_3 \sin\frac{3\pi x}{l} + \ldots\ldots$$

Equating coefficients from both sides, we get

$$\frac{3v_0}{4} = \frac{c\pi}{l} b_1, \quad 0 = \frac{2c\pi}{l} b_2, \quad -\frac{v_0}{4} = \frac{3c\pi}{l} b_3 \ldots\ldots\ldots\ldots$$

$$\therefore \quad y = \frac{lv_0}{12c\pi}\left(9\sin\frac{\pi x}{l}\sin\frac{c\pi t}{l} - \sin\frac{3\pi x}{l}\sin\frac{3c\pi t}{l}\right)$$

is the desired solution.

Example 3: A tightly stretched string with fixed end points $x = 0$ and $x = l$ is initially at rest in its equilibrium position. If it is set vibrating by giving to each of its points a velocity $\lambda x(l - x)$, find the displacement of the string at any distance x from one end at any time t.

Solution: Here the boundary conditions are $y(0, t) = y(l, t) = 0$

The displacement $y(x, t)$ of any point of the string is given by

$$\frac{\partial^2 y}{\partial t^2} = c^2 \frac{\partial^2 y}{\partial x^2} \qquad ...(1)$$

Its solution is

$$y(x, t) = \sum_{n=1}^{\infty} \left(a_n \cos\frac{n\pi ct}{l} + b_n \sin\frac{n\pi ct}{l} \right) \sin\frac{n\pi x}{l} \qquad ...(2)$$

Since the string was at rest initially $y(x, 0) = 0$

$\therefore$ From (2), $0 = \sum_{n=1}^{\infty} a_n \sin\frac{n\pi x}{l}$ implies that $a_n = 0$

$$\therefore \quad y(x, t) = \sum_{n=1}^{\infty} b_n \sin\frac{n\pi ct}{l} \sin\frac{n\pi x}{l} \qquad ...(3)$$

and

$$\frac{\partial y}{\partial t} = \sum_{n=1}^{\infty} \frac{n\pi c}{l} b_n \cos\frac{n\pi ct}{l} \sin\frac{n\pi x}{l}$$

$$= \frac{\pi c}{l} \sum_{n=1}^{\infty} n b_n \cos\frac{n\pi ct}{t} \sin\frac{n\pi x}{l}$$

But given $\frac{\partial y}{\partial t} = \lambda x (l - x)$ when $t = 0$

$$\therefore \quad \lambda x (l - x) = \frac{\pi c}{l} \sum_{n=1}^{\infty} n b_n \sin\frac{n\pi x}{l}$$

$$\therefore \quad \frac{\pi c}{l} n b_n = \frac{2}{l} \int_0^l \lambda x(l - x) \sin\frac{n\pi x}{l} dx$$

$$= \frac{2\lambda}{l} \left[x(l - x) \left(-\frac{l}{n\pi} \cos\frac{n\pi x}{l} \right) - (l - 2x) \left(-\frac{l^2}{n^2\pi^2} \sin\frac{n\pi x}{l} \right) + (-2) \left(\frac{l^3}{n^3\pi^3} \cos\frac{n\pi x}{l} \right) \right]_0^l$$

$$= \frac{4\lambda l^2}{n^3\pi^3} (1 - \cos n\pi)$$

$$= \frac{4\lambda l^2}{n^3\pi^3}\left[1-(-1)^n\right]$$

$= 0$ when n is even.

$$= \frac{8\lambda l^2}{n^3\pi^3} \text{ when } n \text{ is odd}$$

$$= \frac{8\lambda l^2}{\pi^3(2m-1)^3} \text{ taking } n = 2m - 1$$

$$\therefore \quad b_n = \frac{8\lambda l^2}{c\pi^4(2m-1)^4}$$

$\therefore$ From (3) the required solution is

$$y(x, t) = \frac{8\lambda l^3}{c\pi^4}\sum_{n=1}^{\infty}\frac{1}{(2m-1)^4}\sin\frac{(2m-1)\pi ct}{l}\sin\frac{(2m-1)\pi x}{l}.$$

Example 4: Solve completely the equation $\frac{\partial^2 y}{\partial t^2} = c^2\frac{\partial^2 y}{\partial x^2}$ representing the vibrations of a string of length l, fixed at both ends, given that $y(0, t) = 0$; $y(l, t) = 0$, $y(x, 0) = f(x)$ and $\frac{\partial y}{\partial t}(x, 0) = 0$, $0 < x < l$.

Ans.: $y(x, t) = \sum_{n=1}^{\infty} a_n \cos\frac{n\pi ct}{l}\sin\frac{n\pi x}{l}$

where $a_n = \frac{2}{l}\int_0^l f(x)\sin\frac{n\pi x}{l}\,dx$

Example 5: Find the deflection $y(x, t)$ of the vibrating string of length π and ends fixed corresponding to zero initial velocity and initial deflection $f(x) = K(\sin x - \sin 2x)$, given $c^2 = 1$.

Ans.: $y(x, t) = K(\cos t \sin x - \cos 2t \sin 2x)$

Example 6: The end of a tightly stretched string of length l are fixed at

$x = 0$ and $x = l$. The string is at rest with the point $x = b$ drawn aside through a small distance 'd' and released at time $t = 0$. Show that

$$y(x, t) = \frac{2dl^2}{\pi^2 b(l-b)} \sum_{n=1}^{\infty} \frac{1}{n^2} \sin\frac{n\pi b}{l} \cos\frac{n\pi ct}{l}$$

3.3 D'ALEMBERT'S SOLUTION

Let $\frac{\partial^2 u}{\partial t^2} = c^2 \frac{\partial^2 u}{\partial x^2}$ ....(1) where $c^2 = \frac{T}{p}$, be the wave equation.

Transform the independent variables x, t to v and z through the transformation.

$$v = x + ct\,; \qquad z = x - ct$$

Here, we are using $u(x, t)$ as the position function because we will be changing variables from (x, y) plane to a (v, z) plane.

Find $\frac{\partial v}{\partial x} = 1$, $\frac{\partial v}{\partial t} = c$, $\frac{\partial z}{\partial x} = 1$, $\frac{\partial z}{\partial t} = -c$ and $\frac{\partial^2 v}{\partial x^2} = 0$, $\frac{\partial^2 z}{\partial x^2} = 0$

But
$$\frac{\partial u}{\partial x} = \frac{\partial u}{\partial v}\frac{\partial v}{\partial x} + \frac{\partial u}{\partial z}\frac{\partial z}{\partial x}$$

$$= \frac{\partial u}{\partial v}\,.\,1 + \frac{\partial u}{\partial z}\,.\,1 = u_v + u_z$$

Again,
$$\frac{\partial^2 u}{\partial x^2} = \frac{\partial}{\partial x}\left(\frac{\partial u}{\partial x}\right) = \frac{\partial}{\partial v}\left(\frac{\partial u}{\partial x}\right)\frac{\partial v}{\partial x} + \frac{\partial}{\partial z}\left(\frac{\partial u}{\partial x}\right)\frac{\partial z}{\partial x}$$

$$= \frac{\partial}{\partial v}(u_v + u_z)\,.\,1 + \frac{\partial}{\partial z}(u_v + u_z)\,.\,1$$

$$= u_{vv} + u_{zv} + u_{vz} + u_{zz}$$

$$= u_{vv} + 2u_{zv} + u_{zz} \qquad ....(2)$$

And,
$$\frac{\partial u}{\partial t} = \frac{\partial u}{\partial v}\frac{\partial v}{\partial t} + \frac{\partial u}{\partial z}\frac{\partial z}{\partial t}$$

$$= \frac{\partial u}{\partial v}(c) + \frac{\partial u}{\partial z}(-c)$$

$$= cu_v - cu_z$$

Again, $$\frac{\partial^2 u}{\partial t^2} = \frac{\partial}{\partial v}\left(\frac{\partial u}{\partial t}\right)\frac{\partial v}{\partial t}+\frac{\partial}{\partial z}\left(\frac{\partial u}{\partial t}\right)\frac{\partial z}{\partial t}$$

$$= \frac{\partial}{\partial v}(cu_v - cu_z)(c)+\frac{\partial}{\partial z}(cu_v - cu_z)\,(-c)$$

$$= c^2\,[u_{vv} - u_{zv} - u_{vz} + u_{zz}]$$

$$= c^2\,[u_{vv} - 2u_{zv} + u_{zz}] \quad \text{....(3)}$$

Using (2) and (3) in (1) we get

$$c^2\,[u_{vv} - 2u_{zv} + u_{zz}] = c^2\,[u_{vv} + 2u_{zv} + u_{zz}]$$

$$\therefore \quad 4\,u_{zv} = 0$$

$$\therefore \quad u_{zv} = 0$$

i.e., $$\frac{\partial^2 u}{\partial z\,\partial v} = 0$$

Integrating w.r.t. z keeping v constant, we get

$$\frac{\partial u}{\partial v} = F(v)$$

Integrating w.r.t. v keeping z constant

$$u = \int F(v)dv + \psi(z)$$

$$u = \varphi\,(v) + \psi\,(z)$$

$$\therefore \quad u = \varphi\,(x + ct) + \psi\,(x - ct) \quad \text{....(4)}$$

This is known as D' Alembert's solution of wave equation.

To determine φ and ψ

Let us suppose initially $u(x, 0) = f(x)$ and $\frac{\partial y}{\partial t}(x, 0) = 0$

Differentiate (4) w.r.t. 't' we get

$$\frac{\partial u}{\partial t} = c\varphi'(x+ct) - c\psi'(x-ct)$$

At $t = 0$, $\varphi'(x) - \psi'(x) = 0$ ...(5)

and $u\,(x, 0) = \varphi\,(x) + \psi\,(x) = f\,(x)$ ....(6)

$\therefore$ Equation (5) gives $\varphi(x) = \psi(x) + k$ (by integrating (s))

$\therefore$ Equation (6) becomes

$$2\psi(x) + k = f(x)$$

$$\therefore \quad \psi(x) = \frac{1}{2}[f(x) - k] \text{ and } \varphi(x) = \frac{1}{2}[f(x) + k]$$

Hence equation (4) takes the form

$$u(x, t) = \frac{1}{2}[f(x+ct)+k] + \frac{1}{2}[f(x-ct)-k]$$

$$u(x, t) = f(x + ct) + f(x - ct) \qquad \text{....(7)}$$

which is the D' Alembert's solution of the wave equation.

Note: D'Alembert's solution gives a very useful method of solving partial differential equations by change of variables.

Example 1: Find the deflection of a vibrating string of unit length having fixed ends with initial velocity zero and initial deflection $f(x) = k(\sin x - \sin 2x)$.

Solution: By D'Alembert's method, the solution is

$$y(x, t) = \frac{1}{2}\,[f(x + ct) + f(x - ct)]$$

$$= \frac{1}{2}\,[k(\sin(x + ct) - \sin 2(x + ct)$$

$$+ k(\sin x - ct) - \sin 2(x - ct)]$$

$$= k[\sin x \cos ct - \sin 2x \cos 2ct]$$

Also $y(x, 0) = k[\sin x - \sin 2x) = f(x)$

and $\dfrac{\partial y}{\partial t}(x, 0) = k(-c \sin x \sin ct + 2c \sin 2x \sin 2ct)_{t=0} = 0$

Hence the boundary conditions are satisfied.

Example 2: Using D' Alembert's method, find the deflection of a vibrating string of unit length having fixed ends with initial velocity zero and initial deflection $f(x) = a(x - x^3)$.

Ans.: $y(x, t) = ax(1 - x^2 - 3c^2 t^2)$

HEAT EQUATIONS

Heat flow

The fundamental principles involved in the problems of heat conduction are:

i) Heat flows from a higher temperature to the lower temperature.

ii) The quantity of heat in a body is proportional to its mass and temperature.

iii) The rate of heat flow across an area is proportional to the area and to the rate of change of temperature with respect to its distance normal to the area.

If q (cal/sec) be the quantity of heat that flows across a slab of area (α) cm^2 and thickness δx in one second where the difference of the temperature at the face is δT then by (ii)

$$q = -k\,\alpha\,\frac{dT}{dx}$$

where k is the constant depending upon the material of the body and is called thermal conductivity and α is the slab area.

3.4 DERIVATION OF ONE-DIMENSIONAL HEAT EQUATION

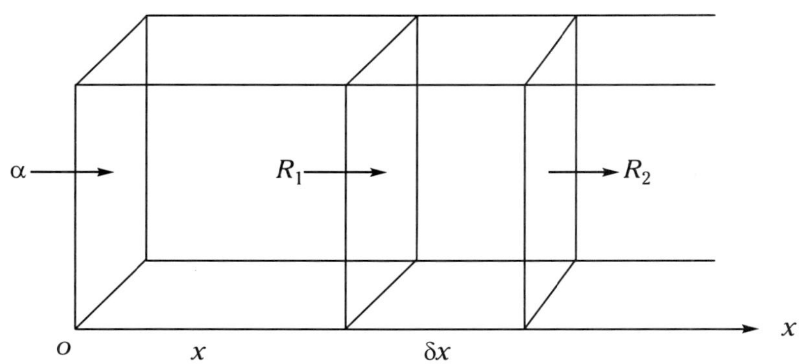

Heat and radiation phenomenena are modeled by a partial differential equation called the heat equation. Now we derive one-dimensional heat equation.

Consider a homogeneous bar of uniform cross-section α (cm^2) suppose that the sides are covered with a material impervious to heat so that the stream lines of heat flow are all parallel and perpendicular to the area α. Take one end of the bar as the origin and the direction of flow as the positive x-axis.

Let ρ be the density (gr/cm²) and k the thermal conductivity (cal/cm). Let $u(x, t)$ be the temperature at a distance x from 0. If δu be the temperature change in a slab of thickness δx, of the bar then the quantity of heat in this slab

$$= s\rho\alpha\delta x\, \delta u.$$

Hence the rate of increase of heat in this slab is

$$s\rho\alpha\delta x \frac{\partial u}{\partial t} = R_1 - R_2$$

where R_1 and R_2 are respectively the rates (cal/sec) of inflow and outflow of heat.

Now, $R_1 = -k\alpha\left(\frac{\partial u}{\partial x}\right)_x$ and $R_2 = -k\alpha\left(\frac{\partial u}{\partial x}\right)_{x+\delta x}$ the –ve sign appearing as a result given because heat flow from a higher temeprature to the lower temperature.

$$\therefore \quad s\rho\alpha\delta x \frac{\partial u}{\partial t} = -k\alpha\left(\frac{\partial u}{\partial x}\right)_x + k\alpha\left(\frac{\partial u}{\partial x}\right)_{x+\delta x}$$

$$\therefore \quad \frac{\partial u}{\partial t} = \frac{k}{s\rho}\left[\frac{\left(\frac{\partial u}{\partial x}\right)_{x+\delta x} - \left(\frac{\partial u}{\partial x}\right)_x}{\delta x}\right]$$

As $\delta x \to 0$, we get

$$\frac{\partial u}{\partial t} = c^2 \frac{\partial^2 u}{\partial x^2}$$ Where $c^2 = \frac{k}{s\rho}$ called the diffusivity of the substance (cm²/sec.)

This is one-dimensional heat flow equation.

3.5 SOLUTION OF HEAT EQUATION

Let $\frac{\partial u}{\partial t} = c^2 \frac{\partial^2 u}{\partial x^2}$ ......(1) be the heat flow equation of one dimension.

Assume that

$$u(x, t) = X(x)\, T(t) \qquad ....(2)$$

Where X is the function of x only and T is the function of t alone, be the solution of equation (1).

From (2) we have, $\frac{\partial u}{\partial t} = XT'$ and $\frac{\partial^2 u}{\partial x^2} = X''T$

Substituting in (1)

$$XT' = c^2 X'' T$$

$$\therefore \quad \frac{X''}{X} = \frac{T'}{c^2 T} = k \text{ (say)}, \text{ variables are separated.}$$

$\therefore$ $X'' - kX = 0$ and $T' - c^2 kT = 0$ are the linear differential equations.

$\therefore$ AEs are $D^2 - k = 0$ and $D_1 - c^2 k = 0$

$$\therefore \quad D = \pm\sqrt{k} \quad \text{and} \quad D_1 = c^2 k$$

The solution of these equations depends on the nature of the roots.

Case i: When $k > 0$ (+ve) say $= p^2$

$\therefore$ $D = \pm p$ and $D_1 = c^2 p^2$

$\therefore$ $X = c_1 e^{px} + c_2 e^{-px}$; $T = c_3 e^{c^2 p^2 t}$

Case ii: When $k < 0$ (–ve) say $= -p^2$

$\therefore$ $D = \pm pi$ and $D_1 = -c^2 p^2$

$\therefore$ $X = c_4 \cos px + c_5 \sin px$; $T = c_6 e^{-c^2 p^2 t}$

Case iii: When $k = 0$

$\therefore$ $X = c_7 x + c_8$; $T = c_9$

Of these three solutions we have to choose that solution which is consistent with the physical nature of the problem. As we are dealing with problems on heat conduction, the suitable solution is

$$u = (c_1 \cos px + c_2 \sin px)\, e^{-c^2 p^2 t}$$

Example 1: A rod of length l with insulated sides is initially at a uniform temperature u_0. Its ends are suddenly cooled to 0°C and are kept at that temperature. Find the temperature function $u(x, t)$.

Solution: The temperature function $u(x, t)$ satisfies the differential equation.

$$\frac{\partial u}{\partial t} = c^2 \frac{\partial^2 u}{\partial x^2} \qquad \text{...(1)}$$

Its solution is

$$u(x, t) = (c_1 \cos px + c_2 \sin px)\, e^{-c^2 p^2 t} \qquad ...(2)$$

Since the ends $x = 0$ and $x = l$ are cooled to 0°C and kept at that temperature throughout, the boundary conditions are $u(0, t) = u(l, t) = 0$ for all t.

Also $u(x, 0) = u_0$ is the initial conditon.

Since $u(0, t) = 0$ we have from (2)

$$0 = c_1\, e^{-c^2 p^2 t} \qquad \therefore\ c_1 = 0$$

$$\therefore \quad u(x, t) = c_2 \sin px\, e^{-c^2 p^2 t} \qquad ...(3)$$

Since $u(l, t) = 0$ we have from (3)

$$0 = c_2 \sin pl\, e^{-c^2 p^2 t} \quad \Rightarrow \sin pl = 0$$

$$\Rightarrow pl = n\pi$$

$$\therefore \quad p = \frac{n\pi}{l}, \ n \text{ is an integer.}$$

$\therefore$ Solution (3) reduces to

$$u(x, t) = b_n \sin \frac{n\pi x}{l}\, e^{-\frac{c^2 n^2 \pi^2 t}{l^2}} \quad \text{where } c_2 = b_n$$

The general solution is obtained by adding all solutions for $n = 1, 2, 3, \ldots$.

$$\therefore \quad u(x, t) = \sum_{n=1}^{\infty} b_n \sin \frac{n\pi x}{l}\, e^{-\frac{c^2 n^2 \pi^2 t}{l^2}}$$

Since $u(x, 0) = u_0$ we have

$$u_0 = \sum_{n=1}^{\infty} b_n \sin \frac{n\pi x}{l}$$

Which is half range sine series for u_0.

$$\therefore \quad b_n = \frac{2}{l} \int_0^l u_0 \sin \frac{n\pi x}{l}\, d_x = \begin{cases} 0 & \text{when } n \text{ is even} \\ \dfrac{4u_0}{n\pi} & \text{when } n \text{ is odd} \end{cases}$$

Hence the temperature function

$$u(x, t) = \frac{4u_0}{\pi} \sum_{n=1,3,5} \frac{1}{n} \sin \frac{n\pi x}{l} e^{-\frac{c^2 n^2 \pi^2 t}{l}}$$

$$= \frac{4u_0}{\pi} \sum_{n=1}^{\infty} \frac{1}{2n-1} \sin \frac{(2n-1)\, x}{l} e^{-\frac{c^2 (2n-1)^2 \pi^2 t}{l^2}}$$

Example 2: An insulated rod of length l has its ends A and B maintained at 0°C and 100°C respectively unit steady state conditions prevail. If B is suddenly reduced to 0°C and maintained at 0°C, find the temperature at a distance x from A at time t.

Solution: The differential equation of the temperature functions

$$\frac{\partial u}{\partial t} = c^2 \frac{\partial^2 u}{\partial x^2} \qquad \text{....(1)}$$

Prior to the temperature change at the end B, when $t = 0$, the heat flow was independent of time (steady state condition) when the temperature u depends only upon x and not on t Equation (1) reduces to

$$\frac{\partial^2 u}{\partial x^2} = 0$$

Its general solution is

$$u = ax + b \qquad \text{....(2)}$$

Where a and b are constants.

Since $u = 0$ for $x = 0$ and $u = 100$ for $x = l$, we get from (2) $b = 0$ and $a = \frac{100}{l}$.

$\therefore$ The initial condition is expressed by $u(x, 0) = \frac{100}{l} x$. Also the boundary conditions for the subsequent flow are $u(0, t) = 0$ and $u(l, t) = 0$ for all values of t.

The general solution of (1) satisfying the boundary conditions is

$$u(x, t) = \sum_{n=1}^{\infty} b_n \sin \frac{n\pi x}{l} e^{-\frac{c^2 n^2 \pi^2 t}{l^2}} \qquad \text{....(3)}$$

(Proved in previous example)

Since $u(x, 0) = \frac{100}{l}x$, we have

$$\frac{100}{l}x = \sum_{n=1}^{\infty} b_n \sin\frac{n\pi x}{l}$$

Which is half range sine series for $\frac{100}{l}x$

$$\therefore \qquad b_n = \frac{2}{l}\int_0^l \frac{100}{l}x \sin\frac{n\pi x}{l}dx$$

$$= \frac{200}{l^2}\left[x\left(\frac{-\cos\frac{n\pi x}{l}}{\frac{n\pi}{l}}\right) - (1)\left(\frac{-\sin\frac{n\pi x}{l}}{\left(\frac{n\pi}{l}\right)^2}\right)\right]_0^l$$

$$= \frac{200}{l^2}\left[-\frac{l^2}{n\pi}\cos n\pi\right] = -\frac{200}{n\pi}(-1)^n$$

$$= \frac{200}{n\pi}(-1)^{n+1}$$

Hence the temperature function is

$$u(x, t) = \frac{200}{\pi}\sum_{n=1}^{\infty}\frac{(-1)^{n+1}}{n}\sin\frac{n\pi x}{l}e^{-\frac{c^2n^2\pi^2 t}{l}}$$

Example 3: Solve the differential equation $\frac{\partial u}{\partial t} = \alpha^2\frac{\partial^2 u}{\partial x^2}$ for the conduction of heat along a rod without radiation, subject to the following conditions.

i) u is not infinite for $t \to \infty$

ii) $\frac{\partial u}{\partial x} = 0$ for $x = 0$ and $x = l$

iii) $u = lx - x^2$ for $t = 0$ between $x = 0$ and $x = l$.

Solution: Given heat flow equation of one dimension is

$$\frac{\partial u}{\partial t} = \alpha^2 \frac{\partial^2 u}{\partial x^2} \qquad ....(1)$$

Put $u = X(x)\,T(t)$ in equation (1)

$$XT' = \alpha^2 X''\,T$$

$$\therefore \quad \frac{X''}{X} = \frac{T'}{\alpha^2 T} = -k^2 \text{ (say)}$$

$$\therefore \quad \frac{d^2X}{dx^2} + k^2 X = 0 \text{ and } \frac{dT}{dt} + k^2\alpha^2 T = 0 \qquad ...(2)$$

Their solutions are

$$X = c_1 \cos kx + c_2 \sin kx, \quad T = c_3 e^{-k^2\alpha^2 t}.$$

It k^2 is changed to $-k^2$, the solutions are

$$X = c_4 e^{kx} + c_5 e^{-kx}, \qquad T = c_6\, e^{k^2\alpha^2 t} \qquad ....(4)$$

If $k^2 = 0$, the solutions are

$$X = c_7 x + c_8, \quad T = c_9 \qquad ....(5)$$

In (4), $T \to \infty$ for $t \to \infty$, therefore u also $\to \infty$

i.e., the given condition (1) is not satisfied. So we reject the solutions (4) while (3) and (5) satisfy this conditon. Applying the condition (ii) to (5) we get $c_7 = 0$.

$$\therefore \quad u = XT = c_8\, c_9 = a_0 \text{ (say)} \qquad ....(6)$$

From (3), $\frac{\partial u}{\partial x} = (-c_1 \sin kx + c_2 \cos kx)\, k\, c_3 e^{-k^2\alpha^2 t}$

Applying the condition (ii) we get $c_2 = 0$ and $-c_1 \sin lk + c_2 \cos kl = 0$

i.e., $c_2 = 0$ and $kl = n\pi$ (n a non-zero integer)

$$\therefore \quad u = c_1 \cos kx \,.\, c_3 e^{-k^2\alpha^2 t}$$

$$= a_n \cos\left(\frac{n\pi x}{l}\right) e^{-\frac{n^2\pi^2\alpha^2 t}{l^2}} \qquad ....(7)$$

Thus the general solution being the sum of (6) and (7) we have

$$u = a_0 + \Sigma a_n \cos\frac{n\pi x}{l} \cdot e^{-\frac{n^2\pi^2\alpha^2 t}{l^2}} \quad ....(8)$$

Now, using the condition (iii), we get

$$lx - x^2 = a_0 + \Sigma\, a_n \cos\frac{n\pi x}{l}$$

This being the expansion of $lx - x^2$ as a half range cosine series in $(0, l)$, we get

$$a_0 = \frac{1}{l}\int_0^l (lx - x^2)\, dx = \frac{1}{l}\left[\frac{lx^2}{2} - \frac{x^3}{3}\right]_0^l = \frac{l^2}{6}$$

$$a_n = \frac{2}{l}\int_0^l (lx - x^2)\cos\frac{n\pi x}{l}\, dx$$

$= 0$ when n id odd

$= \dfrac{-4l^2}{n^2\pi^2}$ when n is even.

Hence by taking $n = 2m$, the required solution is

$$u = \frac{l^2}{6} - \frac{l^2}{\pi^2}\sum_{m=1}^{\infty}\frac{1}{m^2}\cos\left(\frac{2m\pi x}{l}\right)e^{-\frac{4m^2\pi^2\alpha^2 t}{l^2}}$$

Example 4: Solve the following boundary value problem

$$\frac{\partial u}{\partial t} = \alpha^2\frac{\partial^2 u}{\partial x^2}\, 0 < x < l, \quad \frac{\partial(0\ t)}{\partial x} = 0, \quad \frac{\partial u}{\partial x}(l, t) = 0, \ u(x, 0) = x$$

Ans.: $u(x,t) = \dfrac{l}{2} - \dfrac{4l}{\pi^2}\displaystyle\sum_{1}^{\infty}\frac{1}{(2n-1)^2}\cos\frac{(2n-1)\pi x}{l}\, e^{-\frac{(2n-1)^2 c^2\pi^2 t}{l^2}}$

Example 5: Solve the equation $\dfrac{\partial u}{\partial t} = \dfrac{\partial^2 u}{\partial x^2}$ with boundary conditions

$u(x, 0) = 3\sin n\pi x$, $u(0, t) = 0$, $u(1, t) = 0$ where $0 < x < 1$, $t > 0$.

Ans.: $u(x, t) = 3\displaystyle\sum_{1}^{\infty} e^{-n^2\pi^2 t}\sin n\pi x$

Example 6: A homogeneous rod of conducting material of length 100 cm has its ends kept at zero temperature and the temperature initially is

$$u(x, 0) = x \qquad 0 \le x \le 50$$
$$= 100 - x \qquad 50 \le x \le 100$$

Find the temperature $u(x, t)$ at any time.

Ans.: $$-u(x, t) = \frac{400}{\pi^2} \sum_{n=0}^{\infty} \frac{(-1)^n}{(2n+1)^2} e^{-\left[\frac{(2n+1)\,c\pi}{100}\right]^2 t} \sin \frac{(2n+1)\pi x}{100}$$

Example 7: The ends A and B of a rod 20 cm long have the temperature at 30°C and 80°C until steady state prevails. The temperature of the ends are changed to 40°C and 60°C respectively. Find the temperature distribution in the rod at time t.

Ans.: $$u(x, t) = x + 40 - \frac{20}{\pi} \sum_{1}^{\infty} \frac{1}{n} (2\cos n\pi + 1) \sin \frac{n\pi x}{l} e^{-\frac{c^2 n^2 \pi^2 t}{400}}.$$

Example 8: Find the solution of $\frac{\partial \theta}{\partial t} = k \frac{\partial^2 \theta}{\partial x^2}$ such that

i) θ is not infinite when $t \to +\infty$

ii) $\frac{\partial \theta}{\partial x} = 0$ when $x = 0$ for all values of t

iii) $\theta = \theta_0$ when $x = l$ for all values of t

Ans.:

$$\theta = \frac{4\theta_0}{\pi}\left[e^{-\left(\frac{\pi}{2l}\right)^2 kt} \cos \frac{\pi x}{2l} - \frac{1}{3} e^{-\left(\frac{3\pi}{2l}\right)^2 kt} \cos \frac{3\pi}{2l} x + \frac{1}{5} e^{-\left(\frac{5\pi}{2l}\right)^2 kt} \cos \frac{5\pi x}{2l} \ldots\right]$$

3.6 DERIVATION OF TWO-DIMENSIONAL HEAT FLOW EQUATION

Consider the flow of heat in a metal plate in the XOY plane. If the temperature at any point is independent of the z-coordinate and depends on x, y and t only. Then the flow is called two dimensional and the heat flow lies in the plane XOY only and is zero along the normal to the plane XOY.

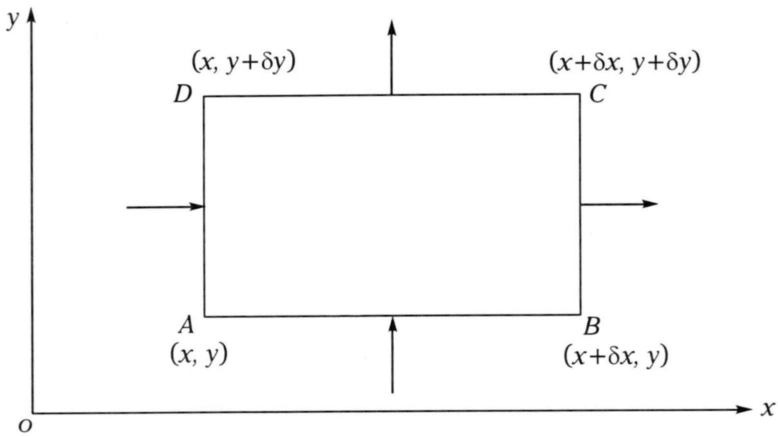

Take a rectangular element of the plate with δx and δy and thickness α. As in the case of the one-dimensional heat flow along a bar, the quantity of heat that entres the plate per second from the sides AB and AD is given by

$-k\alpha\, \delta x \left(\frac{\partial u}{\partial y}\right)_y$ and $-k\alpha \delta y \left(\frac{\partial u}{\partial x}\right)_x$ respectively and that which flows out through the sides CD and BC per second is

$$-k\alpha\, \delta x \left(\frac{\partial u}{\partial y}\right)_{y+\delta y} \quad \text{and} \quad -k\alpha \delta y \left(\frac{\partial u}{\partial x}\right)_{x+\delta x} \quad \text{respectively.}$$

Therefore, the total gain of heat by the rectangular plate $ABCD$ per second

$$= -k\alpha\delta x\left(\frac{\partial u}{\partial y}\right)_y - k\alpha\delta y\left(\frac{\partial u}{\partial x}\right)_x + k\alpha\delta x\left(\frac{\partial u}{\partial y}\right)_{y+\delta y} + k\alpha\delta y\left(\frac{\partial u}{\partial y}\right)_{x+\delta x}$$

$$= k\alpha\,\delta x\delta y\left[\frac{\left(\frac{\partial u}{\partial x}\right)_{x+\delta x} - \left(\frac{\partial u}{\partial y}\right)_x}{\delta x} + \frac{\left(\frac{\partial u}{\partial y}\right)_{y+\delta y} - \left(\frac{\partial y}{\partial y}\right)_y}{\delta y}\right] \qquad ...(1)$$

The rate of gain of heat by the plate is also given by

$$s\rho\delta x\delta y\,\frac{\partial u}{\partial t} \qquad .....(2)$$

Where s = specific heat, ρ = density of the metal plate.

Equating (1) and (2), we have

$$k\alpha\delta x\delta y\left[\frac{\left(\frac{\partial u}{\partial x}\right)_{x+\delta x}-\left(\frac{\partial u}{\partial x}\right)_x}{\delta x}+\frac{\left(\frac{\partial u}{\partial y}\right)_{y+\delta y}-\left(\frac{\partial u}{\partial y}\right)_y}{\delta y}\right]=s\rho\delta x\delta y\frac{\partial u}{\partial t}$$

Dividing both sides by $\alpha\delta x\delta y$ and taking the limit as $\delta x \to 0$, $\delta y \to 0$

We get

$$k\left(\frac{\partial^2 u}{\partial x^2}+\frac{\partial^2 u}{\partial y^2}\right)=s\rho\frac{\partial u}{\partial t}$$

or $$\frac{\partial u}{\partial t} = c^2\left(\frac{\partial^2 u}{\partial x^2}+\frac{\partial^2 u}{\partial y^2}\right) \quad \text{where } c^2=\frac{k}{\delta\rho} \qquad ....(3)$$

Equation (3) gives the temperature distribution of the plate in the transient state.

Note 1: In the state, u is independent of t, so that $\frac{\partial u}{\partial t}=0$ and the above equation reduces to

$$\frac{\partial^2 u}{\partial x^2}+\frac{\partial^2 u}{\partial y^2}=0 \qquad ....(4)$$

which is known as **Laplace's equation in two dimensions.**

Note 2: The equation of heat flow in a solid (3-dimensional heat flow) can be derived as

$$c^2\left(\frac{\partial^2 u}{\partial x^2}+\frac{\partial^2 u}{\partial y^2}+\frac{\partial^2 u}{\partial z^2}\right)=\frac{\partial u}{\partial t}$$

In the steady state, it reduces to

$$\frac{\partial^2 u}{\partial x^2}+\frac{\partial^2 u}{\partial y^2}+\frac{\partial^2 u}{\partial z^2}=0 \qquad ....(5)$$

which is **Laplace's equation in three dimensions.**

2.7 SOLUTION OF LAPLACE EQUATION IN TWO DIMENSIONS

Let the Laplace's equation in two dimensions be

$$\frac{\partial^2 u}{\partial x^2}+\frac{\partial^2 u}{\partial y^2}=0 \qquad ...(1)$$

Assume that

$$u(x, y) = X(x)\, Y(y) \qquad ....(2)$$

where X is a function of x only and Y is a function of y alone, be the solution of equation (1).

$$\therefore \quad \frac{\partial^2 u}{\partial x^2} = X''Y \text{ and } \frac{\partial^2 u}{\partial y^2}=XY''$$

Substituting in (1) we get

$$X''Y+XY''=0$$

or $$\frac{X''}{X}=-\frac{Y''}{Y}=k \text{ say.}$$

$\therefore$ $X' - kX = 0$ and $Y'' + kY = 0$ are the ordinary differential equations.

$\therefore$ AEs are $D^2 - k = 0$ and $D_1^2+k=0$, where $\frac{d}{dx}=D,\ \frac{d}{dy}=D_1$

$\therefore$ $D=\pm\sqrt{k}$ and $D_1=\pm\sqrt{-k}$

The solutions depend upon the nature of the roots.

Case i: When k is positive and $= p^2$ say

$D = \pm\, p \qquad D_1 = \pm\, p_i$

$X = c_1\, e^{px} + c_2\, e^{-px}\,; \qquad y = c_3 \cos py + c_4 \sin py$

Case ii: When k is negative and $= -p^2$ say

$D = \pm\, p_i\,, \qquad D_1 = \pm\, p$

$X = c_5 \cos px + c_6 \sin px\,;\ Y = c_7\, e^{py} + c_8\, e^{-py}$

Case iii: When $k = 0$

$X = c_9\, x + c_{10}\,; \qquad Y = c_{11} y + c_{12}.$

Of these three solutions, we have to choose that solution which is consistant with the physical nature of the problem and the given boundary conditions.

$$\therefore \quad u(x, y) = (c_1 \cos px + c_2 \sin px)\ (c_3\, e^{py} + c_4\, e^{-py})$$

is required solution of Laplace equation in two dimension.

Laplace's Equation

1) We know that the two-dimensional heat flow equation in steady state reduces to

$$\frac{\partial^2 u}{\partial x^2}+\frac{\partial^2 u}{\partial y^2}=0$$

which is the Laplace's equation in two dimensions.

2) The three-dimensional heat flow equation in steady state is

$$\nabla^2 u=\frac{\partial^2 u}{\partial x^2}+\frac{\partial^2 u}{\partial y^2}+\frac{\partial^2 u}{\partial z^2}=0$$

which is Laplace's equation in three dimensions.

3) The Laplace equation in polar co-ordinates is given by

$$\frac{\partial^2 u}{\partial r^2}+\frac{1}{r}\frac{\partial u}{\partial r}+\frac{1}{r^2}\frac{\partial^2 u}{\partial \theta^2}=0$$

This equaiton can be obtained from (1) by putting $x = r\cos\theta$, $y = r\sin\theta$, thus changing the independent variables from (x, y) to (r, θ).

4) Laplace's equation in cylindrical coordinates is given by

$$\frac{\partial^2 u}{\partial r^2}+\frac{1}{r}\frac{\partial u}{\partial r}+\frac{1}{r^2}\frac{\partial^2 u}{\partial \theta^2}+\frac{\partial^2 u}{\partial z^2}=0$$

This equation can be obtained from (2) by putting

$x = r\cos\theta$, $y = r\sin\theta$ and $z = z$, thus changing the independent variables (x, y, z) to (r, θ, z).

5) Laplace's equation in spherical coordinates is given by

$$\frac{\partial^2 u}{\partial r^2}+\frac{2}{r}\frac{\partial u}{\partial r}+\frac{1}{r^2}\frac{\partial^2 u}{\partial \theta^2}+\frac{\cot\theta}{r}\frac{\partial u}{\partial \theta}+\frac{1}{r^2\sin^2\theta}\frac{\partial^2 u}{\partial \varphi^2}=0$$

This equation can be obtained from (2) by putting

$$x = r\sin\theta\cos\varphi,\quad y = r\sin\theta\sin\varphi,\quad z = r\cos\theta$$

Thus changing the independent variables (x, y, z) to (r, θ, φ).

3.8 SOLUTION OF LAPLACE'S EQUATION IN POLAR CO-ORDINATES

Let the Laplace's equation in polar co-ordinates be

$$r^2 \frac{\partial^2 u}{\partial r^2} + r \frac{\partial u}{\partial r} + \frac{\partial^2 u}{\partial \theta^2} = 0 \qquad ...(1)$$

Suppose that

$$u(r, \theta) = R(r)\ F(\theta)$$

where R is the function of r only and F is the function of θ alone, be a solution of equation (1).

$$\therefore \quad \frac{\partial^2 u}{\partial r^2} = R''F, \quad \frac{\partial u}{\partial r} = R'F \text{ and } \frac{\partial^2 u}{\partial \theta^2} = RF''$$

Substituting in (1)

$$(r^2 R'' + rR')\ F + RF'' = 0$$

$$\therefore \quad (r^2 R'' + rR')\ F + RF'' = 0$$

Separating the variables

$$\frac{r^2 R'' + rR'}{R} = -\frac{F''}{F} = k \quad \text{say}$$

$\therefore$ $r^2 R'' + rR' - kR = 0$ and $F'' + kF = 0$ are the two differential equations.

$r^2 R'' + rR' - kR = 0$ ....(2) is a homogeneous linear differential equation and to find the solution put

$$r = e^z \quad \text{or} \quad z = \log r$$

$$\therefore \quad \frac{\partial R}{\partial r} = \frac{\partial R}{\partial z} \cdot \frac{\partial z}{\partial r} = \frac{\partial R}{\partial z} \cdot \frac{1}{r}$$

$$\therefore \quad r \frac{\partial R}{\partial r} = DR \quad \text{where} \quad D = \frac{\partial}{\partial z}$$

$$\text{and} \quad r^2 \frac{\partial^2 R}{\partial r^2} = D(D-1)R$$

Substituting in (2)

$$D(D-1)\ R + DR - kR = 0$$

$\therefore \qquad (D^2 - k) R = 0$

And $\qquad \dfrac{d^2 F}{d\theta^2} + kF = 0$

Hence the AEs are $D^2 - k = 0$ and $D_1^2 + k = 0, \quad \dfrac{d}{d\theta} = D_1$

$\therefore \qquad D = \pm\sqrt{k}, \qquad D_1 = \pm\sqrt{-k}$

The solutions depend upon the nature of the roots of the equations.

Case i: When k is positive and $= p^2$, say

$\therefore \qquad R = c_1 e^{pz} + c_2 e^{-pz} \; ; \qquad F = c_3 \cos p\theta + c_4 \sin p\theta$

Case ii: When k is negative and $= -p^2$, say

$$R = c_5 \cos pz + c_6 \sin pz \; ;$$
$$= c_5 \cos (p \log r) + c_6 \sin (p \log r)$$
$$F = c_7 e^{p\theta} + c_8 e^{-p\theta}$$

Case iii: When $k = 0$

$$R = c_9 z + c_{10} = c_9 \log r + c_{10} \; ; \qquad F = c_{11} \theta + c_{12}$$

Thus the three possible solutions are

$$u = (c_1 r^p + c_2 r^{-p})(c_3 \cos p\theta + c_4 \sin p\theta) \qquad ...(3)$$
$$u = [c_5 \cos (p \log r) + c_6 \sin (p \log r)](c_7 e^{p\theta} + c_8 e^{-p\theta}) \qquad ...(4)$$
$$u = (c_9 \log r + c_{10})(c_{11} \theta + c_{12}) \qquad ...(5)$$

Of these solutions, we choose the one which is consistent with the physical nature of the problem.

Note: Usually we require a solution extending upto the origin since u must be finite at the origin, we reject (4) and (5)

$\therefore$ In this case, the solution may be written as

$u = (A \cos p\theta + B \sin p\theta) r^p.$

The general solution will consist of a sum of similar terms with different (arbitrary) values of A, B, and p.

Example 1: Solve $\dfrac{\partial^2 u}{\partial x^2} + \dfrac{\partial^2 u}{\partial y^2} = 0$ which satisfies the conditions

$u(0, y) = u(l, y) = u(x, 0) = 0$ and $u(x, a) = \sin \dfrac{n\pi x}{l}$.

Solution: The given Laplace's equation is

$$\frac{\partial^2 u}{\partial x^2} + \frac{\partial^2 u}{\partial y^2} = 0 \qquad ...(1)$$

The three possible solutions of (1) are

$$u = (c_1 e^{px} + c_2 e^{-px})(c_3 \cos py + c_4 \sin py) \qquad ....(2)$$

$$u = (c_1 \cos px + c_2 \sin px)(c_3 e^{py} + c_4 e^{-py}) \qquad ...(3)$$

$$u = (c_1 x + c_2)(c_3 y + c_4) \qquad ...(4)$$

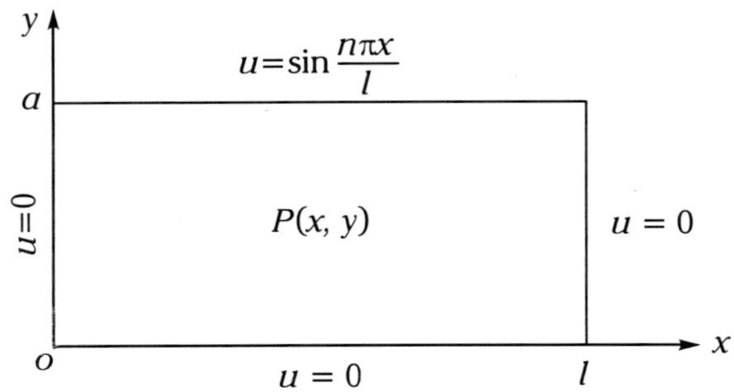

Given boundary conditions

$u(0, y) = 0 \qquad u(l, y) = 0, \quad u(x, 0) = 0$ and

$u(x, a) = \sin\frac{n\pi x}{l}$ are shown in the figure.

The only possible solution is (3)

i.e. $$u(x, y) = (c_1 \cos px + c_2 \sin px)(c_3 e^{py} + c_4 e^{-py})$$

Since $u(0, y) = 0$

$\therefore \qquad 0 = c_1 (c_3 e^{py} + c_4 e^{-py})$

$\therefore \qquad c_1 = 0$

Equation (5) reduces to

$$u(x, y) = c_2 \sin px \, (c_3 e^{py} + c_4 e^{-py}) \qquad ...(6)$$

Since $u(l, y) = 0, \quad 0 = c_2 \sin pl \, (c_3 e^{py} + c_4 e^{-py})$

$\therefore \qquad \sin pl = 0$ i.e., $pl = n\pi$

or $p = \frac{n\pi}{l}$, n being an integer.

Also $u(x, 0) = 0 \;\therefore\; 0 = c_2 \sin\frac{n\pi x}{l}(c_3 + c_4)$

$\therefore \quad c_3 + c_4 = 0 \qquad \therefore \quad c_3 = -c_4$

$\therefore$ Equation (6) becomes

$$u(x, y) = c_2 \sin\frac{n\pi x}{l}\left(c_3 e^{\frac{n\pi y}{l}} - c_3 e^{-\frac{n\pi y}{l}}\right)$$

$$= c_2 c_3 \sin\frac{n\pi x}{l}\left(e^{\frac{n\pi y}{l}} - e^{-\frac{n\pi y}{l}}\right)$$

Replacing $c_2\, c_3$ by b_n we have

$$u(x, y) = 2\, b_n \sin\frac{n\pi x}{l}\sinh\frac{n\pi y}{l} \qquad ...(7)$$

Putting $y = a$ we have $u(x, a) = \sin\frac{n\pi x}{l}$, by condition (4)

$$\therefore \quad \sin\frac{n\pi x}{l} = 2\, b_n \sin\frac{n\pi x}{l}\sinh\frac{n\pi a}{l}$$

$$\therefore \quad 2b_n \sinh\frac{n\pi a}{l} = 1 \qquad \therefore \quad b_n = \frac{1}{2\sinh\frac{n\pi a}{l}}$$

Hence (7) reduces to

$$u(x, y) = \sin\frac{n\pi x}{l} \cdot \frac{\sinh\frac{n\pi y}{l}}{\sinh\frac{n\pi a}{l}}$$

$$= \left(\frac{1}{\sinh\frac{n\pi a}{l}}\right)\sin\frac{n\pi x}{l}\sinh\frac{n\pi y}{l}$$

Example 2: Solve the Laplace's equation $\frac{\partial^2 u}{\partial x^2} + \frac{\partial^2 u}{\partial y^2} = 0$ with the boundary conditions. $u(0, y) = 0$ for all values of y, $u(\pi, y) = 0$ for all values of y, $u(x, \infty) = 0$ in $0 < x < \infty$ and $u(x, 0) = u_0$ in $0 < x < \pi$.

Solution: Given Laplace's equation is

$$\frac{\partial^2 u}{\partial x^2} + \frac{\partial^2 u}{\partial y^2} = 0 \qquad ...(i)$$

and the conditions

$$u(0, y) = 0 \quad \text{....(ii)}, \qquad u(\pi, y) = 0 \quad \text{...(iii)}$$

$$u(x, \infty) = 0 \quad \text{....(iv) and} \quad u(x, 0) = u_0 \quad \text{...(v)}$$

in $0 < x < \pi$.

The solution of equation (1) is

$$u(x, y) = (c_1 \cos px + c_2 \sin px)(c_3 e^{py} + c_4 e^{-py}) \quad \text{...(vi)}$$

By (ii) $\quad u(0, y) = 0 \quad \therefore \quad 0 = c_1 (c_3 e^{py} + c_4 e^{-py})$

$\therefore \quad c_1 = 0$ and equation (vi) reduces to

$$u(x, y) = c_2 \sin px (c_3 e^{py} + c_4 e^{-py}) \quad \text{...(vii)}$$

By (iii), $u(\pi\ y) = 0 \quad \therefore 0 = c_2 \sin p\pi (c_3 e^{py} + c_4 e^{-py})$

$\therefore \quad \sin p\pi = 0$ i.e., $p\pi = n\pi$ as $c_2 \neq 0 \quad \therefore p = n$ is an integer.

Also to satisfy the condition (iv)

i.e., $\quad u = 0$ as $y \to \infty, \ c_3 = 0$

Hence (vii) takes the form

$$u(x, y) = b_n \sin nx\, e^{-ny} \quad \text{...(viii)}$$

$$\therefore \quad u(x, y) = \sum_{n=1}^{\infty} b_n \sin nx\, e^{-ny}$$

Putting $y = 0$

$$u(x, 0) = \sum_{n=1}^{\infty} b_n \sin nx \quad \text{...(ix)}$$

From (v) $u(x, 0) = u_0$

In order that the condition (v) may be satisfied, the results (v) and (ix) must be the same.

This requires the expansion of u_0 in a half range Fourier series in $(0, \pi)$ Thus

$$u = \sum_{n=1}^{\infty} b_n \sin nx$$

where, $\quad b_n = \dfrac{2}{\pi} \displaystyle\int_0^{\pi} u_0 \sin nx\, dx \quad$ and $\quad u = u_0$

$$= \frac{2u_0}{n\pi}\left[1-(1)^n\right]$$

$\therefore \quad b_n = 0 \qquad$ if n is even

$$= \frac{4u_0}{n\pi} \qquad \text{if } n \text{ in odd.}$$

Hence (viii) becomes

$$u(x, y) = \frac{4u_0}{\pi}\left[e^{-y}\sin x + \frac{1}{3}e^{-3y}\sin 3y + \frac{1}{5}e^{-5y}\sin 5y + \ldots\ldots\ldots\right]$$

Example 3: A rectangular plate with insulated surface is 10 cm wide and so long compared to its width that it may be considered infinite in length without introducing an appreciable error. If the temperature of the short edge $y = 0$ is given by

$$u = 20\,x \qquad \text{for} \quad 0 \le x \le 5$$

and

$$u = 20\,(10 - x) \quad \text{for} \quad 0 \le x \le 10$$

and the two long edges $x = 0$, $x = 10$ as well as the other short edge are kept at 0°C. Prove that the temperature u at any point (x, y) is given by

$$u = \frac{800}{\pi^2}\sum_{1}^{\infty}\frac{(-1)^{n+1}}{(2n-1)^2}\sin\frac{(2n-1)\pi x}{10}\,.\,e^{-\frac{(2n-1)}{10}\pi y}$$

Solution: The temperature $u(x, y)$ at any point $p(x, y)$ satisfies the equation

$$\frac{\partial^2 u}{\partial x^2} + \frac{\partial^2 u}{\partial y^2} = 0 \qquad \ldots(1)$$

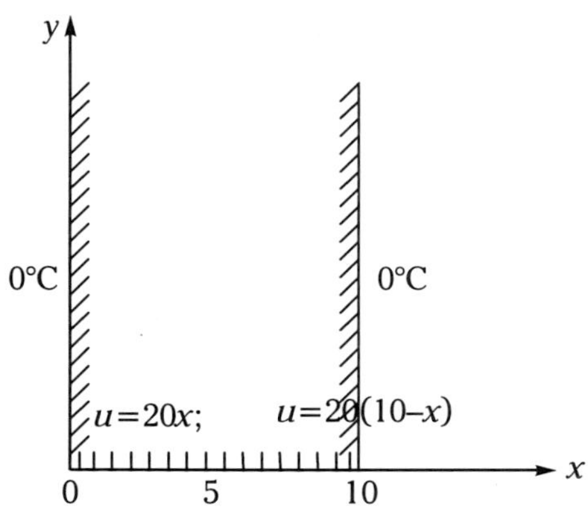

The boundary conditions are

$$u(0, y) = 0 \quad \text{for all values of } y \ge 0 \quad ...(2)$$

$$u(10, y) = 0 \quad \text{for all values of } y \ge 0 \quad ...(3)$$

$$u(x, \infty) = 0 \text{ in } 0 \le x \le 10 \quad ...(4)$$

$$u(x, 0) = \begin{cases} 20x & 0 \le x \le 5 \\ 20(1-x) & 5 \le x \le 10 \end{cases} \quad ....(5)$$

Now the possible solutions of (1) are

$$u = (c_1 e^{px} + c_2 e^{-px})(c_3 \cos py + c_4 \sin py) \quad ...(6)$$

$$u = (c_1 \cos px + c_2 \sin px)(c_3 e^{py} + c_4 e^{-py}) \quad ...(7)$$

$$u = (c_1 x + c_2)(c_3 y + c_4) \quad ....(8)$$

The solution (6) cannot satisfy the condition (2) since we get $u \ne 0$ for $x = 0$ for all values of y and solution (8) cannot satisfy the condition (4). Thus the only possible solution is (7).

$$\therefore \quad u(x, y) = (c_1 \cos px + c_2 \sin px)(c_3 e^{py} + c_4 e^{-py}) \quad ...(9)$$

Since $u(0, y) = 0 \quad \therefore \quad 0 = c_1 (c_3 e^{py} + c_4 e^{-py}) \quad \Rightarrow \quad c_1 = 0$

$\therefore$ Equation (9) reduces to

$$u(x, y) = c_2 \sin px\,(c_3 e^{py} + c_4 e^{-py}) \quad ....(10)$$

Since $u(10, y) = 0 \quad \therefore \quad 0 = c_2 \sin 10p\,(c_3 c^{py} + c_4 e^{-py})$

$\Rightarrow \quad \sin 10p = 0$ i.e., $10p = n\pi$ or $p = \dfrac{n\pi}{10}$, n being an integer.

Also $u(x, \infty) = 0 \quad \therefore \quad c_3 = 0$

Hence equation (10) reduces to

$$u(x, y) = c_2 c_4 \sin \frac{n\pi x}{10} e^{-\frac{n\pi y}{10}}$$

Replacing $c_2\, c_4$ by b_n, the most general solution is

$$u(x, y) = \sum_{n=1}^{\infty} b_n \sin \frac{n\pi x}{10} e^{-\frac{n\pi y}{10}} \quad ...(ii)$$

putting $y = 0$, we have $u(x, 0) = \sum_{n=1}^{\infty} b_n \sin \dfrac{n\pi x}{10}$

Where $$b_n = \frac{2}{10}\int_0^{10} u(x,0)\sin\frac{n\pi x}{10}dx$$

$$= \frac{2}{10}\left[\int_0^{5} 20x\sin\frac{n\pi x}{10}dx + \int_5^{10} 20(10-x)\sin\frac{n\pi x}{10}dx\right]$$

$$= \frac{800}{n^2\pi^2}(-1)^{\frac{n-1}{2}} \quad \text{when } n \text{ is odd}$$

$$= 0 \quad \text{when } n \text{ is even.}$$

Hence from (1) the required solution is

$$u(x, y) = \frac{800}{\pi^2}\sum_{\pi^2} \frac{(-1)^{\frac{n-1}{2}}}{n^2}\sin\frac{n\pi x}{10}e^{-\frac{(2n-1)\pi y}{10}}$$

$$\left(\because (-1)^{n+1} = (-1)^{n-1}(-1)^2 = (-1)^{n-1}\right)$$

Example 4: Solve the Laplace equation $\frac{\partial^2 u}{\partial x^2} + \frac{\partial^2 u}{\partial y^2} = 0$ within the rectangle $0 \le x \le a,\ 0 \le y \le b$ given that $u(0, y) = u(a, y) = u(x, b) = 0$ and $u(x, 0) = x(a - x)$.

Ans.: $$u(x, y) = \frac{8a^2}{\pi^3}\sum_{n=0}^{\infty} \frac{1}{(2n+1)^3}\sin\frac{(2n+1)\pi x}{a} \cdot \frac{\sinh\frac{(2n+1)\pi(b-u)}{a}}{\sin\frac{(2n+1)\pi b}{a}}$$

Example 5: Solve $\frac{\partial^2 u}{\partial x^2} + \frac{\partial^2 u}{\partial y^2} = 0$ for $0 < x < \pi,\ 0 < y < \pi$ given that $u(0, y) = u(\pi, y) = u(x, \pi) = 0$ and $u(x, 0) = \sin^2 x$.

Ans.: $$u(x, y) = -\frac{8}{\pi}\sum_{n=1,3,5}^{\infty} \frac{\sin nx \sinh n(\pi - y)}{n(n^2-4)\sinh \pi x}$$

Example 6: A long rectangular plate of width 'a' cm with insulated surface has its temperature v equal to zero on both the long sides and one of the short sides so that $v(0, y) = 0$, $v(a, y) = v(x, \infty) = 0$, $v(x, 0) = kx$. Show that the steady state temperature within the plate is

$$v(x, y) = \frac{2ak}{\pi} \sum_{n=1}^{\infty} \frac{(-1)^{n+1}}{n} e^{-\frac{n\pi y}{a}} \sin \frac{n\pi x}{a}$$

Example 7: A rectangular plate with insulated surface is 8 cm wide and so long compared to its width that it may be considered infinite in length without introducing an appreciable error. If the temperature along one short edge $y = 0$ is given by $u(x, 0) = 100 \sin \frac{\pi x}{8}$, $0 < x < 8$ with two long edges $x = 0$ and $x = 8$ as well as the other short edge are kept at 0°C. Show that the steady-state temperature at any point of the plate is given by $u(x, y) = 100\, e^{-\frac{\pi y}{8}} \sin \frac{\pi x}{8}$.

Example 8: Solve $\frac{\partial^2 u}{\partial x^2} + \frac{\partial^2 u}{\partial y^2} = 0$ subject to the conditions

$u(0, y) = u(l, y) = u(x, 0) = 0$ and $u(x, a) = \sin \frac{n\pi x}{l}$.

Ans.: $u(x, y) = \sin \frac{n\pi x}{l} \dfrac{\sinh \frac{n\pi y}{l}}{\sinh \frac{n\pi a}{l}}$

Example 9: Solve $\frac{\partial^2 u}{\partial x^2} + \frac{\partial^2 u}{\partial y^2} = 0$ given a square plate bounded by the lines $x = 0$, $y = 0$, $x = 20$ and $y = 20$. Its faces are insulated the temperature along the upper horizontal edge is given by $u(x, 20) = x(20 - x)$ when $0 < x < 20$ while other three edges are kept at 0°C.

Ans.: $u(x, y) = \dfrac{3200}{\pi^3} \sum_{1}^{\infty} \dfrac{\sin \frac{(2n-1)\pi x}{20} \sinh \frac{(n-1)\pi y}{20}}{(2n-1)^2 \sinh (2n-1)\pi}$

Example 10: A rectangular plate has sides a and b. Taking the sides of length a as OX and that of length b as OY and other sides to be $x = 0$ and $y = b$, the sides $x = 0$, $x = a$, $y = b$ are insulated and the edges $y = 0$ is kept at temperature $u_0 \cos \frac{\pi x}{a}$. Find the temperature $u(x, y)$ in the steady state.

Ans.: $u(x, y) = u_0 \cos\frac{\pi x}{a} \cosh\frac{\pi}{a}(b-y) \operatorname{sech}\frac{\pi b}{a}$.

3.9 QUESTION BANK

1. Solve the wave equation $c^2 \frac{\partial^2 u}{\partial x^2} = \frac{\partial^2 u}{\partial t^2}$, $0 < x < \pi$ given

   $u(0, t) = u(\pi, t) = 0$; $u(x, 0)=0$ and $\frac{\partial u}{\partial t}(x, 0) = A(\sin x - \sin 2x)$ $A \neq 0$.

2. Obtain the D' Alembert's solution of the wave equation $u_{tt} = c^2 u_{xx}$ subject to the conditions $u(x, 0) = f(x)$ and $\frac{\partial u}{\partial t}(x, 0) = 0$.

3. Solve the wave equation $\frac{\partial^2 u}{\partial t^2} = c^2 \frac{\partial^2 u}{\partial x^2}$ given $u(0, t) = 0$ $u(l, t) = 0$;

   $\frac{\partial u}{\partial t} = 0$ when $t = 0$ and $u(x, 0) = u_0 \sin \frac{\pi x}{l}$.

4. Obtain the various possible solutions of the Laplace's equation $u_{xx} + u_{yy} = 0$ by the method of separation of variable.

5. Derive one-dimensional wave equation $\frac{\partial^2 u}{\partial t^2} = c^2 \frac{\partial^2 u}{\partial x^2}$.

6. Derive one-dimensional heat equation $\frac{\partial u}{\partial t} = c^2 \frac{\partial^2 u}{\partial x^2}$.

7. Obtain D'Alembert's solution of wave equation $\frac{\partial^2 u}{\partial t^2} = c^2 \frac{\partial^2 u}{\partial x^2}$.

8. A tightly stretched string of length l with fixed ends is initially in equilibrium position. It is set to vibrate by giving each point a velocity $v_0 \sin^3 \dfrac{\pi x}{l}$. Find the displacement $y(x, t)$.

9. A string stretched between the two fixed points (0, 0) and (1, 0) and released at rest from the position $y = \lambda \sin(\pi x)$ show that the formula for its subsequent displacement $y(x, t)$ is $\lambda \cos(c\pi t) \sin(\pi x)$.

10. Obtain the solutions of the Laplace's equation $\dfrac{\partial^2 u}{\partial x^2} + \dfrac{\partial^2 u}{\partial y^2} = 0$ by the method of separation of variables.

11. Solve $\dfrac{\partial^2 u}{\partial x^2} + \dfrac{\partial^2 u}{\partial y^2} = 0$ which satisfies the conditions $u(0, y) = u(l, y) = u(x, 0) = 0$ and $u(x, a) = \sin \dfrac{n\pi x}{l}$.

12. Find the solution of Laplace's equation in polar co-ordinates $r^2 \dfrac{\partial^2 u}{\partial r^2} + r \dfrac{\partial u}{\partial r} + \dfrac{\partial^2 u}{\partial \theta^2} = 0$.

UNIT 4

CURVE FITTING AND OPTIMIZATION

4.1 METHOD OF LEAST SQUARES

This method is most widely used. It is a mathematical method and with its help a trend line is fitted to the data in such a manner that the following two conditions are satisfied.

1) $\Sigma (y_i - y_c) = 0$ i.e., the sum of deviations of the actual values of y and the computed values of y is zero.

2) $\Sigma (y_i - y_c)^2$ is least. i.e., the sum of the squares of the deviations of the actual and computed values is least from this line and hence the name. **Method of Least Squares.** The line so obtained by this method is known as "the line of best fit". The method of least Squares may be used either to fit a straight line trend or a parabolic trend.

A) Fitting of a Straight line

Let $y = a + bx$ — (1) be the line of best fit where a and b are parameters to be determined Let y_i be the value of y corresponding to the value x_i of x as determined by equation (1)

ie $$y_i = a + b\,x_i \qquad \dots (2)$$

The value of Y_i is called the estimated value of the given value y_i corresponding to x_i

Let $S = \Sigma (y_i - Y_i)^2 \qquad \dots (3)$

Using (2) $S = \Sigma (y_i - a - bx_i)^2 \qquad \dots (4)$

We determine a and b so that S is minimum

Two necessary conditions are $\dfrac{\partial S}{\partial a} = 0$ and $\dfrac{\partial S}{\partial b} = 0$

From (4) $\Sigma (y_i - a - bx_i) = 0$

$$\therefore \qquad \Sigma y_i = na + b\,\Sigma x_i \qquad \dots (5)$$

And $\quad \Sigma (y_i - a - bx_i)\, x_i = 0$

$\therefore \quad \Sigma x_i y_i = a\Sigma x_i + b\,\Sigma x_i^2 \qquad \ldots (6)$

The equations, $\Sigma y_i = na + b\,\Sigma x_i$

$$\Sigma x_i y_i = a\,\Sigma x_i + b\,\Sigma x_i^2$$

are called Normal equations. Solving these equations we get a and b. Putting these values in (1) we get the line of best fit.

Working Procedure

1) Let $y = a + bx$ be the line of best fit.
2) Form the normal equations

   $\Sigma y = na + b\,\Sigma x$

   $\Sigma xy = a\,\Sigma x + \mathrm{b}\,\Sigma x^2$
3) Solve the normal equations for a and b
4) Substitute these values in (1).

   We get the line of best fit.

Example 1: By the method of least squares fit a straight line to the following data:

x_i:	1	3	4	6	8	9	11	14
y_i:	1	2	4	4	5	7	8	9

Solution: Let $y = a + bx$ ... (1) be the line of best fit and the normal equation are

$\Sigma y = na + b\,\Sigma x$

$\Sigma xy = a\,\Sigma x + b\,\Sigma x^2 \qquad \ldots(2)$

To find the values of a and b we form the table as follows:

x	y	x^2	xy
1	1	1	1
3	2	9	6
4	4	16	16
6	4	36	24
8	5	64	40
9	7	81	63
11	8	121	88
14	9	196	126
$\Sigma x = 56$	$\Sigma y = 40$	$\Sigma x^2 = 524$	364

Here $n = 8$

Substituting these values in (2) we get

$$40 = 8a + 56b$$

$$364 = 56a + 524b$$

Solving these equations we get $a = 0.545$ and $b = 0.64$.

$\therefore \quad y = a + bx$

$y = 0.545 + 0.64x$. which is the line of best fit.

Example 2: Using the method of least squares fit a linear relation of the form $p = a + bw$ to the following data where p is the pull required to lift the weight w.

W (kg wt) :	50	70	10	120
P(kg wt) :	12	15	21	25

Estimate P when W is 150 kg

Solution: Here $n = 4$, denote $w = x$, $P = y$ then the linear law to be fitted is

$$y = a + bx \qquad \text{.....(1)}$$

The normal equations are

$$\Sigma y = na + b\,\Sigma x$$

$$\Sigma xy = a\,\Sigma x + b\,\Sigma x^2$$

To find the values of a and b we form the table as follows.

$W = x$	$P = y$	x^2	xy
50	12	2500	600
70	15	4900	1050
100	21	10,000	2,000
120	25	14,400	3,000
$\Sigma x = 340$	$\Sigma y = 73$	$\Sigma x^2 = 31800$	$\Sigma xy = 6750$

Substituting in the normal equations (2)

$$73 = 4a + 340b$$

$$6750 = 340a + 3180b$$

Solving we get $a = 2.2785$ and $b = 0.1879$.

$\therefore \quad P = a + bw$

$P = 2.2785 + 0.187\, w$ is the linear relation between P and W.

When $W = 150$ kg, $P = 2.2785 + 0.1879 \times (150)$

$= 30.4635$ kg.

Example 3: The following table gives the production (in 000 units) of certain commodity in different years.

Year x	1958	1968	1978	1988	1998
Prod: y :	8	10	12	10	16

Fit a straight line trend to the data and estimate the production in the year 2005.

Solution: Here $n = 5$

Let $y = a + bX$ .... (1) be the linear trend

Where $X = x - 1957$ then the normal equations are

$$\Sigma y = na + b\,\Sigma X$$

$$\Sigma Xy = a\,\Sigma X + b\,\Sigma X^2$$

We prepare a table as follows

$X = x_i - 1957$	y	X^2	Xy
1	8	1	8
11	10	121	110
21	12	441	252
31	10	961	310
41	16	1681	656
$\Sigma = 105$	$\Sigma y = 56$	$\Sigma X^2 = 3205$	$\Sigma Xy = 1336$

Substituting in the normal equations

$$56 = 5a + 105b$$

$$1336 = 105a + 3205\, b$$

Solving $a = 7.84$ and $b = 0.16$

Now, $y = a + bx$

$= 7.84 + 0.16X$

$y = 7.84 + 0.16\,(x - 1957)$

When $x = 2005$

$$y = 7.85 + 0.16\,(2005 - 1957)$$
$$= 15.52$$

$\therefore$ when $x = 2005$, production is 15.52 thousand units

Example 4: Fit a curve of the form $y = a + \frac{b}{x}$ to the following data and find y when $x = 8$.

x	2	3	4	6	10
y	50	47	46	45	44.

Solution: Here $n = 5$

The curve is of the form $y = a + \frac{b}{x}$

or $\quad y = a + bX \quad \ldots(1)$ where $X = \frac{1}{x}$

The normal equations are

$$\Sigma y = na + b\,\Sigma X$$
$$\Sigma Xy = a\,\Sigma X + b\,\Sigma X^2$$

To find a and b we prepare a table as follows:

X	$X = \frac{1}{x}$	y	X^2	Xy
2	$\frac{1}{2}$	50	$\frac{1}{4}$	25
3	$\frac{1}{3}$	47	$\frac{1}{9}$	$\frac{47}{3}$
4	$\frac{1}{4}$	46	$\frac{1}{16}$	$\frac{23}{2}$
6	$\frac{1}{6}$	45	$\frac{1}{36}$	$\frac{45}{6}$
10	$\frac{1}{10}$	44	$\frac{1}{100}$	$\frac{22}{5}$
	$\Sigma X = 1.35$	$\Sigma y = 232$	$\Sigma X^2 = 0.4614$	$\Sigma Xy = 64.0667$

Substituting in the normal equations (11)

$$232 = 5a + 1.35b$$

$$64.067 = 1.35a + 0.4614b$$

Solving, $a = 42.4247,\ b = 14.7234.$

$\therefore$ $y = a + bX$

$$= 42.4247 + \frac{14.7234}{x} \qquad \because\ X = \frac{1}{x}$$

When $x = 8$, $y = 42.4247 + \dfrac{14.7234}{x}$

$$= 4.2651$$

Example 5: Fit straight line trend for the following series. Estimate the value for the year 1997.

Year	x:	1990	1991	1992	1993	1994	1995	1996
Prod of Steel in M.tons	y:	60	72	75	65	80	85	95

Solution: Let $y = a + bx$ ... (1) be the straight line trend.

Let $X = x - 1993$ and from the following table.

Year	Prod	$X = x - 1993$	Xy	X^2
1990	60	–3	–180	9
1991	72	–2	–144	4
1992	75	–1	–75	1
1993	65	0	0	0
1994	80	1	80	1
1995	85	2	170	4
1996	95	3	285	9
	532	0	136	28

Here $n = 7$, since $\Sigma X = 0$,

$$a = \frac{\Sigma y}{n} = \frac{532}{7} = 76$$

$$b = \frac{\Sigma XY}{\Sigma X^2} = \frac{136}{28} = 4.857$$

∴ The straight line trend is

$$y = 76 + 4.857X$$
$$= 76 + 4.857\,(x - 1993)$$

When $x = 1997$

$$y = 76 + 4.857\,(1997 - 1993)$$
$$= 95.428 \text{ M. tons.}$$

4.2 FITTING OF A PARABOLA

Let the parabola of best fit be

$$y = a + bx + cx^2 \qquad \ldots (1)$$

Where a, b, c are parameters to be determined.

Let Y_i be the value of y corresponding to the value of x_i of x.

i.e., $Y_i = a + bx_i + cx_i^2$ ...(2)

also let $S = \Sigma\,(y_i - Y_i)^2$ ... (3)

using (2) in (3) we get

$$S = \Sigma\,(Y_i - a - bxi - cx_i^2)^2 \qquad \ldots (4)$$

We determine $a, b, c,$ So that S is minimum.

∴ The necessary conditions are $\frac{\partial S}{\partial a} = 0$, $\frac{\partial S}{\partial b} = 0$ and $\frac{\partial S}{\partial c} = 0$.

From (4) we get

$$\Sigma y_i = na + b\,\Sigma x_i + c\,\Sigma x_i^2$$
$$\Sigma x_i y_i = a\,\Sigma x_i + b\,\Sigma x_i^2 + c\,\Sigma x_i^3$$
$$\Sigma x_i^2 y_i = a\,\Sigma x_i^2 + b\,\Sigma x_i^3 + c\,\Sigma x_i^4.$$

These three equations are called the normal equations. Solving these simultaneous equations for $a, b,$ and c and substituting in (1) we get the parabola of best fit.

Working Procedure

1) Let $y = a + bx + cx^2$ be the parabola of best fit.

2) Form the normal equations

$$\Sigma y_i = na + b\,\Sigma x_i + c\,\Sigma xi^2$$

$$\Sigma x_i y_i = a \Sigma x_i + b \Sigma x_i^2 + c \Sigma x_i^3$$
$$\Sigma x_i^2 y_i = a \Sigma x_i^2 + b \Sigma x_i^3 + c \Sigma x_i^4.$$

3) Solve these simultaneous equations for *a, b,* and *c*.
4) Substitute these values in equation (1)

We get the parabola of best fit.

Example 1: Fit a parabola to the following data.

x:	1	2	3	4	5	6	7	8	9
y:	2	6	7	8	10	11	11	10	9

Estimate y when $x = 4.5$.

Solution: Here $n = 9$

Let $y = a + bx + cx^2$ ... (1) be the parabola of best fit.

The normal equations are

$$\Sigma y = na + b \Sigma x + c \Sigma x^2$$
$$\Sigma xy = a \Sigma x + b \Sigma x^2 + c \Sigma x^3$$
$$\Sigma x^2 y = a \Sigma x^2 + b \Sigma x^3 + c \Sigma x^4$$

To determine *a, b, c,* we form the table as follows:

x	y	x^2	x^3	x^4	xy	x^2y
1	2	1	1	1	2	2
2	6	4	8	16	12	24
3	7	9	27	81	21	63
4	8	16	64	256	32	128
5	10	25	125	625	50	250
6	11	36	216	1296	66	396
7	11	49	343	2401	77	539
8	10	64	512	4096	80	640
9	9	81	729	6561	81	729
45	74	285	2025	15333	421	2771

Substituting in the normal equations (2) we get

$$74 = 9a + 45b + 285c$$
$$421 = 45a + 285b + 2025c$$
$$2771 = 285b + 2025c + 15333c$$

Solving these equations we get $a = -0.9282$, $b = 3.523$ and $c = -0.2673$

$\therefore$ The parabola of best fit is

$$y = -0.9282 + 3.523x - 0.2673x^2$$

When $x = 4.5$

$$y = -0.9282 + 3.523\,(4.5) - 0.2673\,(4.5)^2$$
$$= 9.5721$$

Thus when $x = 4.5$, $y = 9.5721$.

Example 2 : Fit a parabola $y = a + bx + cx^2$ to the following data.

x :	-3	-2	-1	0	1	2	3
y :	4.63	2.11	0.67	0.09	0.63	2.15	4.58

Solution: Here $n = 7$ and the parabola of best fit is

$$y = a + bx + cx^2 \qquad \text{..... (1)}$$

The normal equations are

$$\Sigma\, y = na + b\,\Sigma\, x + c\,\Sigma\, x^2$$
$$\Sigma\, xy = a\,\Sigma\, x + b\,\Sigma\, x^2 + c\,\Sigma\, x^3$$
$$\Sigma\, x^2y = a\,\Sigma\, x^2 + b\,\Sigma\, x^3 + c\,\Sigma\, x^4$$

To determine the parameters a, b, c we from the table as follows.

x	y	x^2	x^3	x^4	xy	x^2y
-3	4.63	9	-27	81	-13.89	41.67
-2	2.11	4	-8	16	-4.22	8.44
-1	0.67	1	-1	1	-0.67	0.67
0	0.09	0	0	0	0	0
1	0.63	1	1	1	0.63	0.63
2	2.15	4	8	16	4.5	8.6
3	4.58	9	+27	81	13.34	41.22
0	14.86	28	0	196	-0.11	101.23

Substituting in the normal equations (2)

$$14.86 = 7a + 0.b + 28c$$
$$-0.11 = 0.a + 28b + 0.c$$
$$101.23 = 28a + 0b + 196c.$$

Solving these equations $a = 0.1329$, $b = -0.0039$ and $c = 0.4975$.

∴ The parabola of best fit is

$$y = 0.1329 - 0.0039x + 0.4975x^2$$

Example 3: The prices of a commodity during 1994 – 99 are given below. Fit a parabola $y = a + bx + cx^2$ to these data. Also estimate the price for the year 2000.

Year:	1994	1995	1996	1997	1998	1999
Prices (in Rs)	100	107	128	140	181	192

Solution: Here $n = 6$ and let $X = x - 1996$ then the parabola of best fit is

$$y = a + bx + cx^2 \quad \ldots (1)$$

And the normal equations are

$$\begin{aligned} \Sigma y &= na + b\,\Sigma x + c\,\Sigma x^2 \\ \Sigma xy &= a\,\Sigma x + b\,\Sigma x^2 + c\,\Sigma x^3 \\ \Sigma x^2y &= a\,\Sigma x^2 + b\,\Sigma x^3 + c\,\Sigma x^4 \end{aligned} \quad \ldots (2)$$

We prepare a table as follows:

Year x	Price y	$X = x - 1996$	X^2	X^3	X^4	Xy	X^2y
1994	100	–2	4	–8	16	–200	400
1995	107	–1	1	–1	1	–107	107
1996	128	.0	0	0	0	0	0
1997	140	1	1	1	1	140	140
1998	181	2	4	8	16	362	724
1999	192	3	9	27	81	576	1728
Total	848	3	19	27	115	771	3099

Substituting in (2)

$$\begin{aligned} 848 &= 6a + 3b + 19c \\ 771 &= 3a + 19b + 27c \\ 3099 &= 19a + 27b + 115c \end{aligned}$$

Solving $a = 126.657$, $b = 18.042$ and $c = 1.786$

∴ The parabola of best fit is

$$y = 126.657 + 18.042X + 1.786X^2$$

where $X = x - 1996$

when $x = 2000, X = 2000 - 1996 = 4.$

$\therefore$ $y = 126.657 + (18.042)(4) + (1.786)(4)^2$

$= 227.41$

Thus the likely price for the year 2000 is Rs. 227.41.

4.3 FITTING OF THE CURVE OF THE FORM $y = ab^x$

Let $y = ab^x$... (1) be the curve to be fitted.

Taking log both sides

$$\log y = \log^a + x\log^b$$

Or $u = A + Bx$ ...(2)

Where $u = \log^y$, $A = \log^a$ and $B = \log^b$

The normal equations are

$$\Sigma\, ui = nA + B\,\Sigma\, xi$$

$$\Sigma\, xui = A\,\Sigma\, xi + B\,\Sigma\, x_i^2 \quad ...(3)$$

Solving the simultaneous equations (3) we get A and B.

Where $a = e^A$ and $b = e^B$.

Substituting a and b in (1) we get the curve of best fit.

Example 1: Fit a curve of the form $y = ab^x$ for the following data.

x :	1	2	3	4	5	6	7	8
y :	1.0	1.2	1.8	2.5	3.6	4.7	6.6	9.1

Solution: Here $n = 8$.

Let $y = ab^x$ ...(1) be the curve to be fitted.

Taking log both sides

$$\text{Log}^y = \log^a + x\log^b$$

Or $u = A + Bx$ ....(2)

The normal equations are

$$\Sigma u_i = nA + B\,\Sigma\, x_i$$

$$\Sigma x_i u_i = A\,\Sigma\, x_i + B\,\Sigma\, x_i^2$$

To find A and B we form the table as follows

x	y	$u_i = \log y$	x^2	xu_i
1	1.0	0	1	0
2	1.2	0.1825	4	0.3646
3	1.8	0.5878	9	1.7634
4	2.5	0.9163	16	3.6652
5	3.6	1.209	25	6.4045
6	4.7	1.5475	36	9.2856
7	6.6	1.8871	49	13.2097
8	9.1	2.2083	64	17.6764
36	30.5	8.6103	204	52.3594

Substituting in (3)

$$8.6103 = 8A + 36B$$

$$52.3594 = 36A + 204B$$

Solving we get $A = -0.3822$ and $B = 0.3241$

Now $a = e^A = e^{0.3241} = 0.6824$

$b = e^B = e^{0.3241} = 1.3828$

$\therefore$ The curve of best fit is

$$y = ab^x$$

$$= (0.6824)\,(1.3828)^x$$

Example 2: Fit a curve of the form $y = ab^x$ in the least square sense to the following data.

x :	0	2	4	5	7	10
y:	100	120	256	390	710	1600.

Solution: Here $n = 6$.

Let $y = ab^x$ ... (1) be the curve of best fit.

Taking log both sides

$$\text{Log } y = \log a + x \log b$$

Or $u = A + Bx$ ... (2)

The normal equations are

$$\Sigma u_i = nA + B\,\Sigma xi$$

$$\Sigma x_i u_i = A\,\Sigma xi + B\,\Sigma xi^2$$

To determine A and B we form the table as follows.

X	y	$u_i = \log y_i$	x^2	$x_i u_i$
0	100	4.6051	0	0
2	120	4.7875	4	9.757
4	256	5.5452	16	22.1808
5	390	6.5653	25	29.8305
7	710	6.5653	49	45.9571
10	1600	7.3778	100	73.778
28	3175	34.847	194	181.3214

Substituting in the normal equations (3)

$$34.847 = 6A + 28B \qquad \text{Solving } A = 4.4298,\ B = 0.2953$$

$$181.324 = 28A + 194B \ \therefore \quad a = e^A = 83.9146 : b = e^B = 1.3435$$

$\therefore$ The curve of best fit is

$$y = (83.9146)\,(1.3435)^x.$$

Example 3: The population of a certain city at 10 years intervals is given below.

Year	x:	1931	1941	1951	1961	1971	1981	1991
Population: in lakhs	y:	3.9	5.3	7.3	9.6	12.9	17.1	23.1

Fit a curve of the form $y = ab^x$ and estimate the population of the city in the year 2006.

Solution: Here $n = 7$. Take 1931 as the base year and set $X_i = x_i - 1931$

X_i	y	$u_i = \log y_i$	X_i^2	$X_i u_i$
0	3.9	1.3610	0	0
10	5.3	1.6677	100	16.667
20	7.3	1.9879	400	39.758
30	9.6	2.2618	900	67.854
40	12.9	2.5572	1600	102.288
50	17.1	2.8391	2500	141.955
60	23.2	3.1442	3600	188.652
210	79.3	15.8189	9100	557.142

Let $y = ab^X$ .... (1) be the curve of best fit and $X = x_i - 1931$

Normal equations are

$$\Sigma u = nA + B \Sigma X_i$$
$$\Sigma X_i u = A \Sigma X_i + B \Sigma X_i^2$$

$\therefore$ $15.8189 = 7A + 210B$ Solving $A = 1.3745$

$557.184 = 210A + 9100B$ $B = 0.02951$

$\therefore$ $a = e^A = 3.9530$ and $b = e^A = 0.0299$

$\therefore$ The curve of best fit is

$$y = ab^X$$
$$= (3.9530)\,(0.0299)^x$$

For $x = 2006$ which corresponds to $X = 75$ we get $y = 36.02$

$\therefore$ Population in 2006 will be 36 lakhs.

4.4 EXPONENTIAL CURVE $y = ae^{bx}$

Let $y = ae^{bx}$ .... (1) be the curve of best fit

Taking $\log_{10}$ both sides of the equation (1)

$$\text{Log}^y = \log^a{}_{10} + bx \log^e{}_{10}$$
$$Y = A + Bx \qquad \ldots(2)$$

Where $Y = \log^y{}_{10}$ $A = \log^a{}_{10}$ $B = b \log^e{}_{10}$

The normal equations are

$$\Sigma y = nA + B \Sigma x$$
$$\Sigma xy = A \Sigma x + B \Sigma x^2$$

Solve for A and B these equations and calculate a and b.

Example 4 : A constant temperature the pressure p and the volume v of a gas are connected by the relation Pv^T = constant. Find the best fitting equation of this form to the following data and estimate v when $p = 4$.

P.kg sq cm:	0.5	1.00	1.5	2.0	2.5	3.0
V cc:	1620	1000	750	620	520	460

Solution: Consider $Pv^T = k$ where k is constant

Taking log both sides

$$\log P + T \log^v = \log^k$$
$$\log P = \log^k - T \log^v$$

Let $\log^P = Y$, $\log^k = a$, $-T = b$, $\log^v = x$

Then $y = a + bx$ which is the straight line.

Table for determining a and b.

P	v	$x = \log^v{}_e$	$y = \log^p$	xy	x^2
0.5	1620	7.39	–0.69	–5.0991	54.6121
1.0	1000	6.91	0	0	47.7481
1.5	750	6.62	0.40	2.648	43.8244
2.0	620	6.48	0.69	4.4367	41.3449
2.5	520	6.25	0.92	5.75	39.062
3.0	460	6.13	1.1	6.743	37.5769
		39.73	2.42	14.4786	264.1689

Substituting in the normal equations

$$6a + 39.73b = 2.42$$

$$39.73a + 264.1689b = 14.4789$$

Solving $a = 9.8061 \quad b = -1.42$

$$a = \log_e^k \quad \therefore \quad k = e^a = 18144$$

$$\text{And } -T = b \quad \therefore \quad T = -b = 1.42.$$

$$\therefore \quad Pv^T = k.$$

$$Pv^{1.42} = 18144$$

Put $P = 4$, $\quad 4v^{1.42} = 18144$

$$\therefore \quad v^{1.42} = 4536$$

$$\therefore \quad v = 375.94288 = 376$$

$$\therefore \quad \text{when } p = 4,\ v = 376.$$

Example 5 : Fit the curve of the form $y = ae^{bx}$ to the following data by least squares method.

x:	1	5	7	9	12
y:	10	15	12	15	21

Solution: Let $y = ae^{bx}$...(1) be the curve to be fitted.

Then $y = A + Bx$ where $Y = \log^y{}_{10}$, $A = \log^a{}_{10}$, and $B = b \log^e{}_{10}$

The normal equations are

$$\Sigma y = nA + B \Sigma x$$

$$\Sigma xy = A \Sigma x + B \Sigma x^2$$

We form the table as follows.

X	Y	$Y = \log^y_{10}$	x^2	xy
1	10	1.0000	1	1
5	15	1.1761	25	5.8805
7	12	1.0792	49	7.5544
9	15	1.1761	81	10.5849
12	21	1.3222	144	15.8664
$\Sigma x = 34$		$\Sigma y = 5.7536$	$\Sigma x^2 = 300$	$\Sigma xy = 40.8862$

Substituting in (2) we get

$$5.7536 = 5A + 34B \quad \text{Solving we get } A = 0.9766, \; B = 0.02561$$

$$40.8862 = 34A + 300B$$

$$\therefore \quad a = \text{anti}\log A = 9.4754$$

$$b = \frac{B}{\log^e_{10}} = 0.059.$$

Hence the required curve is $y = (9.4754)\, e^{0.059x}$.

Example 6 : By the method of least squares, obtain a relation of the form $y = ab^x$ for the following data.

X:	2	3	4	5	6
Y:	8.3	15.4	33.1	65.2	127.4

Solution: Let $y = ab^x$ ... (1) be the curve to be fitted.

$$\therefore\ Y = A + bx \qquad \text{...(2)}$$

Where $A = \log^a_{10}$, $B = \log^b_{10}$ and $Y = \log^y_{10}$

The normal equations are

$$\Sigma Y = nA + B\,\Sigma x$$

$$\Sigma xY = A\,\Sigma x + B\,\Sigma x^2 \qquad \text{.....(3)}$$

To determine A and B we form the table as follows.

X	y	$Y = \log^y_{10}$	x^2	xY
2	8.3	0.9191	4	1.8382
3	15.4	1.1872	9	3.5616
4	33.1	1.5198	16	6.0792
5	65.2	1.8142	25	9.0710
6	127.4	2.1052	36	12.6312
20		7.5444	90	33.1812

Substituting these values in (3)

$$7.5455 = 5A + 20B$$

$$33.1812 = 20A + 90B$$

Solving we get $A = 0.31$ and $B = 0.3$

$\therefore \quad a = \text{antilog } A = 2.04$ and $b = \text{antilog } B = 1.995$

Hence the required curve is

$$y = (2.04)\,(1.1995)^x.$$

4.5 QUESTION BANK

1. The results of measurements of electric resistance R of a copper bar at various temperatures θ are listed below

θ	19	25	30	36	40	45	50
R:	76	77	79	80	82	83	85

Fit a relation of the form $R = a + b\theta$ that fits the data.

Ans.: $R = 70.052 + (.292)\,\theta$.

2. Fit a straight line of the form $y = a + bx$ to the following data by the method of least squares.

x:	0	5	10	15	20	25
y:	12	15	17	22	24	30

Ans.: $y = 11.1 + 0.71x$

3. The weights of a calf taken at weekly intervals are given below:

Age :	1	2	3	4	5	6	7	8	9	10
Weight :	52.5	58.2	55.0	70.2	75.4	81.1	87.2	95.5	102.2	108.4

Fit a straight line of best fit.

Ans.: $y = 46.05 + 6.1x$

4. A simply supported beam carries a concentrated load P (lb) at its mid point. Corresponding to various values of P the maximum deflection Y (in) is measured. The data is given below. Find a law of the type $Y = a + bp$.

P:	100	120	140	160	180	200
Y:	0.45	0.55	0.60	0.70	0.80	0.85

Ans.: $Y = 0.048 + 0.0041P$.

5. Using the method of averages, fit a parabola $y = ax^2 + bx + c$ to the following data

x:	20	40	60	80	100	120
y:	5.5	9.1	14.9	22.8	33.3	46.0

Ans.: $y^2 = 4.18 + 0.01x + 0.0028x^2$.

6. By the method of least squares fit a second degree parabola $y = a + bx + cx^2$ to the following data.

x:	1	2	3	4	5	6	7	8	9
y:	2	6	7	8	10	11	11	10	9

Ans.: $y = -1 + 3.55x - 0.27x^2$

7. Fit an equation of the form $y = ae^{bx}$ to the following data by the method of least squares

x:	1	2	3	4
y:	1.65	2.7	4.5	7.35

Ans.: $y = e^{0.5x}$

8. Fit a curve of the form $y = ax^b$ to the following data

x:	1	2	3	4	5
y:	7.1	27.8	62.1	110	161

Ans.: $y = 7.173\, x^{1.952}$

9. Fit a curve of the form $y = ab^x$ in least square sense to the data given below.

x :	2	3	4	5	6
y :	144	172.8	207.4	248.8	298.5

Ans.: $y = 99.86\, (1.2)^x$.

10. The voltage v cross a capacitor at time t seconds is given by the following table. Use the principle of least squares to fit a curve of the form $v = ae^{kt}$ to the data.

t :	0	2	4	6	8
V :	150	63	28	12	5.6

Ans.: $v = 146.3^{-0.4118t}$

4.6 OPTIMIZATION

Introduction

Linear programming deals with the optimization (maximization or minimization) of linear functions subject to linear constraints. This technique has found its applications to important areas. The followings are some of the linear programming problems.

1) Production allocation is a linear programming problem.
2) Production planning problem.
3) Advertising media selection problem.
4) Product mix problem of a chemical industry.
5) Blending problem of allow in production engineering.
6) Quality – control inspection problem.
7) Fluid blending problem of oil company.
8) Diet – constituent requirement problem.
9) War – strategy problem where the strategy bomber has to interrupt the enemy tank production.
10) Production – scheduling problem.
11) Warehouse problem – selling identical items.
12) Trim – loss problem in paper mill production.
13) Caterer problem in organizing parties.
14) Transportation problem.

4.7 FORMULATION OF LPP

To formulate an LPP it is necessary to specify i) decision variables ii) objective function and iii) constraints. The decision variables are the variables for which a decision is required to be taken. Using these decision variables the decision maker shall then formulate the objective function and the constraints involved in making the decision. We illustrate this through the following examples.

Example 1: A manufacturer produces two types of models M_1 and M_2. Each M_1 Model requires 4 hours of grinding and 2 hours of polishing where as each M_2 model requires 2 hours of grinding and 5 hours of polishing. The manufacturer has 2 grinders and 3 polishers. Each grinder works for 40 hours a week and each polisher works for 60 hours a week. Profit on an M_1 model is Rs 3 and on an M_2 model is Rs 4. Whatever is produced in a week 13 sold in the market. How should the manufacturer allocate his production capacity to the two types of models, so that he may make the maximum profit in a week.

Formation : Let x_1 be the number of M_1 models and x_2 the number of M_2 models produced per week.

Then the weekly profit in Rs is

$$Z = 3x_1 + 4x_2 \qquad \dots \text{(i)}$$

To produce these number of models, the total number of grinding hours needed per week $= 4x_1 + 2x_2$ and the total number of polishing hours required per week $= 2x_1 + 5x_1$.

Since the number of grinding hours available is not more than 80 hours and the number of polishing hours is not more than 180 therefore.

$$4x_1 + 2x_2 \leq 80 \qquad \dots \text{(ii)}$$

$$2x_1 + 5x_2 \leq 180 \qquad \dots \text{(iii)}$$

Also, since the negative number of models are not produced obviously we must have $x_1 \geq 0$ and $x_2 \geq 0$. Hence this allocation problem is to find x_1, x_2 which maximize $Z = 3x_1 + 4x_2$

Subject to $4x_1 + 2x_1 \leq 80$

$2x_1 + 5x_2 \leq 180$

And $x_{1,}\ x_2 \geq 0.$

***Note* :** The variables that enter into the problem are called **decision variables.** The expression (1) showing the relationship between the manufacturers goal and the decision is called the **objective function.** The inequalities (ii), (iii), and (iv) are called the **constraints.**

The objective function and the constraints being all linear, it is a linear programming problem (LPP). This is an example of a real situation from industry.

Example 2: The Ashok Fasteners, a small scale industry in Belgaum, produces two types of screw namely, male (or external) screw called product A and female (or internal) screw called product B. The profit per unit from the product A is Rs 10 and from the product B is Rs 20.

The available machine hours are 20 and the labour hours are 30. If the products *A* and *B* respectively require 5 and 2 machine hours and 3 and 4 labour hours for the production of each unit formulate the LPP to decide how many units of each product shall be produced in order to have the maximum profit.

Formation : Let x_1 and x_2 be the number of *A* and *B* to be produced.

$\therefore$ x_1 and x_2 are decision variables. The objective of the decision maker is to increase the profit.

Since the profits Rs 10 and Rs 20 for each unit of products *A* and *B* respectively, the objective is to maximize the profit function.

$$z = 10x_1 + 20x_2$$

The constraints of the problem are the available labour and machine hours.

The constraint with regard to machine hours is

$$5x_1 + 2x_1 \leq 20$$

Similarly, the constraint connected to labour hours is

$$3x_1 + 4x_2 \leq 30$$

In addition to these constraints we also have another constraint that the decision variables shall be non – negative

Thus we have the LPP is

Maximize $\quad z = 10x_1 + 20x_2 \qquad \ldots$(i)

Subject to the constraints

$$5x_1 + 2x_2 \leq 20 \qquad \ldots\text{(ii)}$$

$$3x_1 + 4x_2 \leq 30 \qquad \ldots\text{(iii)}$$

And $\quad x_1 \geq 0,\ x_2 \geq 0 \qquad \ldots$ (iv)

Example 3: A firm making castings uses electric furnace to melt iron with the following specifications :

	Minimum	Maximum
Carbon	3.20%	3.40%
Silicon	2.25%	2.35%

Specifications and costs of various raw materials used for this purpose are given below :

Materials	Carbon %	Silicon %	Costs (Rs)
Steel scrap	0.4	0.15	850/ton
Cast iron scrap	3.80	2.40	900/ton
Remelt from foundry	3.50	2.30	500/ton

If the total charge of iron metal required is 4 tons find the weight in kg of each raw material that must be used in the optimal mix at minimum cost.

Formation

Let x_1, x_2, x_3 be the amounts (in kg) of these raw materials.

The objective is to minimize the cost.

That is minimize $Z = \dfrac{850}{1000}x_1 + \dfrac{900}{1000}x_2 + \dfrac{500}{1000}x_3$ ....(i)

For iron melt to have a minimum of 3.2% carbon

$$0.4x_1 + 3.80\,x_2 + 3.5\,x_3 \geq 3.2 \times 4{,}000 \quad \text{...(ii)}$$

For iron melt to have a maximum of 3.4% carbon

$$0.4\,x_1 + 3.8x_2 + 3.5\,x_3 \leq 3.4 \times 4000 \quad \text{...(iii)}$$

For iron melt to have a minimum of 2.25% silicon,

$$0.15x_1 + 2.41x_2 + 2.35x_3 \geq 2.25 \times 4000 \quad \text{...(iv)}$$

For iron melt to have a maximum of 2.35% silicon

$$0.15x_1 + 2.41x_2 + 2.35x_3 \leq 2.35 \times 4000 \quad \text{... (v)}$$

Also, since the materials added up must be equal to the full charge weight of 4 tons.

$$\therefore \quad x_1 + x_2 + x_3 = 4{,}000 \quad \text{...(vi)}$$

In addition to these we have another constraint that the decision variables shall be non-negative.

That is $x_1 \geq 0,\ x_2 \geq 0,\ x_3 \geq 0$ ... (vii)

Thus the LPP is to find $x_1,\ x_2\, x_3$ which minimize

$$z = 0.85x_1 + 0.9x_2 + 0.5x_3$$

Subject to the constraints

$$0.4x_1 + 3.8x_2 + 3.5x_3 \geq 12{,}800$$
$$0.4x_1 + 3.8x_2 + 3.5x_3 \leq 13{,}600$$
$$0.15x_1 + 2.41\,x_2 + 2.35x_3 \geq 9{,}000$$

$$0.15x_1 + 2.41x_2 + 2.35\, x_3 \leq 9{,}400$$

And $x_1 + x_2 + x_3 = 4{,}000$

And $x_1 \geq 0, x_2 \geq 0, x_3 \geq 0.$

Example 4: A company has three operational departments (weaving, processing and packing) with capacity to produce three different types of clothes namely, suitings, shirtings and woolens yielding a profit of Rs 2, Rs 4 and Rs 3 per meter respectively. One meter suiting requires 3 minutes in weaving, 2 minutes in processing and 1 minute in packing. Similarly one meter of shirting requires 4 minutes in weaving, 3 minutes in each department. In a week total run time of each department is 60, 40 and 80 hours weaving, processing and packing departments respectively.

Formulate the linear programming problem to find the product mix to maximize the profit.

Formulation:

The data of the problem is summarized below

	Departments Weaving in minutes	 Processing in minutes	 Packing in minutes	Profit Rs per meter
Suitings	3	2	1	2
Shirtings	4	1	3	4
Woolens	3	3	3	3
Availability (min)	60 × 60	40 × 60	80 × 60	

Let x_1, x_2, x_3 be the weekly production of suitings, shirtings and woolens.

$\therefore\ x_1, x_2, x_3$ are decision variables.

The objective of the decision maker is to increase the profit. Since the profits are Rs 2, Rs 4 and Rs 3 for each units of the products, the objective is maximize the profit function.

$$Z = 2x_1 + 4x_2 + 3x_3.$$

The constraints of the problem are, the available weaving time.

$$3x + 4x_2 + 3x_3 \leq 3600$$

The available processing time

$$2x_1 + x_2 + 3x_3 \leq 2400$$

The available packing time

$$x_1 + 3x_2 + 3x_3 \leq 4800$$

In addition to these constraints we also have another constraint that the decision variables shall be non-negative.

Thus the linear programming problem can be put in the algebraic format as,

Maximize $Z = 2x_1 + 4x_2 + 3x_3$

Subject to the constraints

$$3x_1 + 4x_2 + 3x_3 \leq 3600$$

$$2x_1 + x_2 + 3x_3 \leq 2400$$

$$x_1 + 3x_2 + 3x_3 \leq 4800$$

And $\quad x_1 \geq 0, x_2 \geq 0, x_3 \geq 0.$

4.8 GRAPHICAL METHOD

Linear programming problems involving only two variables can be effectively solved by a graphical technique which provides a pictorial representation of the solution and one gets insight into the basic concepts used in solving large linear programming problems.

Working Procedure

Working procedure to solve a linear programming problem graphically involves the following steps.

Step 1 : Formulate the given problem as a linear programming problem.

Step 2: Plot the given constraints as equalities on $x_1 - x_2$ co-ordinate plane and determine the convex region formed by them.

Step 3: Determine the vertices of the convex region and find the value of the objective function at each vertex. The vertex which gives the optimal (maximum or minimum) value of the objective function gives the desired optimal solution to the problem.

Note : A region or a set of points is said to be **convex** if the line joining any two of its points lies completely in the region.

Example : A company makes two kinds of leather belts. Belt A is high quality and belt B is of lower quality. The respective profits are Rs 4.00

and Rs. 3.00 per belt. Each belt of type A requires twice as much time as a belt of type B, and if all belts were of type B, the company could make 1000 per day. The supply of leather is sufficient for only 800 belts per day (both A and B combined). Belt A requires a fancy buckle and only 400 per day are available. There are only 700 buckles a day available for belt B. Determine the optimal product mix.

Solution: The mathematical formulation of the given LPP is

Maximize $z = 4x_1 + 3x_2$ ...(i)

Subject to the constraints

$2x_1 + x_2 \leq 1000$ ...(ii)

$x_1 + x_2 \leq 800$ ... (iii)

$x_1 \leq 400$ and $x_2 \leq 700$ ... (iv)

And $x_1 \geq 0$ and $x_2 \geq 0$ ... (v)

Where x_1= number of belts of type A and x_2 = number of belts of type B.

Consider $x_1 - x_2$ co-ordinate system. The non–negativity restrictions imply that the values of $x_1 - x_2$ lie in the first quadrant only.

Now $2x_1 + x_2 = 1000$ and $x_1 + x_2 = 800$.

x_1	0	500
x_2	1000	0

x_1	0	800
x_2	800	0

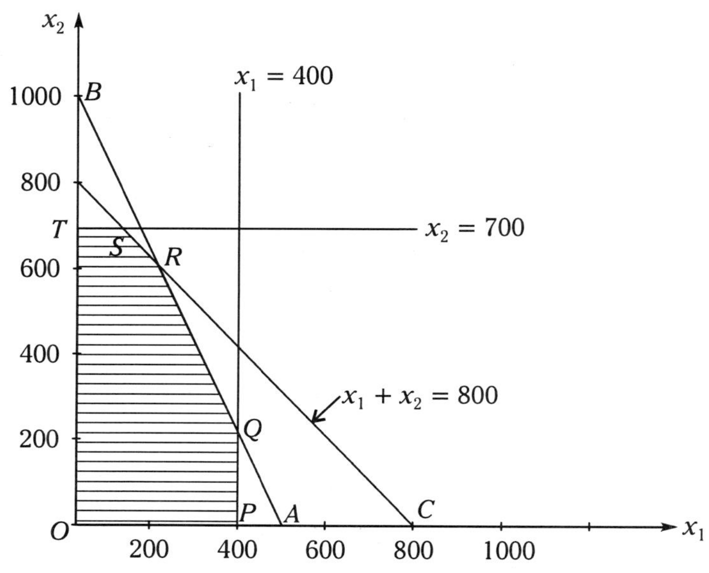

The optimum value of the function occurs at one of the extreme (corner) points of the feasible region. The co-ordinates of the extreme points are

$O(0,0)$, P (400,0) Q (400 200), R (200, 600), S (100, 700) and T (0,700)

Now we compute the Z value corresponding to the extreme points.

Extreme points	(x_1, x_2)	$Z = 4x_1 + 3x_2$	
O	(0,0)	0	
P	(400,0)	1600	
Q	(400 200)	2200	
R	(200,600)	2600	maximum value,
S	(100,700)	2500	
T	(0,700)	2100	

The optimum solution occurs at the point R that is, $x_1 = 200$ and $x_2 = 600$ with the objective function value of Rs 2600.

Hence to maximize profit, the company should produce 200 belts of type A and 600 belts of type B per day.

Example 2 : Solve the LPP graphically, maximize $Z = 3x_1 + 4x_2.$

Subject to $4x_1 + 2x_2 \le 80$

$2x_1 + 5x_2 \le 180$

And $x_1 \ge 0, x_2 \ge 0$

Solution: The mathematical formulation of the given LPP is

Maximize, $Z = 3x_1 + 4x_2$ ...(i)

Subject to $4x_1 + 2x_2 \le 80$ ...(ii)

$2x_1 + 5x_2 \le 180$ ...(iii)

And $x_1 \ge 0$, $x_2 \ge 0$.

Consider $x_1 - x_2$ co-ordinate system. The non-negativity restrictions imply that the values of $x_1 - x_2$ lie in the first quadrant only.

Now $4x_1 + 2x_2 = 80$ and $2x_1 + 5x_2 = 180$

x_1	0	20
x_2	40	0

x_1	0	90
x_2	36	0

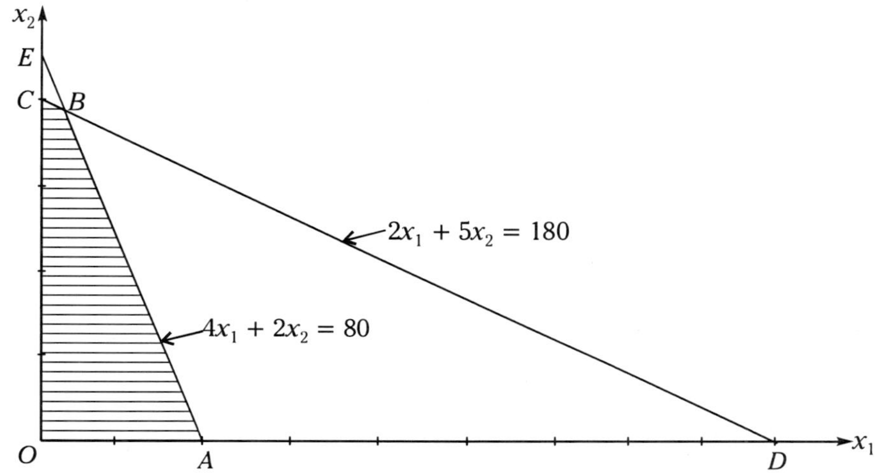

The optimum value of the function occurs at one of the extreme (corner) points of the feasible region. The co-ordinates of the extreme points are

$O(0.0)$ $A\,(20.0)$ $B\,(2.5, 35)$ $C\,(0, 36)$

The value of the objective function Z corresponding to the extreme points computed as

Extreme points	$(x_1\ x_2)$	$Z = 3x_1 + 4x_2$	
O	(0 ,0)	0	
A	(20,0)	60	
B	(2.5,35)	147.5	maximum value.
C	(0.36)	144	

Thus the maximum value of Z is 147.5 and it occurs at B. Hence the optimal solution to the problem is

$$x_1 = 2.5,\quad x_2 = 35 \text{ and } z_{\max} = 147.5.$$

Example 3: Using graphical method, solve the LPP

Maximize $z = 5x_1 + 7x_2$

Subject to constraints

$$x_1 + x_2 \le 4\ ;\quad 3x_1 + 8x_2 \le 24; 10x_1 + 7x_2 \le 35$$

and $x_1, x_2, \geq 0$.

Solution: The mathematical formulation of the given LPP is

maximize $z = 5x_1 + 7x_2$ ... (i)

Subject to constraints

$x_1 + x_2 \leq +4$ ... (ii)

$3x_1 + 8x_2 \leq 24$ ... (iii)

$10x_1 + 7x_2 \leq 35$ ...(iv)

And $x_1 \geq 0, x_2 \geq 0$ ...(v)

Now $x_1 + x_2 = 4$ $3x_1 + 8x_2 = 24$ $10x_1 + 7x_2 = 35$

x_1	0	4
x_2	4	0

x_1	0	8
x_2	3	0

x_1	0	3.5
x_2	5	0

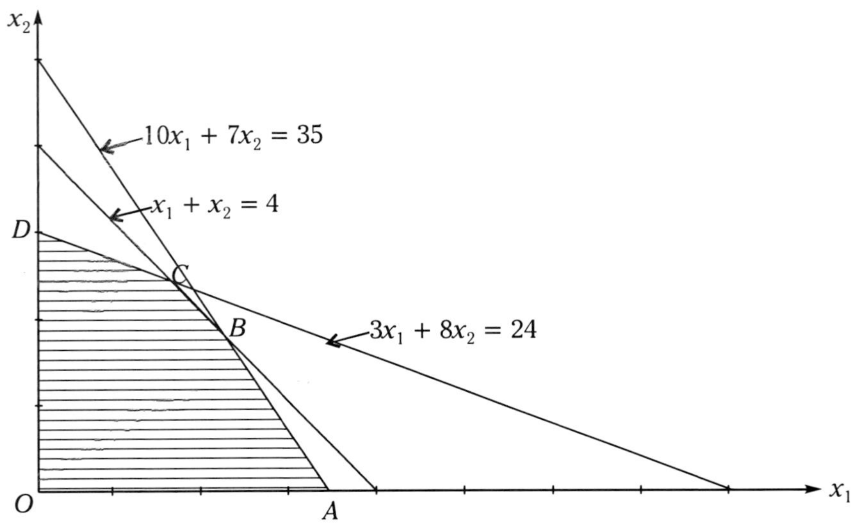

The desired points (x_1, x_2) lies within the convex region $O\ A\ B\ C\ D$. Its vertices are O (0.0), A (3.5,0) $B\left(\frac{5}{3}, \frac{7}{3}\right)$ $C\left(\frac{8}{5}, \frac{12}{5}\right)$ and D (0.3)

Now, we compute the Z value corresponding to the extreme points.

Extreme points	$(x_{1,} y_2)$	$z = 5x_1 + 7x_2$	
O	(0,0)	0	
A	(3.5, 0)	17.5	
B	$\left(\frac{5}{3}, \frac{8}{3}\right)$	24.6	
C	$\left(\frac{8}{5}, \frac{12}{5}\right)$	$\boxed{24.8}$	maximum value.
D	(0.3)	21.	

The maximum value of Z is 24.8 and it occurs at D. Hence the optimal solution to the problem is $x_1 = \frac{8}{5}, x_2 = \frac{12}{5}$ and z max = 24.8.

Example 4 : A company making cold drinks has two bottling plants located at towns B_1 and $B_{2.}$ Each plant produces three drinks $D_{1,}$ D_2 and D_3 and their production capacity per day is shown below.

	Cold drinks A	Plant at B
D_1	6000	2000
D_2	1000	2500
D_3	3000	3000.

The marketing department of the company forecasts a demand of 80,000 bottles of D_1 22,000 bottles of D_2 and 40,000 bottles of D_3 during the month of June. The operating costs per day of plants at B_1 and B_2 are Rs 6000 and Rs 4000 respectively. Find graphically the number of days for which each plant must be run in June. So as to minimize the operating costs while meeting the market demand.

Solution: Let the plants B_1 and B_2 be run for x_1 and x_2 days. Then the objective is to minimize the operation costs.

i.e., minimize $Z = 6000x_1 + 4000x_2$ ...(i)

Constraints on the demand for the three cold drinks are

For $D_{1,}$ $6000x_1 + 2000x_2 \geq 80{,}000$ or $3x_1 + x_2 \geq 40$ ...(ii)

For D_2 $1000x_1 + 2500x_2 \geq 22{,}000$ or $x_1 + 2.5x_2 \geq 22$ ...(iii)

For D_3 $\quad 3000x_1 + 3000x_2 \geq 40{,}000$ or $x_1 + x_2 \geq \dfrac{40}{3}$ ...(iv)

Also $x_1 \geq 0,\ x_2 \geq 0$ ....(v)

Thus the LPP is to minimize (i) subject to constraints (ii) to (v).

Now,

$3x_1 + x_2 = 40$ $\qquad$ $x_1 + 2.5x_2 = 22$ $\qquad$ $x_1 + x_2 = \dfrac{40}{3} = 13.3$

x_1	0	$\frac{40}{3}$
x_2	40	0

x_1	0	22
x_2	8.8	0

x_1	0	$\frac{40}{3}$
x_2	$\frac{40}{3}$	0

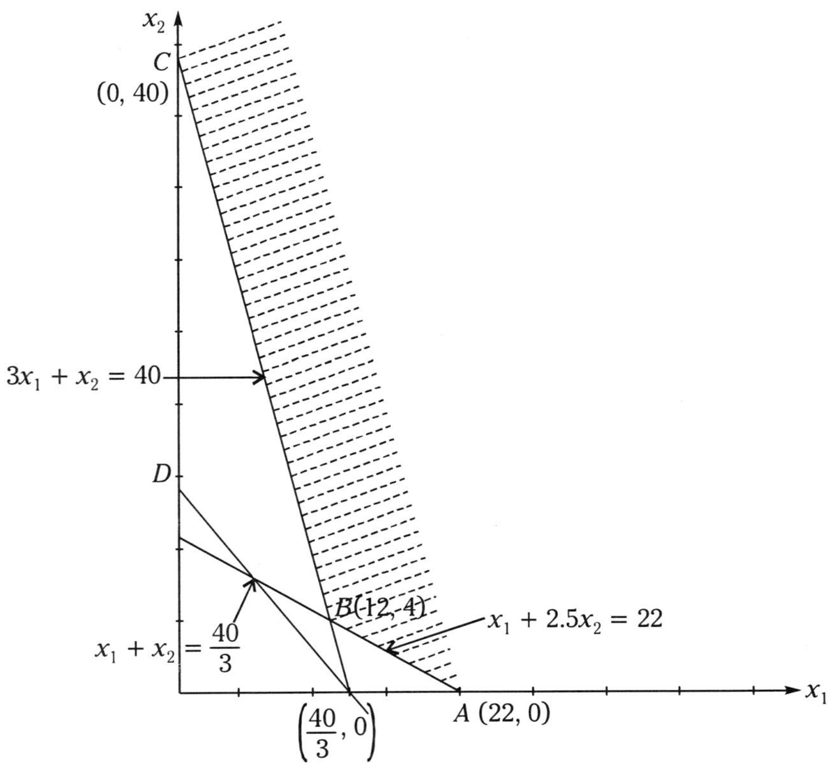

The vertices of the convex region $A\ B\ C$ are A (22, 0), B (12, 4) and C (40,0).

The value of the objective function (i) at these points are

Extreme points	(x_1, x_2)	$Z = 6000x_1 + 4000x_2$	
A	(22,0)	$Z = 132{,}000$	
B	(12,4)	$\boxed{Z = 88{,}000}$	minimum
C	(0,40)	$Z = 160{,}000.$	

Thus the minimum value of Z is Rs. 88,000 and it occurs at B. Hence the solution to the problem is $x_1 = 12$ days, $x_2 = 4$ days and Z min = 88,000.

Example 5: Find the maximum value of $Z = 2x + 3y$ subject to the

Constraints : $x + y \le 30 \quad y \ge 3.$

$O \le y \le 12, \; x - y \ge O$ and $O \le x \le 20.$

Solution: The mathematical formulation of the given LPP is

Maximize $Z = 2x + 3y.$

Subject to the constraints

$x + y \le 30 \qquad y \ge 3$

$O \le y \le 12, \qquad x - y \ge O$ and $O \le x \le 20.$

Now $\quad x + y = 30$

x	0	30
y	30	0

$y = 3, \; y = 12, x = 20$ and $x = y.$

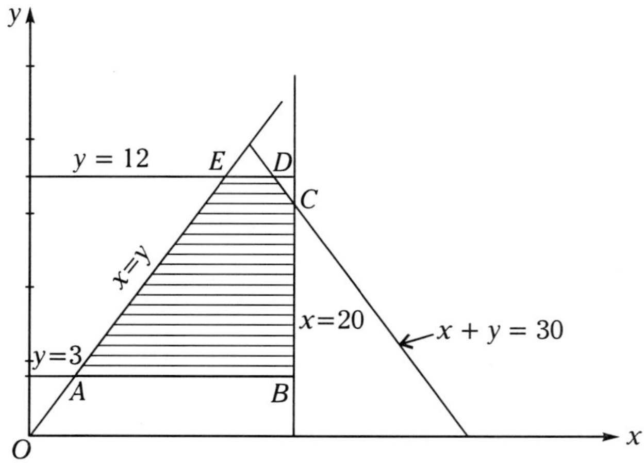

The optimal value of the function occurs at one of the extreme points of

the feasible region. The co-ordinate of the extreme points are $A(3,3)$, $B(20,3)$ $C(20,10)$ $D(18,12)$ and $E(12,12)$.

The value of the objective function Z corresponding to the extreme points computed as follows.

Extreme points	$(x\ y)$	$Z = 2x + 3y$	
A	(3,3)	15	
B	(20,3)	49	
C	(20,10)	70	
D	(18,12)	72	Maximum value
E	(12,12)	60	

Thus the maximum value of Z is 72 and it occurs at D. Hence the optimal solution to the problem is $x = 18$, $y = 12$ and Z max = 72.

Exceptional cases

The constraints generally give region of feasible solution which may be bounded or unbounded. In problems involving two variables and having a finite solution we observed that the optimal solution existed at a vertex of the feasible region. In fact this is true for all LPPs for which solution exists. Thus it may be stated as , "If there exists an optimal solution of an LPP, it will be at one of the vertices of the solution space".

LPP may have

i) a unique optimal solution

ii) an infinite number of optimal solutions

iii) an unbounded solution.

iv) no solution.

Example 6: Use graphical method to solve the LPP

Maximize $Z = 2x_1 + 4x_2$ subject to the constraints
$x_1 + 2x_2 \leq 5$, $x_1 + x_2 \leq 4$, … $x_1, x_2 \geq 0$.

Solution: The mathematical formulation of the given LPP is

Maximize $\quad Z = 2x_1 + 4x_2$

Subject to the constraints

$$x_1 + 2x_2 \leq 5,\ x_1 + x_2 \leq 4 \text{ and } x_1, x_2 \geq 0$$

$x_1 + 2x_2 = 5$

x_1	0	5
x_2	$5/2$	0

$x_1 + x_2 = 4$

x_1	0	4
x_2	4	0

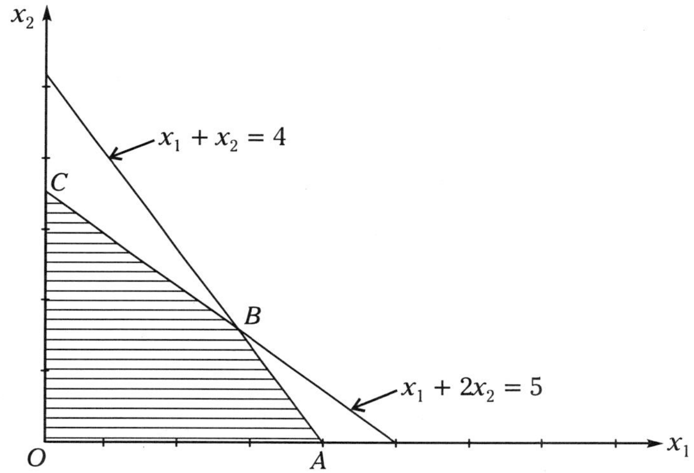

The optimal value of the function occurs at one of the extreme points of the feasible region. The co-ordinates of the extreme points are $O(0,0)$ $A(4,0)$ $B(3.1)$ and $C(0,2.5)$.

The value of the objective function Z corresponding to the extreme points computed as follows,

Extreme points	$(x_1\, x_2)$	$Z = 2x_1 + 4x_2$	
O	(0,0)	$Z = 0$	
A	(4,0)	$Z = 8$	
B	(3,1)	$\boxed{z = 10}$	←maximum
C	(0,2.5)	$\boxed{z = 10}$	←maximum.

Thus the maximum value of Z (= 10) occurs at two vertices, B and C. Therefore, any point on the line joining B and C will also give the same maximum value of Z. i.e., there are infinite number of feasible solutions which yield the same maximum value of Z. Thus there is no unique optimal solution to the problem.

Obs : An LPP having more than one optimal solution is said to have an alternative or multiple optimal solutions.

Example 7: A firm uses milling machines, grinding machines and lathes to produce two motor parts. The machining time required for each part, the machining times available on different machines and the profit on each motor part are given below.

Types of Machine	Machining time reqd. for the Motor Part (mtrs)		Maxi. time available Per week (minutes)
	I	II	
Milling Machines	10	4	2000
Grinding Machines	3	2	900
Lathes	6	12	3000
Profit / unit (Rs.)	100	40	

Determine the number of parts I and II to be manufactured per week to maximize the profit.

Solution: Let x_1 , x_2 be the number of parts I and II manufactured per week. Then objective being to maximize the profit. We have to maximize

$$Z = 100x_1 + 40x_2 \qquad \ldots \text{(i)}$$

Constraints being on the time available on each machine, we get

For milling machines $\quad 10x_1 + 4x_2 \le 2{,}000 \qquad \ldots$ (ii)

For grinding machines $\quad 3x_1 + 2x_2 \le 900 \qquad \ldots$ (iii)

For lathes $\quad 6x_1 + 12x_2 \le 3{,}000 \qquad \ldots$ (iv)

Also $\quad x_1 , x_2 \ge 0 \qquad \ldots$ (v)

Now,

$10x_1 + 4x_2 = 2000$

x_1	0	200
x_2	500	0

$3x_1 + 2x_2 = 900$

x_1	0	300
x_2	450	0

$6x_1 + 12x_2 = 3000$

x_1	0	500
x_2	250	0

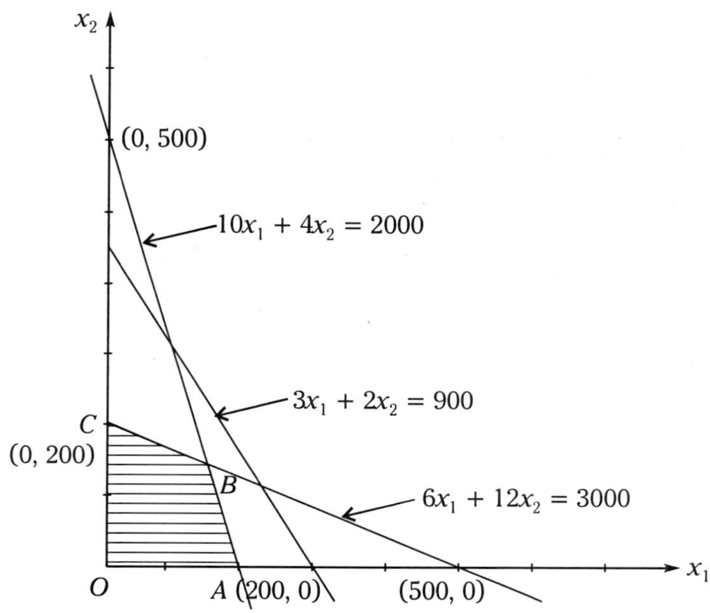

The constraint (iii) is a redundant so it does not affect the solution space. The vertices of the convex region $OABC$ are O (0,0) A (200,0) B (125, 187.5) and C (0,250).

We compute the Z – value corresponding to the extreme points.

Extreme points	$(x_1\ x_2)$	$Z = 100x_1 + 40x_2$
O	(0,0)	$Z = 0$
A	(200,0)	$Z = 20{,}000$ ←maximum
B	(125, 187.5)	$Z = 20{,}000$ ←maximum
C	(0,250)	$Z = 10{,}000$.

The maximum value of Z occurs at two points A and B. ∴ Any point on the line joining A and B will also give the same maximum value of Z. i.e., there are infinite number of feasible solutions which yield the same maximum value of Z.

Obs : An LPP having more than one optimal solution is said to have alternative or multiple optimal solution. It implies that the resources can be combined in more than one way to maximize the profit.

Example 8: Using graphical method solve the following LPP maximize $Z = 2x_1 + 3x_2$

Subject to the constraints

$x_1 - x_2 \le 2 : x_1 + x_2 \ge 4$ and $x_1, x_2 \ge 0$.

Solution: Consider the co-ordinate system $x_1 - x_2$. Any point (x_1, x_2) satisfying $x_1, x_2 \ge 0$ lies in the first quadrant.

The mathematical formulation of the given LPP is

Maximize $\quad Z = 2x_1 + 3x_2 \quad$ ...(i)

Subject to $\quad x_1 - x_2 \le 2 \quad$ ... (ii)

$\quad x_1 + x_2 \ge 4 \quad$ ... (iii)

And $\quad x_1, x_2 \ge 0 \quad$ ... (iv)

Now,

$x_1 - x_2 = 2$

x_1	0	2
x_2	–2	0

$x_1 + x_2 = 4$

x_1	0	4
x_2	4	0

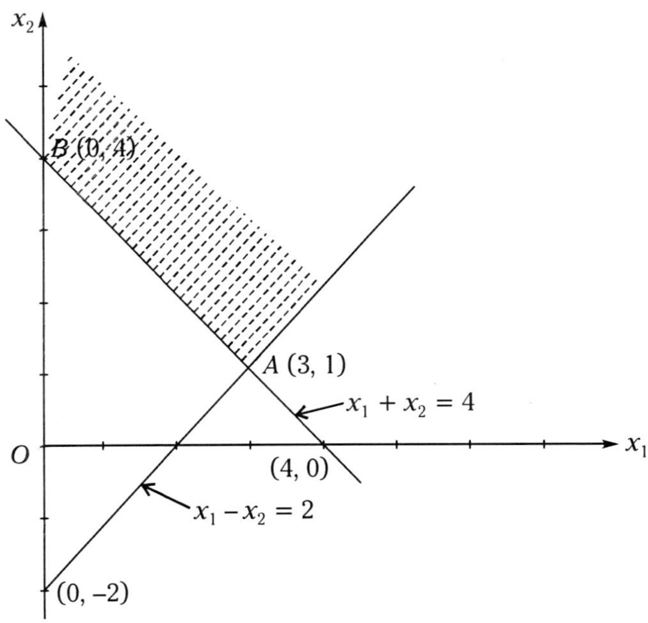

The optimum value of the function occurs at one of the extreme points of the feasible region. **But** here solution space is unbounded. The vertices of the feasible region are A (3, 1) and B (0, 4).

The value of the objective function (i) at these points are z_{at} $A = 9$ and Z_{at} $B = 12$. But there are points in this region for which Z will have much higher values. For instance, the point (5.5) lies in the shaded region and the value of Z is 25. In fact the maximum value of Z occurs at infinity.

Thus the problem has an unbounded solution.

Example 9: Solve graphically the following LPP

Maximize $z = 4x_1 + 3x_2$

Subject to the constraints

$$x_1 - x_2 \leq -1 \; ; \quad -x_1 + x_2 \leq 0 \; \text{ and } \; x_{1,} x_2 \geq 0.$$

Solution: The mathematical formulation of the given LPP is

Maximize $\quad Z = 4x_1 + 3x_2$ ...(i)

Subject to $\quad x_1 - x_2 \leq -1$ ... (ii)

$\quad -x_1 + x_2 \leq 0$ ... (iii)

And $\quad x_1 , x_2 \geq 0$ ... (iv)

Now $\quad x_1 - x_2 = -1 \qquad -x_1 + x_2 = 0$

$\therefore \quad x_2 = x_1 + 1 \qquad x_2 = x_1$

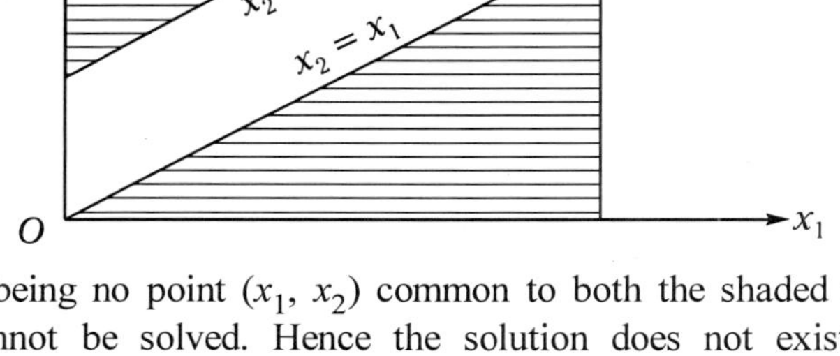

There being no point (x_1, x_2) common to both the shaded regions; the problem cannot be solved. Hence the solution does not exist, since the constraints are inconsistent.

4.9 GENERAL LINEAR PROGRAMMING PROBLEM

This is a problem of optimization of a linear function Z (profit, sales, costs etc) subject to a set of linear constraints. Here, function Z is called **objective function.**

Thus the general LPP is

Optimize $Z = c_1 x_1 + c_2x_2 + c_3x_3 + \ldots\ldots + c_nx_n$ .... (i)

Subject to the constraints (or restrictions)

$$a_{11} x_1 + a_{12} x_2 + \ldots\ldots\ldots + a_{1n} x_n \le b_1 \text{ (or } \ge b_1)$$
$$a_{21} x_1 + a_{22} x_2 + \ldots\ldots\ a_{2n} x_n \le b_2 \text{ (or } \ge b_2)$$
$$\ldots\ldots\ldots\ldots\ldots\ldots\ldots\ldots\ldots\ldots\ldots\ldots\ldots\ldots$$
$$\ldots\ldots\ldots\ldots\ldots\ldots\ldots\ldots\ldots\ldots\ldots\ldots\ldots\ldots$$
$$am_1 x_1 + am_2 x_2 + \ldots\ldots + a_{mn} x_n \le b_m \text{ (or } \ge b_m)$$

and non-negative constraints

$$x_j \ge 0 \text{ where } j = 1,2,3 \ldots\ldots\ldots\ldots n$$

where aij, bj and cj are all constants.

Let LPP be represented in the matrix form as

Optimize $Z = CX$

Subject to the constraint

$$AX \le B \text{ or } (=, \ge B)$$

And $X \ge O$

Here $C = C_1, C_2, C_3 \ldots\ldots\ldots\ldots\ldots C_n.$

$$X = \begin{bmatrix} x_1 \\ x_2 \\ \vdots \\ x_n \end{bmatrix} \quad B = \begin{bmatrix} b_1 \\ b_2 \\ \vdots \\ b_m \end{bmatrix} \quad \text{and } A = \begin{bmatrix} a_{11} & a_{12} & \cdots & a_{1n} \\ a_{21} & a_{22} & \cdots & a_{2n} \\ \vdots & & & \\ a_{m1} & a_{m2} & \cdots & a_{mn} \end{bmatrix}$$

The function $Z = CX$ which is to be optimized is called **objective function.**

The variables $x_1, x_2 \ldots\ldots x_n$ whose values are to be determined are called **decision variables.**

1) Feasible solution : A set of real variables

$X = (x_1, x_2 \ldots\ldots x_n)$ which

i) Satisfies the constraints $AX \le B$ (or $\ge B$)

ii) Satisfies the non-negativity constraints $X \ge O$ is called feasible solution.

2) Optimal solution : A set of real values

$X = (x_1, x_2, \ldots\ldots\ldots x_n)$ which

i) Satisfies the constraints $AX \leq B$ (or $= \geq B$)

ii) Satisfies the non-negativity restriction $X \geq O$ and

iii) Optimizes the objective function $Z = CX$ is called optimal solution.

Note:

a) If an LPP has many (two or more) optimal solutions, it is said to have multiple solutions.

b) If an LPP has one optimal solution, it is said to have unique solution.

c) For some LPP the optimal value of Z may be infinity. In this case the LPP is said to have unbounded solution.

d) There may be case where the LPP may not have any feasible solution at all.

Slack variables

If the constraints of a general LPP be

$$\sum_{j=1}^{n} a_{ij} x_i \leq b_i \, (i = 1,2,3, \,\ldots\ldots\ldots\, k)$$

Then the non-negative variable Si which satisfy

$$\sum_{j=1}^{n} a_{ij} x_i + S_i = b_i \, (i = 1,2,3, \,\ldots\ldots\ldots k)$$

are called slack variables.

Surplus variables

If the constraints of a general LPP be

$$\sum_{j=1}^{n} a_{ij} x_i \geq b_i \qquad i = k,\, k+1 \,\ldots\ldots$$

Then the non-negative variable Si which satisfies

$$\sum_{j=1}^{n} a_{ij} x_i - S_i = b_i \qquad i = k,\, k+1 \,\ldots\ldots\ldots$$

are called surplus variables.

4.10 CANONICAL AND STANDARD FORMS OF LPP

After the formulation of LPP the next step is to obtain its solution. But before any method is used to find its solution, the problem must be presented in a suitable form. As such there are two forms.

Canonical form

Maximize $Z = c_1 x_1 + \ldots x_2 + \ldots\ldots\ldots c_n x_n$

Subject to constraints

$ai_1 x_1 + ai_2 x_2 + \ldots + a_{in} x_n \leq b_i$, $i = 1\ 2 \ldots\ldots n$ and $x_1, x_2, x_3 \ldots x_n \geq 0$.

by making some elementary transformations.

This form of LPP is called its canonical form and has the following characteristics.

i) Objective function is of maximization type

ii) All constraints are of ($\leq$) type.

iii) All variables x_i are non-negative.

The canonical form is a format for an LPP which finds its use in Duality theory.

Standard form

The general LPP can also be put in the form

Maximize $Z = c_1 x_1 + c_2 x_2 + \ldots\ldots\ldots c_n x_n$

Subject to the constraints

$$ai_1 x_1 + ai_2 x_2 + \ldots\ldots\ldots + ai_n x_n = b_i$$

$$i = 1,2,3,\ldots\ldots\ldots\ldots\ldots n$$

$$x_1, x_2, x_3 \ldots\ldots\ldots x_n \geq 0.$$

This form of LPP is called its standard form and has the following characteristics.

1) Objective function is of maximization type.

2) All constraints are expressed as equations.

3) Right hand side of each constraint is non-negative.

4) All variables are non-negative.

Example 1: Convert the following LPP to the standard form.

Maximize $z = 3x_1 + 5x_2 + 7x_3$

Subject to the constraints

$$6x_1 - 4x_2 \leq 5\ ;\ 3x_1 + 2x_2 + 5x_3 \geq 11\ ;$$

$$4x_1 + 3x_3 \leq 2 \text{ and } x_1\, x_2 \geq 0.$$

Solution: As x_3 is unrestricted let $x_3 = x_3' - x_3''$

Where $x_3', x_3'' \geq 0$. Now the given constraints can be expressed as

$$6x_1 - 4x_2 \leq 5$$
$$3x_1 + 2x_2 + 5x_3' - 5x_3'' \geq 11$$
$$4x_1 + 3x_3' - 3x_3'' \leq 2$$
$$x_1, x_2, x_3', x_3'' \geq 0$$

Introducing the slack / surplus variables the problem in standard form becomes

Maximize $z = 3x_1 + 5x_2 + 7x_3' - 7x_3''$

Subject to
$$6x_1 - 4x_2 + S_1 = 5$$
$$3x_1 + 2x_2 + 5x_3' - 5x_3'' - s_2 = 11$$
$$4x_1 + 3x_3' - 3x_3'' + S_3 = 2$$

And $x_1, x_2, x_3', x_3'', S_1, S_2, S_3 \geq 0.$

Example 2: Express the following problem in the standard form

Minimize $Z = 3x_1 + 4x_2$

Subject to
$$2x_1 - x_2 - 3x_3 = -4$$
$$3x_1 + 5x_2 + x_4 = 10$$
$$x_1 - 4x_2 = 12.$$

And $x_1, x_3, x_4 \geq 0.$

Solution: Here x_3, x_4 are the slack / surplus variables and x_1, x_2 are the decision variables. As x_2 is unrestricted

Let $x_2 = x_2' - x_2''$ where $x_2', x_2'' \geq 0.$

$\therefore$ The problem in the standard form is

Maximize $Z' (= -z) = -3x_1 - 4x_2' + 4x_2''.$

Subject to
$$-2x_1 + x_2' - x_2'' + 3x_3 = 4$$
$$3x_1 + 5x_2' - 5x_2'' + x_4 = 10$$

$$x_1 - 4x_2' + 4x_2'' = 12$$

And $\quad x_1\, x_2',\, x_2'',\, x_3,\, x_4 \geq 0.$

Example 3: Rewrite in standard form the following LPP

Minimize $\quad z = 2x_1 + x_2 + 4x_3$ subject to the constraints

$$-2x_1 + 4x_2 \leq 4\ ;\ x_1 + 2x_2 + x_3 \geq 5,\ 2x_1 + 3x_3 \leq 2$$

$x_1, x_2 \geq 0$ and x_3 is unrestricted.

Solution: Since x_3 is unrestricted in sign we write $x_3 = x_3' - x_3''$

We introduce slack / surplus variables the problem in the standard form is

Maximize $\quad Z' (= -z) = -2x_1 - x_2 - 4x_3' - 4x_3''$

Subject to $\quad -2x_1 + 4x_2 + S_1 = 4$

$$x_1 + 2x_2 + x_3' - x_3'' - S_2 = 5$$

$$2x_1 + 3x_3' - 3x_3'' + S_3 = 2$$

And $\quad x_1, x_2, x_3', x_3'', S_1, S_2, S_3 \geq 0.$

4.11 SIMPLEX METHOD

The best known and the most widely used procedure for locating the optimal vertex is the **Simplex Method**. The simplex method is also called the "Simplex Technique" or "Simplex Algorithm". It is an iterative procedure for solving a linear programming problem in a finite number of steps. The method provides an algorithm which consists in moving from one vertex of the region of feasible solutions to another in such a manner that the value of the objective function at the succeeding vertex is less (or more as the case may be) than at the preceding vertex. This procedure of jumping from one vertex to another is then repeated. Since the number of vertices is finite, the method leads to an optimal vertex in a finite number of steps or indicates the existence of an unbounded solution.

In simplex method an infinite number of solutions is reduced to a finite number of promising solutions by the following facts.

a) When there are m constraints and $(m + n)$ variables (m being $\leq n$), the starting solution is found by setting n variables equal to zero and then solving the remaining m equations, provided the solution exists and is unique. The n zero variables are known as **non-basic variables** while the remaining m variables are called **basic variables** and they form a **basic solution.**

b) In an LPP the variables must always be non-negative some of the basic solutions may contain negative variables. Such solutions are called **basic infeasible solutions** and should not be considered. To achieve this, we start with a basic solution which is non-negative. The next basic solution must always be non-negative. This is ensured by the feasibility condition. Such a solution is known as **basic feasible solution**.

If all the variables in the basic feasible solution are positive, then it is called **non-degenerate solution** and if some of the variables are zero, it is called **degenerate solution.**

c) A new basic feasible solution may be obtained from the previous one by equating one of the basic variables to zero and placing it by a new non-basic variable. The eliminated variable is called the **out-going variable** while the new variable is known as the **incoming variable.** The incoming variable must improve the value of the objective function which is ensured by the optimality condition. This process is repeated till no further improvement is possible. The resulting solution is called the optimal basic feasible solution or simply **optimal solution.**

d) The simplex method is, therefore based on the following two conditions:

   i) Feasibility condition : It ensures that if the starting solution is basic feasible the subsequent will also be basic feasible.

   ii) Optimality condition: It ensures that only improved solutions will be obtained.

Example 1: Obtain all the basic solutions to the following system of linear equations

$$x_1 + 2x_2 + x_3 = 4$$

$$2x_1 + x_2 + 5x_3 = 5$$

Solution: The given system of equations can be written in the matrix form as $AX = b$

where $$A = \begin{bmatrix} 1 & 2 & 1 \\ 2 & 1 & 5 \end{bmatrix} \quad X = \begin{bmatrix} x_1 \\ x_2 \\ x_3 \end{bmatrix} \text{ and } \quad b = \begin{bmatrix} 4 \\ 5 \end{bmatrix}$$

Since the rank of the matrix A is 2, the maximum number of linearly independent columns of A is 2. Thus we can take any of the following 2×2 sub-matrices as basic matrix B.

$$\begin{bmatrix} 1 & 2 \\ 2 & 1 \end{bmatrix}, \begin{bmatrix} 1 & 1 \\ 2 & 5 \end{bmatrix} \text{ and } \begin{bmatrix} 2 & 1 \\ 1 & 5 \end{bmatrix}$$

The variables not associated with the columns of B are x_3, x_2 and x_1 respectively, in the three different cases.

Let us take first $B = \begin{bmatrix} 1 & 2 \\ 2 & 1 \end{bmatrix}$, A basic solution to the given.

System is now obtained by setting $x_3 = 0$ and solving the system

$$\begin{bmatrix} 1 & 2 \\ 2 & 1 \end{bmatrix} \begin{bmatrix} x_1 \\ x_2 \end{bmatrix} = \begin{bmatrix} 4 \\ 5 \end{bmatrix}$$

Thus a basic (non-basic) solution to the given system is

(Basic) $x_1 = 2$, $x_2 = 1$ (Non-basic) $x_3 = 0$

Similarly, the other two basic and non basic solutions are

(Basic) $x_1 = 5$ $x_3 = -1$ (Non-basic) $x_2 = 0$

and (Basic) $x_2 = \dfrac{5}{2}$ $x_3 = \dfrac{2}{3}$ (Non-basic) $x_1 = 0$

We observe that all the above three basic solutions are non-degenerate solution.

Example 2: Show that the following system of linear equation has a degenerate solution.

$$2x_1 + x_2 - x_3 = 2$$
$$3x_1 + 2x_2 + x_3 = 3$$

Solution: The given system of equation can be written as

$AX = b$

where $A = \begin{bmatrix} 2 & 1 & -1 \\ 3 & 2 & 1 \end{bmatrix}$ $X = \begin{bmatrix} x_1 \\ x_2 \\ x_3 \end{bmatrix}$ $b = \begin{bmatrix} 2 \\ 3 \end{bmatrix}$

Since the rank of A is 2, the maximum number of linearly independent columns of A is 2. Thus we can take any of the following 2 × sub-matrices of A, as basic matrix B.

$$\begin{bmatrix} 2 & 1 \\ 3 & 2 \end{bmatrix} \quad \begin{bmatrix} 1 & -1 \\ 2 & 1 \end{bmatrix} \text{ and } \begin{bmatrix} 2 & -1 \\ 3 & 1 \end{bmatrix}$$

The variables not associated with the columns of these sub-matrices are respectively x_3 , x_1 and x_2.

Consider $B = \begin{bmatrix} 2 & 1 \\ 3 & 2 \end{bmatrix}$, a basic solution to the given system is obtained by setting $x_3 = 0$ and solving the system

$$\begin{bmatrix} 2 & 1 \\ 3 & 2 \end{bmatrix} \begin{bmatrix} x_1 \\ x_2 \end{bmatrix} = \begin{bmatrix} 2 \\ 3 \end{bmatrix}$$

Thus the basic solution to the given problem is

(Basic) $x_1 = 1$, $x_2 = 0$ (Non-basic) $x_3 = 0$.

Similarly the other two solutions are

(Basic) $x_2 = \frac{5}{3}$, $x_3 = \frac{-1}{3}$ (Non-basic) $x_1 = 0$

And (Basic) $x_1 = 1$, $x_3 = 0$ (Non-basic) $x_2 = 0$.

In each of the two basic solutions at least one of the basic variables is zero. Hence two of the basic solutions are degenerate solutions.

Definition: A feasible solution to an LPP which is also a basic solution to the problem is called a basic feasible solution to LPP.

1) In the above example (i) observe that [5,0,–1] is not a feasible solution only basic feasible solutions are

 i) $x_2 = \frac{5}{3}$ and $x_3 = \frac{2}{3}$ ii) $x_1 = 2$ and $x_2 = 1$

2) In the above example (2) the non-degenerate solutions $\left[0, \frac{5}{3}, \frac{-1}{3}\right]$ is not feasible. Only basic feasible degenerate solutions are

 i) $x_1 = 1$ and $x_2 = 0$

 ii) $x_2 = 1$ and $x_3 = 0$.

Example 3: Find all the basic solutions of the following system of equations identifying in each case the basic and non-basic variables.

$$2x_1 + x_2 \, 4x_3 = 11$$
$$3x_1 + x_2 + 5x_3 = 14.$$

Solution: The given system of equations can be written in the matrix form as $AX = b$.

$$A = \begin{bmatrix} 2 & 1 & 4 \\ 3 & 1 & 5 \end{bmatrix} \qquad X = \begin{bmatrix} x_1 \\ x_2 \\ x_3 \end{bmatrix} \text{ and } \quad b = \begin{bmatrix} 11 \\ 14 \end{bmatrix}$$

Since the rank of A is 2, the maximum number of linearly independent columns of A is 2. Thus we can take any of the following 2×2 sub-matrices as basic matrix B.

$$\begin{bmatrix} 2 & 1 \\ 3 & 1 \end{bmatrix} \quad \begin{bmatrix} 1 & 4 \\ 1 & 5 \end{bmatrix} \text{ and } \begin{bmatrix} 2 & 4 \\ 3 & 5 \end{bmatrix}$$

The variables not associated with columns of B are x_3, x_1 and x_2 respectively in three different cases.

Let us take first $B = \begin{bmatrix} 2 & 1 \\ 3 & 1 \end{bmatrix}$

A basic solution to the given system is now obtained by setting $x_3 = 0$ and solving the system of equations.

$$\begin{bmatrix} 2 & 1 \\ 3 & 1 \end{bmatrix} \begin{bmatrix} x_1 \\ x_2 \end{bmatrix} = \begin{bmatrix} 11 \\ 14 \end{bmatrix}$$

Thus a basic (non–basic) solution to the given system is

(Basic) $x_1 = 3,$ $\quad x_2 = 5$ $\quad$ (non basic) $x_3 = 0$

Similarly, the other two basic and non–basic solutions are

(Basic) $x_2 = 3,$ $\quad x_3 = -1$ $\quad$ (non basic) $x_1 = 0$

And (basic) $x_1 = \frac{1}{2}$ $\quad x_3 = \frac{5}{2}$ $\quad$ (non basic) $x_2 = 0$

We observe that all the above three basic solutions are non-degenerate solution.

Example 4: Find all the basic solutions to the following problem

Maximize $z = x_1 + 3x_2 + 3x_3$ subject to the constraints

$$x_1 + 2x_2 + 3x_3 = 4 \text{ and } 2x_1 + 3x_2 + 5x_3 = 7.$$

Solution: The total number of basic solutions are 3. Out of these solutions, solutions in which all basic variables $(xj) \geq 0$ will be basic feasible solutions in which all basic variables are ≥ 0 will be non–degenerate basic feasible. The basic feasible solution which optimizes the objective function will be the optimal basic feasible solution.

Now, maximum $z = x_1 + 3x_2 + 3x_3$

Subject to $x_1 + 2x_2 + 3x_3 = 4$

$2x_1 + 3x_2 + 5x_3 = 7.$

The given system of equations can be written as

$$AX = b$$

where $A = \begin{bmatrix} 1 & 2 & 3 \\ 2 & 3 & 5 \end{bmatrix}$ $X = \begin{bmatrix} x_1 \\ x_2 \\ x_3 \end{bmatrix}$ and $b = \begin{bmatrix} 4 \\ 7 \end{bmatrix}$

Since the rank of A is 2, the maximum number of linearly independent columns of A is 2. Thus we can take any of the following 2×2 sub-matrices of A, as a basic matrix B: $\begin{bmatrix} 1 & 2 \\ 2 & 3 \end{bmatrix}$ $\begin{bmatrix} 2 & 3 \\ 3 & 5 \end{bmatrix}$ and $\begin{bmatrix} 1 & 3 \\ 2 & 5 \end{bmatrix}$

The variables not associated with the columns of these matrices are respectively x_3, x_1 and x_2

Consider $B = \begin{bmatrix} 1 & 2 \\ 2 & 3 \end{bmatrix}$ a basic solution to the given system is obtained by setting $x_3 = 0$ and solving the system

The basic solution to the given problem is

(Basic) $x_1 = 2$, $x_2 = 1$ (non-basic) $x_3 = 0$

Similarly the other two solutions are

(Basic) $x_2 = -1$, $x_3 = 2$ (non-basic) $x_1 = 0$

And (Basic) $x_1 = 1$, $x_3 = 1$ (non-basic) $x_2 = 0$.

It may be seen that the first and third solutions are basic feasible. They are also non-degenerate basic feasible solutions.

When $x_1 = 2, \quad x_2 = 1, \quad x_3 = 0$

$$z = x_1 + 3x_2 + 3x_3$$
$$= 2 + 3 + 0 = 5$$

$\therefore \quad z_{max} = 5$. And other two are 4 and 3.

Working procedure for simplex Method

Assuming the existence of an initial basic feasible solution, an optimal solution to any LPP by simplex method is found as follows.

Step 1: i) Check whether the objective function is to be maximized or minimized.

If $Z = c_1 x_1 + c_2 x_2 + c_3 x_3 + \ldots\ldots + c_n x_n$

is to be minimized, then convert it into a problem of maximization, by writing

minimize Z = maximize $(-z)$

ii) Check whether all b's are positive.

Step 2: Express the problem in the standard form.

Convert all inequalities of constraints into equations by introducing slack / surplus variables.

Step 3: Find an initial basic feasible solution, and form the simplex table.

	c_j	c_1	c_2	c_3	...	0	0	0	...	0
c_B	Basis	x_1	x_2	x_3	...	s_1	s_2	s_3	...	b
0	s_1	a_{11}	a_{12}	a_{13}	...	1	0	0	...	b_1
0	s_2	a_{21}	a_{22}	a_{23}	...	0	1	0	...	b_2
0	s_3	a_{31}	a_{32}	a_{33}	...	0	0	1	...	b_3
⋮	⋮	⋮	⋮	⋮	⋮	⋮	⋮	⋮	⋮	⋮

Body matrix Unit matrix

a) The variables s_1, s_2, s_3 .... are called basic variables.

b) The variables x_1, x_2, x_3 .... are called non-basic variables.

c) Cj row denotes the coefficient of the variables in the objective function.

d) C_B – column denotes the coefficient of the basic variables only in the objective function.

e) b – column denotes the values of the basic variables while remaining variables will always be zero.

f) The coefficient of x's (decision variables) in the constraint equations represents the body matrix and the coefficient of slack variables constitute the unit matrix.

Step 4 : Apply optimality test.

Compute $C_j = C_j - Z_j$ where $z_j = C_B\, a_{ij}$

If all C_j are negative then the initial basic solution is optimal.

If even one C_j is positive then the current feasible solution is not optimal and proceed to the next step.

Step 5 : i) Identify the incoming and outgoing variables.

ii) Iterate towards the optimal solution.

Step 6 : Go to step (4) and repeat the computational procedure until either an optimal (or an unbounded) solution is obtained.

Example 1 : Using Simplex method maximize $z = 5x_1 + 3x_2$

Subject to $x_1 + x_2 \le 2$, $5x_1 + 2x_2 \le 10$, $3x_1 + 8x_2 \le 12$, and $x_{1,}x_2 \ge 0$.

Solution:

Step 1: The objective function is to be maximised and all b's are positive.

Step 2: The problem in the standard form is

Max $Z = 5x_1 + 3x_2 + 0S_1\ 0S_2 + 0S_3$

where S_1, S_2, S_3 are slack variables subject to

$$x_1 + x_2 + S_1 + 0S_2 + 0S_3 = 2$$

$$5x_1 + 2x_2 + 0S_1 + S_2 + 0S_3 = 10$$

$$3x_1 + 8x_2 + 0S_1 + 0S_2 + S_3 = 12$$

and $x_1, x_2, S_1, S_2, S_3 \ge 0.$

Step 3: Find initial basic solutions

Three equations and five variables.

Set $x_1 = 0$, $x_2 = 0$, we get the basic solutions $S_1 = 2$, $S_2 = 10$ and $S_3 = 12$

Since all $S_{1,}\ S_{2,}\ S_3$ are positive, the basic solution is also feasible and non-degenerate.

∴ The basic feasible solution is

$x_1 = 0, x_2 = 0$ (non-basic) and $S_1 = 2, S_2 = 10, S_3 = 12$ (basic)

∴ Initial basic feasible solution is given by the following table.

	C_j	5	3	0	0	0		
c_B	Basis	x_1	x_2	s_1	s_2	s_3	b	R
0	s_1	1	1	1	0	0	2	$\frac{2}{1}$ ⇐
0	s_2	5	2	0	1	0	10	$\frac{10}{5}$
0	s_3	3	8	0	0	1	12	$\frac{12}{3}$
	$z_j = \Sigma C_B a_{ij}$	0	0	0	0	0		
	$C_j = c_j - z_j$	5	3	0	0			
		↑						

As C_j is positive under some columns the initial basic feasible solution is not optimal (can be improved) and we proceed to the next step.

Step 4 : C_j (= 5) is maximum ∴ x_1 is incoming variable and the column in which it appears is the key column. Dividing the elements under b – column by the corresponding elements of key–column we find the minimum +ve ratio R is 2.

∴ S_1 as the key row and S_1 is outgoing basic variable.

∴ Removing S_1, the new basic will contain x_1, S_2 and S_3 as the basic variables.

Step 5 : Iteration towards the optimal solution.

	C_j	5	3	0	0	0	
c_B	Basis	x_1	x_2	s_1	s_2	s_3	b
5	x_1	1	1	1	0	0	2 $R_2 - 5R_1$
0	S_2	0	–3	–5	1	0	0 $R_3 - 3R_1$
0	S_3	0	5	–3	0	1	6
0	$z_j = \Sigma C_B a_{ij}$	5	5	5	0	0	10
	$C_j = c_j - z_j$	0	–2	–5	0	0	

As C_j is either zero or negative under all columns, the above table gives the optimal basic feasible solution. Thus optimal solution is $x_1 = 2, x_2 = 0$ and $Z_{max} = 10$.

Example 2: Solve by simplex method the following LPP

Maximize $z = 6x_1 + 8x_2$ subject to the constraints

$$5x_1 + 10x_2 \leq 60 \qquad 4x_1 + 4x_2 \leq 40, x_1, x_2 \geq 0.$$

Solution:

Step 1 : The given L.P.P is maximization problem and all b's are positive.

Step 2 : Express the problem in standard form

Maximum $\quad z = 6x_1 + 8x_2 + 0s_1 + 0s_2$

Subject to $\quad 5x_1 + 10x_2 + s_1 + 0s_2 = 60$

$\quad 4x_1 + 4x_2 + 0s_1 + s_2 = 40$

And $\quad x_1, x_2, s_1, s_2 \geq 0.$

Where s_1 and s_2 are slack variables.

Step 3 : To find an optimal basic feasible solution form the simplex table.

	C_j	6	8	0	0			
c_B	Basis	x_1	x_2	s_1	s_2	b	R	
0	s_1	5	10	1	0	60	$\frac{60}{10} = 6$	←
0	s_2	4	4	0	1	40		
0	z_j	0	0	0	0	0		
	$C_j = c_j - z_j$	6	8	0	0			
			↑					

Step 4 : All the values of $c_j - z_j$ are not less than or equal to zero. Hence the initial solutions are not optimal. The variable x_2 is the entering variable because $c_j - z_j$ is 8 which is highest. The variable S_1 is outgoing variable. Since the ratio w.r.t. this row is the least ratio. Hence the key column corresponding to the variable x_2.

1) **Iteration :** The basic variable S_1 is replaced by x_2 and the table is as follows:

	C_j	6	8	0	0		
C_B	Basis	x_1	x_2	s_1	s_2	b	R
8	x_2	$\frac{1}{2}$	1	$\frac{1}{10}$	0	6	12
0	s_2	2	0	$\frac{-2}{5}$	1	16	8
0	z_j	4	8	$\frac{4}{5}$	0	48	
	$C_j = c_j - z_j$	2	0	$\frac{-4}{5}$	0		

IInd Iteration

The basic variable s_2 is replaced by x_1 and it is not optimal solution $\therefore$ the next iteration.

	C_j	6	8	0	0		
C_B	Basis	x_1	x_2	s_1	s_2	b	R
8	x_2	$\frac{1}{2}$	1	$\frac{1}{5}$	$\frac{-1}{4}$	2	
6	x_1	1	0	$\frac{-1}{5}$	$\frac{1}{2}$	8	
	z_j	6	8	$\frac{2}{5}$	1	64	
	$C_j - z_j$	0	0	$\frac{-2}{5}$	-1		

All the values of $C_j - z_j$ are either zero or negative. Hence the optimal solution is reached

$\therefore \quad x_1 = 8 \qquad x_2 = 2 \qquad Z_{max} = 64.$

Example 3: Solve the LPP by simplex method given that

Maximize $Z = 4x_1 + 3x_2 + 6x_3$ subject to the constraints

$$2x_1 + 3x_2 + 2x_3 \leq 440,$$
$$4x_1 + 3x_3 \leq 470, \quad 2x_1 + 5x_2 \leq 430 \text{ and } x_1, x_2, x_3 \geq 0.$$

Solution: Here, the objective function is to be maximized and all b's are non-negative.

Introduce the slack variables S_1, S_2, S_3 and the problem in the standard form is

Max $\quad Z = 4x_1 + 3x_2 + 6x_3 + 0S_1 + 0S_2 + 0S_3$

Subject to

$$2x_1 + 3x_2 + 2x_3 + s_1 + 0s_2 + 0s_3 = 440.$$
$$4x_1 + 0x_2 + 3x_3 + 0s_1 + s_2 + 0s_3 = 470.$$
$$2x_1 + 5x_2 + 0x_3 + 0s_1 + 0s_2 + s_3 = 430.$$

Setting $x_1 = x_2 = x_3 = 0$ (non-basic) we get $s_1 = 440$, $s_2 = 470$, $s_3 = 430$ (basic).

$\therefore$ The initial basic feasible solution is given by the following table.

	C_j	4	3	6	0	0	0		
c_B	Basis	x_1	x_2	x_3	s_1	s_2	s_3	b	R
0	s_1	2	3	2	1	0	0	440	$\frac{440}{2}$
0	s_2	4	0	3	0	1	0	470	$\frac{470}{3}$ ←
0	s_2	2	5	0	0	0	1	430	$\frac{430}{0}$
$z_j =$	$\Sigma c_B a_{ij}$	0	0	0	0	0	0		
$C_j =$	$c_j - z_j$	4	3	6	0	0	0		
				↑					

As C_j is positive under some columns, the initial basic feasible solution is not optimal and we proceed to the next step.

From the above table x_3 is the incoming variable and s_2 is the outgoing variable and (3) is the key element. Next drop s_2 and introduce x_3 with its associated value 6 under C_B column. Convert the key element to unity and make all other elements of key column zero by row operations. Then the second feasible solution is given by the following table.

c_B	c_j Basis	4 x_1	3 x_2	6 x_3	0 s_1	0 s_2	0 s_3	b	R
0	s_1	$\frac{-2}{3}$	(3)	0	1	$\frac{-2}{3}$	0	$\frac{380}{3}$	$\frac{380}{9}$ ←
6	x_3	$\frac{4}{3}$	0	1	0	$\frac{1}{3}$	0	$\frac{470}{3}$	∞
0	s_3	2	5	0	0	0	1	430	86
	z_j	8	0	6	0	2	0	940	
	C_j	–4	3	0	0	–2	0		
			↑						

As C_j is positive under the second column, the solution is not optimal and we proceed further. Now x_2 is incoming variable and s_1 is outgoing variable and (3) is the key element for the next iteration.

Drop s_1 and introduce x_2 with associated value 3 under, C_B. Convert the key element to unity and make all of the elements of the key column zero. Then the third basic feasible solution is given by the following table.

c_B	c_j Basis	4 x_1	3 x_2	6 x_3	0 s_1	0 s_2	0 s_3	b	R
3	x_2	$\frac{-2}{9}$	1	0	$\frac{1}{3}$	$\frac{-2}{9}$	0	$\frac{380}{9}$	
6	x_3	$\frac{4}{3}$	0	1	0	$\frac{1}{3}$	0	$\frac{470}{3}$	
0	s_3	$\frac{28}{9}$	0	0	$\frac{-5}{3}$	$\frac{10}{9}$	1	$\frac{1970}{3}$	
	z_j	$\frac{22}{3}$	3	6	1	$\frac{4}{3}$	0	$\frac{3200}{3}$	
	C_j	$\frac{-10}{3}$	0	0	–1	$\frac{-4}{3}$	0		

Each $c_f \leq 0$, therefore it gives the optimal solution.

$$x_1 = 0, \quad x_2 = \frac{380}{9}, \quad x_3 = \frac{370}{3} \text{ and } Z_{\max} = \frac{3200}{3} = 1066.67.$$

Example 4: Solve the following LPP by simplex method

Minimize $z = x_1 - 3x_2 + 3x_3$

Subject to $3x_1 - x_2 + 2x_3 \leq 7,\ 2x_1 + 4x_2 \geq -12,$

$-4x_1 + 3x_2\ 8x_3 \leq 10,\ x_1, x_2, x_3 \geq 0.$

Solution: The problem is that of minimizing the objective function convert it to the maximization problem.

i.e., we have Max $Z' = (-Z) = -x_1 + 3x_2 - 3x_3$.

The right hand side of the second constraints is negative we write it as $-2x_1 - 4x_2 \leq 12$.

Now, introduce the slack variables s_1, s_2, s_3 and write LPP in the standard form.

Max $Z' = -x_1 + 3x_2 - 3x_3 + 0s_1 + 0s_2 + 0s_3$

Subject to $3x_1 - x_2 + 2x_3 + s_1 + 0S_2 + 0s_3 = 7$

$-2x_1 - 4x_2 + 0x_3 + 0s_1 + s_2 + 0s_3 = 12$

$-4x_1 + 3x_2 + 8x_3 + 0s_1 + 0s_2 + s_3 = 10$

And $x_1, x_2, x_3, s_1, s_2, s_3 \geq 0.$

Next setting $x_1 = x_2 = x_3 = 0$ (non-basic) we find

$s_1 = 7, \quad s_2 = 12, \quad s_3 = 10$ (basic).

$\therefore$ Initial basic feasible solution is given by the following table.

	c_j	-1	3	-3	0	0	0		
c_B	Basis	x_1	x_2	x_3	s_1	s_2	s_3	b	R
0	s_1	3	-1	2	1	0	0	7	$\frac{7}{-1}$
0	s_2	-2	-4	0	0	1	0	12	$\frac{12}{4}$
0	s_3	-4	(3)	8	0	0	1	10	$\frac{10}{3}$
$z_j =$	$\Sigma c_B a_{ij}$	0	0	0	0	0	0	0	
$C_j =$	$c_j - z_j$	-1	3	-3	0	0	0		

As c_j is positive under second column, the initial basic feasible solution is not optimal and we proceed to the next step.

Now x_2 is incoming variable, s_3 is outgoing variable and (3) is the key element.

$\therefore$ drop s_3 and introduce x_2 with its associated value 3 under c_B column. Convert the key element to unity and make all other elements of the key column zero. Then the second basic feasible solution is given in the following table.

	c_j	–1	3	–3	0	0	0		
c_B	Basis	x_1	x_2	x_3	s_1	s_2	s_3	b	R
0	s_1	$\frac{5}{3}$	0	$\frac{14}{3}$	1	0	$\frac{1}{3}$	$\frac{31}{3}$	$\frac{31}{5}$
0	s_2	$-\frac{22}{3}$	0	$\frac{32}{3}$	0	1	$\frac{4}{3}$	$\frac{76}{3}$	$\frac{-38}{11}$
3	x_2	$\frac{-4}{3}$	1	8.3	0	0	$\frac{1}{3}$	$\frac{10}{3}$	$\frac{-5}{2}$
	z_j	–4	3	8	0	0	1	10	
	C_j	3	0	–11	0	0	–1		

As C_j is positive under first column, the solution is not optimal and we proceed further.

x_1 is incoming variable, s_1 is outgoing variable and $\left(5/3\right)$ is the key element.

$\therefore$ Drop s_1 and introduce x_1 with associated value –1 under C_B column. Convert the key element to unity and make all other elements of the key column zero.

Then the third basic feasible solution is given by the following table.

c_B	c_j / Basis	1 / x_1	3 / x_2	–3 / x_3	0 / s_1	0 / s_2	0 / s_3	b	R
–1	x_1	1	0	$\frac{14}{5}$	$\frac{3}{5}$	0	$\frac{1}{5}$	$\frac{31}{5}$	
0	s_2	0	0	$\frac{156}{5}$	$\frac{22}{5}$	1	$\frac{14}{5}$	$\frac{354}{5}$	
3	x_2	0	1	$\frac{32}{5}$	$\frac{4}{5}$	0	$\frac{3}{5}$	$\frac{58}{5}$	
	z_j	–1	3	$\frac{82}{5}$	$\frac{9}{5}$	0	$\frac{8}{5}$	$\frac{143}{5}$	
	c_j	0	0	$\frac{-97}{5}$	$\frac{-9}{5}$	0	$\frac{-8}{5}$		

Now, since each $C_j \leq 0$, therefore it gives the optimal solution. $x_1 = \frac{31}{5}$,

$x_2 = \frac{58}{5}$, $x_3 = 0$ (non-basic)

And $z'_{\max} = \frac{143}{5}$

Hence $z_{\min} = \frac{-143}{5}$

Example 5 : Maximize $z = 107x_1 + x_2 + 2x_3$

Subject to the constraints $14x_1 + x_2 - 6x_3 + 3x_4 = 7$.

$16x_1 + \frac{1}{2}x_2 - 6x_3 \leq 5$, $3x_1 - x_2 - x_3 \leq 0$, $x_1, x_2, x_3, x_4 \geq 0$.

Solution: The objective function is to be maximized and all b^1s are non–negative.

∴ we express the LPP in the standard form

Max $Z = 107x_1 + x_2 + 2x_3 + 0x_4 + 0s_1 + 0s_2$

Subject to $\quad \frac{14}{3}x_1 + \frac{1}{3}x_2 - 2x_3 + x_4 + 0s_1 + 0s_2 = \frac{7}{3}$.

$$16x_1 + \frac{1}{2}x_2 - 6x_3 + 0x_4 + s_1 + 0s_2 = 5$$

$$3x_1 - x_2 - x_3 + 0x_4 + 0s_1 + s_2 = 0$$

And $\quad x_1, x_2, x_3, x_4, s_1, s_2 \geq 0.$

To find the initial basic feasible solution put $x_1 = x_2 = x_3 = 0$ (non-basic)

$x_4 = \frac{7}{3} \quad s_1 = 5$ and $s_2 = 0$ (basic).

∴ The initial basic solution is given in the following table.

	c_j	107	1	2	0	0	0		
c_B	Basis	x_1	x_2	x_3	x_4	s_1	s_2	b	R
0	x_4	$\frac{14}{3}$	$\frac{1}{3}$	–2	1	0	0	$\frac{7}{3}$	$\frac{7/3}{14/3}$
0	s_1	16	$\frac{2}{3}$	–6	0	1	0	5	$\frac{5}{16}$
0	s_2	(3)	–1	–1	0	0	1	0	$\frac{0}{3}$ ←
$z_j =$	$\Sigma c_B a_{ij}$	0	0	0	0	0	0	0	
$C_j =$	$c_j - z_j$	107 ↑	1	2	0	0	0		

As C_j is positive under some columns, the initial basic feasible solution is not optimal and we proceed further.

Here, x_1 is incoming variable, s_2 is outgoing variable (3) is the key element.

Drop s_2 and introduce x_1 with associated value 107 under c_B convert key element to unity and make all other elements of the key column zeros. (by row operations). Then the second basic feasible solution is given by the following table.

	c_j	107	1	2	0	0	0		
c_B	Basis	x_1	x_2	x_3	x_4	s_1	s_2	b	R
0	x_4	0	$\frac{17}{9}$	$\frac{-4}{9}$	1	0	$\frac{14}{9}$	$\frac{7}{3}$	$\frac{-21}{4}$
0	s_1	0	$\frac{35}{6}$	$\frac{-2}{3}$	0	1	$\frac{-16}{3}$	5	$\frac{-15}{2}$
1.7	x_1	1	$\frac{-1}{3}$	$\frac{-1}{3}$	0	0	$\frac{1}{3}$	0	0
	z_j	107	$\frac{-107}{3}$	$\frac{-107}{3}$	0	0	$\frac{107}{3}$		
	C_j	0	$\frac{110}{3}$	$\frac{113}{3}$ ↑	0	0	$\frac{-107}{3}$		

As C_j is positive under some columns, the solution is not optimal. Here $\frac{113}{3}$ being the largest positive value of C_j, x_3 is the incoming variable. But all the values of R being ≤ 0, x_3 will not enter the basis. This indicates that the solution to the problem is unbounded.

4.12 QUESTION BANK

Formation of LPP

1. A firm produces three products. These products are processed on three different machines. The time required to manufacture one unit of each of the three products and the daily capacity of the three machines are given below.

Machines	Time per unit (minutes)			Machine Capacity
	Product I	Product II	Product III	(Minutes/day)
M_1	2	3	2	440
M_2	4	–	3	470
M_3	2	5	–	430

It is required to determine the daily number of units to be manufactured for each product. The profit per unit for products I, II, III is Rs 4, Rs 3 and Rs 6 respectively. It is assumed that all the amounts produced are consumed. Formulate the mathematical LPP that will maximize the daily profit.

Ans.: $z = 4x_1 + 3x_2 + 6x_3$

and constraints are

$2x_1 + 3x_2 + 2x_3 \leq 440,\ 4x_1 + 0x_2 + 3x_3 \leq 470,\ 2x_1 + 5x_2 + 0x_3 \leq 430.$

2. A person wants to decide the constituents of a diet which will fulfill his daily requirements of proteins, fats and carbohydrates at the minimum cost. The choice is to be made from four different types of foods. The yields per unit of these foods are given below.

Food type	Yield per unit			Cost per unit
	Proteins	Fats	Carbohydrates	(Rs.)
1	3	2	6	45
2	4	2	4	40
3	8	7	7	85
4	6	5	4	65
Minimum requirement	800	200	700	

Formulate linear programming model for the problem.

Ans.: Minimize $z = 45x_1 + 40x_2 + 85x_3 + 65x_4$

And constraints are, for proteins $3x_1 + 4x_2 + 8x_3 + 6x_4 \geq 800$

For fats : $2x_1 + 2x_2 + 7x_3 + 5x_4 \geq$ for carbohydrates

$$6x_1 + 4x_2 + 7x_3 + 4x_4 \geq 700.$$

3. A firm manufactures 3 products *A, B* and *C*. The profits are Rs 3, Rs 2 and Rs 4 respectively. The firm has two machines M_1 and M_2 and below is the required processing time in minutes for each machine on each product.

Machine	Product		
	A	*B*	*C*
M_1	4	3	5
M_2	2	2	4

Machines M_1 and M_2 have 2000 and 2500 machine minutes respectively. The firm must manufacture 100 A^1s, 200 B^1s and 50 C^1s but not more than 150 A^1s set up an LPP to maximize profit.

Ans.: Maximize $z = 3x_1 + 2x_2 + 4x_3$

Subject to $4x_1 + 3x_2 + 5x_3 \leq 2000$; $2x_1 + 2x_2 + 4x_3 \leq 2500$,

$100 \leq x_1 \leq 150$; $200 \leq x_2$ and $50 \leq x_3$.

4. A firm manufactures headache pills in two sizes A and B, size A contains 2 grains of aspirin, 5 grains of bicarbonate and 1 grain of codeine. Size B contains 1 grain of aspirin, 8 grains of bicarbonate and 6 grains of codeine. It is found by users that it requires at least 12 grains of codeine for providing immediate effect. It is required to determine the least number of pills a patient should take to get immediate relief. Formulate the problem as a standard LPP.

Ans.: Minimize $Z = x_1 + x_2$ subject to $2x_1 + x_2 \geq 12$,

$5x_1 + 8x_2 \geq 74$, $x_1 + 6x_2 \geq 24$ and $x_1, x_2 \geq 0$.

5. A factory engaged in the manufacturing of pistons, rings and valves for which the profits per unit are Rs 10, 6 and 4 respectively wants to decide the most profitable mix. It takes one hour of preparatory work, ten hours of machining and two hours of packing and allied formalities for a piston. Corresponding time requirements for rings and valves are 1, 4 and 2 and 1, 5, and 6 hours respectively. The total number of hours available for preparatory work, machining and packing and allied formalities are 100, 600 and 300 respectively. Determine the most profitable mix, assuming that what all produced can be sold, formulate the LPP.

Ans.: Maximize $z = 10x_1 + 6x_2 + 4x_3$ subject to $x_1 + x_2 + x_3 \leq 100$

$10x_1 + 4x_2 + 5x_3 \leq 600$; $2x_1 + 2x_2 + 6x_3 \leq 300$, $x_1, x_2, x_3 \geq 0$.

B. Solve the following examples by graphical method

6. Find the maximum value of $z = 10x_1 + 6x_2$ subject to

$5x_1 + 3x_2 \leq 30$, $x_1 + 2x_2 \leq 18$ and $x_1, x_2 \geq 0$.

Ans.: $z \max = 60\ x_1 = \frac{6}{7}\ x_2 = \frac{60}{7}\ x_1 = 6, x_2 = 0.$

$\therefore$ infinite solutions lying between $\left(\frac{6}{7}, \frac{60}{7}\right)$ and (6,0).

7. Maximize $z = 50x_1 + 60x_2$ subject to the constraints

   $2x_1 + 3x_2 \leq 1500$, $3x_1 + 2x_2 \leq 1500$, $0 \leq x_1 \leq 400$, $0 \leq x_2 \leq 400$.

Ans.: $x_1 = 300$ and $x_2 = 300$ maxi $z = 33000$.

8. Show graphically that the maximum or minimum values of the objective function for the problem are same.

   Maximize (or minimize) $z = 5x_1 + 3x_2$ subject to the constraints.

   $x_1 + x_2 \leq 6$, $2x_1 + 3x_2 \leq 3$, 0 63> $x_1 \geq 3$, 0 63> $x_2 \geq 3$.

Ans.: $x_1 = 3$, and $x_2 = 3$ maxi $z = 24$.

9. Find the maximum value of $z = 5x_1 + 3x_2$ subject to the constraints

   $x_1 + x_2 \leq 6$, $2x_1 + 3x_2 \leq 6$, $0 \leq X_1 \leq 4$, $0 \leq x_2 \leq 3$.

Ans: $x_1 = 6$ and $x_2 = 0$, max $z = 30.323$.

10. Find the maximum value of $Z = 3x_1 + 2x_2$ subject to $-2x_1 + 3x_2 \leq 9$; $x_1 - 5x_2 \leq -20$ and $x_1, x_2 \geq 0$.

Ans.: Infeasible.

C. Solve the following examples by simplex method

11. Using Simplex method, solve the following LPP

    Maximize $z = 4x_1 + 10x_2$

    Subject to $2x_1 + x_2 \leq 50$, $2x_1 + 5x_2 \leq 100$, $2x_1 + 3x_2 \leq 90$, $x_1, x_2 \geq 0$.

Ans.: $x_1 = 0$, $x_2 = 20$, maximize $z = 200$

12. Solve by simplex method the LPP

    Maximize $z = 10x_1 + x_2 + 2x_3$

    Subject to $x_1 + x_2 - 2x_3 \leq 10$, $4x_1 + x_2 + x_3 \leq 20$, $x_1 x_2 x_3 \geq 0$

Ans.: $x_1 = 5$, $x_2 = 0$, $x_3 = 0$ z max $= 50$.

13. Solve by Simplex method the LPP

    Minimize $z = 3x_1 + 5x_2 + 4x_3$

    Subject to $2x_1 + 3x_2 \leq 8$, $2x_2 + 5x_3 \leq 10$, $3x_1 + 2x_2 + 4x_3 \leq 15$, $x_1 x_2\ x_3 \geq 0$.

PART B

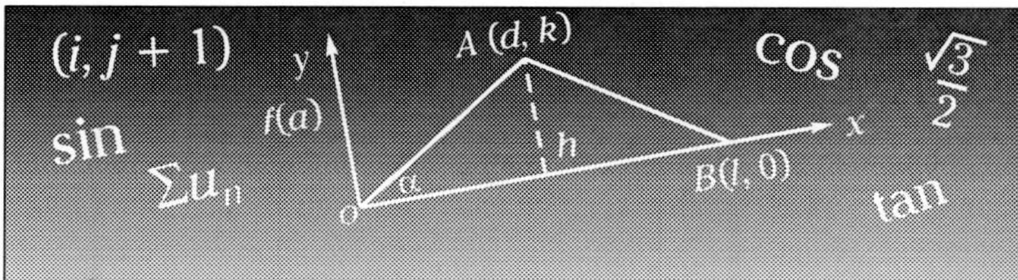

UNIT 5

NUMERICAL METHODS – 1

5.1 SOLUTION OF ALGEBRAIC AND TRANSCENDENTAL EQUATIONS

An expression of the form

$f_n(x) = a_0\, x^n + a_1\, x^{n-1} + a_2 x^{n-2} + \ldots. + a_{n-1}\, x + a_n$, where $a_0, a_1, a_2 \ldots.. a_n$ are constants and n is a positive integer, is said to be a polynomial in x of degree n. The values of x which make $f_n(x) = 0$ are called the roots or the zeroes of the polynomial $f_n(x)$ and every polynomial of nth degree has n-roots.

The equations of the form $f_n(x) = 0$ are called Algebraic or Transcendental according as $f_n(x)$ is purely a polynomial in x or contains some other functions such as logarithmic, exponential, trigonometric functions etc.

For example, the equations $x^3 - 9x - 1 = 0$ and $e^x \sin x + \log^x + 4x^3 = 0$ are called algebraic and transcendental respectively. Solution of an equation $f_n(x) = 0$ means we have to find its roots or zeroes.

For example if, $x^2 - 7x + 12 = 0$ then $x = 3, 4$ are the roots of the equation.

Properties

Let the algebraic equation of nth degree be $a_0\, x^n + a_1\, x^{n-1} + a_2\, x^{n-2} + \ldots. + a_n = 0$, where $a_0 \neq 0$. This equation has n-roots.

1) If two real quantities a and b be substituted for x in any polynomial $f_n(x)$ and if $f_n(a)$ and $f_n(b)$ have apposite signs then one root of $f_n(x) = 0$ lies between a and b.

2) Every equation of an odd degree has at least one real root whose sign is opposite to that of its last term.

3) Every equation of an even degree with last term negative has at least a pair of real roots which are positive and other negative.

4) To find an equation whose roots are with opposite signs to those of the given equation change x to $-x$.

5) To find an equation whose roots are reciprocals of the roots of the given equation change x to $\frac{1}{x}$.

6) The largest root of the equation $f_n(x) = 0$ may be obtained approximately by the root of the equation $a_0x + a_1 = 0$ or by the root of the quadratic expression such that it is larger in absolute value. We will discuss few methods which are applicable to both algebraic and transcendental equations.

5.2 NEWTON-RAPHSON METHOD

Let $f(x) = 0$ be a polynomial equation in x and let x_0 be an approximate value of a root of this equation. Let $x_0 + h$ be the exact value of the corresponding root where h is a very small quantity. Then $f(x_0 + h) = 0$. Since $x_0 + h$ is the root of the equation $f(x) = 0$ .... (1)

Expanding (1) by Taylor's theorem we get

$$f(x_0 + h) = f(x_0) + hf'(x_0) + \frac{h^2}{2!}f''(x_0) + \ldots\ldots = 0$$

Since h is very small, neglecting second and higher order terms and taking the first approximation we have

$$f(x_0) + hf'(x_0) = 0$$

$$\therefore \quad h = -\frac{f(x_0)}{f'(x_0)} \quad \text{provided } f'(x_0) \neq 0$$

$$\therefore \quad x_1 = x_0 + h = x_0 - \frac{f(x_0)}{f'(x_0)} \qquad \ldots.(2)$$

Relation (2) gives the improved value of the root over the previous one. Now, substituting x_1 for x_0 and x_2 for x_1 in (2)

we get $$x_2 = x_1 - \frac{f(x_1)}{f'(x_1)} \qquad \ldots.(3)$$

In general, we can get an approximation on the iteration formula

$$x_{n+1} = x_n - \frac{f(x_n)}{f'(x_n)} \qquad \text{..... (4)}$$

From this formula, we can conclude successive better values of the root, Formula (4) is known as **Newton-Raphson Method.**

Note: i) This method is applicable only when 'h' is very small and $f'(x)$ is large.

ii) This method can be used for both the algebraic and transcendental equations and complex coefficients of an equation.

Example 1: Find the smallest positive root of the equation $x^3 - 2x + 0.5 = 0$ by Newton-Raphson method.

Solution: Let $f(x) = x^3 - 2x + 0.5$

and $f'(x) = 3x^2 - 2$

$x = 0, f(0) = 0.5$ (+ve) and $x = 1,\ f(1) = -0.5$ (–ve)

Since $f(0) f(1) < 0$, the real root lies between 0 and 1.

The formula for Newton-Raphson method is

$$x_{n+1} = x_n - \frac{f(x_n)}{f'(x_n)}$$

Take $x_0 = 0$ and put $n = 0$

$$x_1 = x_0 - \frac{f(x_0)}{f'(x_0)} = 0 - \frac{0.5}{-2} = 0.25$$

$$x_2 = x_1 - \frac{f(x_1)}{f'(x_1)} = 0.25 - \frac{0.0156}{-1.8125} = 0.2586$$

$$x_3 = x_2 - \frac{f(x_2)}{f'(x_2)} = 0.2586 - 0 = 0.2586.$$

Since $x_2 = x_3 = 0.2586$, we stop here. Hence the smallest positive root is 0.2586.

Example 2: Compute the real root of the equation $x \log_{10}^{x} = 1.2$ correct to three decimal places, by using Newton-Raphson Method.

Solution: Given $f(x) = x \log_{10}^{x} - 1.2.$

$$\therefore \qquad f^1(x) = \log_{10}^{x} . 1 + x . \frac{1}{x} \log_{10}^{e}$$

$$= \log_{10}^{x} + .43429$$

$x = 2, \qquad f(2) = 2 \log_{10}^{2} - 1.2 = -0.5980$ (–ve)

$x = 3, \qquad f(3) = 3 \log_{10}^{3} - 1.2 = 0.2313$ (+ve)

So that $f(2) f(3) < 0$. Hence the real root lies between 2 and 3.

Since $| f(3) | < | f(2) |$ we take $x_0 = 3$.

Newton-Raphson method is given by

$$x_{n+1} = x_n - \frac{f(x_n)}{f'(x_n)}$$

The first approximation to the root is

$$x_1 = x_0 - \frac{f(x_0)}{f'(x_0)} = 3 - \frac{3\log_{10}^{3} - 1.2}{\log_{10}^{3} + .43429} = 2.746.$$

$$x_2 = x_1 - \frac{f(x_1)}{f'(x_1)} = 2.746 - \frac{2.746 \log_{10}^{2.746} - 1.2}{\log_{10}^{2.746} + 0.43429} = 2.741$$

$$x_3 = x_2 - \frac{f(x_2)}{f'(x_2)} = 2.741 - \frac{2.741 \log_{10}^{2.741} - 1.2}{\log_{10}^{2.741} + 0.43429} = 2.741.$$

Since $x_2 = x_3 = 2.741$, we stop here and the real root is 2.741.

Example 3: Find the real root of $3x - \cos x - 1 = 0$ by Newton-Raphson method.

Solution: Given $f(x) = 3x - \cos x - 1$

$\therefore f^1(x) = 3 + \sin x$.

$$x = \frac{\pi}{3}, \; f\left(\frac{\pi}{3}\right) = 3 . \frac{\pi}{3} - \cos\frac{\pi}{3} - 1 = 1.64 \text{ (+ve)}$$

$$x = 0, \quad f(0) = 0 - 1 - 1 = -2 \text{ (–ve)}$$

So that $f\left(\pi/3\right) f(0) < 0$. Hence the real root lies between 0 and $\pi/3$ (= 1.047 radians). Also $f\left(\pi/3\right) < f(0)$. We take $x_0 = 1.047$.

Now, by Newton-Raphson Method

$$x_1 = x_0 - \frac{f(x_1)}{f'(x_1)}$$

$$= 1.047 - \frac{3(1.047) - \cos(1.047) - 1}{3 + \sin(1.047)}$$

$$= 0.622379.$$

$$x_2 = x_1 - \frac{f(x_1)}{f'(x_1)} = 0.6071$$

$$x_3 = x_2 - \frac{f(x_2)}{f'(x_2)} = 0.6071$$

Since $x_2 = x_3 = 0.6071$, the real root is 0.6071.

Example 4: Evaluate $\sqrt{12}$ to four decimal places by Newton-Raphson Method.

Solution: Let $x = \sqrt{12}$

$\therefore$ $f(x) = x^2 - 12$ and $f'(x) = 2x$.

Now, $x = 3$, $f(3) = -3$ (–ve)

$x = 4$, $f(4) = 4$ (+ve) and $f(3)\,f(4) < 0$

$\therefore$ the real root lies between 3 and 4.

Take $x_0 = 3$

By Newton-Raphson method

$$x_1 = x_0 - \frac{f(x_0)}{f'(x_0)} = 3 - \left(-\frac{3}{6}\right) = 3.5$$

$$x_2 = x_1 - \frac{f(x_1)}{f'(x_1)} = 3.4642$$

$$x_3 = x_2 - \frac{f(x_2)}{f'(x_2)} = 3.4641$$

$$x_4 = x_3 - \frac{f(x_3)}{f'(x_3)} = 3.4641$$

Since $x_3 = x_4$, the value of $\sqrt{12} = 3.4641$.

Example 5: Find the root of the equation $\sin x = 1 + x^3$ between (–2, –1) to 3 decimal places by Newton-Raphson method.

Solution: Given $f(x) = x^3 - \sin x + 1$

$\therefore \quad f'(x) = 3x^2 - \cos x.$

$x = -1\ f(-1) = 0.8415$ and $x = -2, f(-2) = -60907$

and $|f(-1)| < |f(-2)|$ Hence the real root lies between –1 and –2

We take $x_0 = -1$

By Newton-Raphson method

$$x_1 = x_0 - \frac{f(x_0)}{f'(x_0)} = -1.3421$$

$$x_2 = x_1 - \frac{f(x_1)}{f'(x_1)} = -1.2565$$

$$x_3 = x_2 - \frac{f(x_2)}{f'(x_2)} = -1.2490$$

$$x_4 = x_3 - \frac{f(x_3)}{f'(x_3)} = -1.2491$$

Since $x_3 = x_4 = -1.249$ upto 3 decimal places we take the root of the equation as –1.249.

Example 6: Find to four places of decimals, the smallest root of the equation $e^{-x} = \sin x$ by Newton-Raphson Method.

Solution: Given $f(x) = e^{-x} - \sin x$

$$f'(x) = -e^{-x} - \cos\ x$$

By Newton-Raphson Method, the first approximation to the root is

$$x_1 = x_0 = x_0 - \frac{f(x_0)}{f'(x_0)}$$

Take $x_0 = 0.6$

$$\therefore \quad x_1 = 0.6 - \frac{(e^{-.6}\sin(.6))}{-e^{-.6} - \cos(.6)} = 0.58848$$

$$x_2 = x_1 - \frac{f(x_1)}{f'(x_1)} = 0.588559287$$

Thus the desired root is 0.5885.

Example 7: By using Newton-Raphson Method find the root of $x^4 - x - 10 = 0$ which is near to $x = 2$, correct to 3 decimal places.

Ans.: $x = 1.856$.

Example 8: Find the real root of the equation $x^3 + 1.2\,x^2 - 4x - 4.8 = 0$ to 3 decimal places by N-R Method.

Take $x_0 = -1$

Ans.: $x = -1.200$.

Example 9: Using the Newton-Raphson method find the real root of the equation $x \sin x + \cos x = 0$ near $x = \pi$ carry out 4 iterations (x is in radians)

Ans.: $x = 2.7985$.

Example 10 Using Newton-Raphson method find the root of the equation $e^x = x^3 + \cos 25x$ which is near to 4.5.

Ans.: $x = 4.545$.

Example 12: Develop a recurrence formula for finding $\sqrt{N}$ using Newton-Raphson method and hence compute $\sqrt{32}$ to four decimals.

Ans.: $x = 5.6569$.

5.3 REGULA FALSI METHOD

(Method of false position)

This is the oldest method for finding the real root of a numerical equation $f(x) = 0$. To find the real root of the equation $f(x) = 0$.

Let $x = a$ and $x = b$ be the two numbers such that $f(a) = +\text{ve}$ and $f(b) = -\text{ve}$ and $f(a)\,f(b) < 0$.

Since $f(a)$ and $f(b)$ are of opposite signs the real root of $f(x) = 0$ lies between a and b. Also if $|\,f(b)\,| < |\,f(a)\,|$ we take $x_0 = b$. The value of first approximation by this method is given by

$$x_1 = \frac{a\,f(b) - bf(a)}{f(b) - f(a)}$$

Now, if $f(x_1)$ and $f(a)$ are of opposite signs then the actual root lies between a and x_1

Now, replacing b by x_1 and keeping a as it is we get the next closer approximation.

$$x_2 = \frac{af(x_1) - x_1 f(a)}{f(x_1) - f(a)} \text{ to the actual root.}$$

This process is repeated till the root is found to the desired degree of accuracy.

Example 1: Find the real root of the equation $x\log_{10}^{x} - 1.2 = 0$ by using Regula Falsi method, correct to five places of decimals.

Solution: Let $f(x) = x\log_{10}^{x} - 1.2$

Let $x = 2,\ f(2) = 2\ \log_{10}^{2} - 1.2 = -0.59794$

$x = 3,\ f(3) = 3\ \log_{10}^{3} - 1.2 = +\ 0.23136$

Hence there exists at least one real root of the equation $f(x) = 0$ in the interval (2,3) and it is closer to 3.

Let $x_1 = 2,\ f(x_1) = -0.59794.$

$x_2 = 3,\ f(x_2) = +\ 0.23136.$

By, Regula Falsi method

$$x_3 = \frac{x_1 f(x_2) - x_2 f(x_1)}{f(x_2) - f(x_1)}$$

$$= \frac{2\times .23136 - 3\times .59794}{.23136 - (-.59794)}$$

$$= 2.746$$

Also $f(x_3) = +\ 0.004.$

Since $f(x_1)\,f(x_3) < 0$ the root lies between 2.746 and 3.

$$\therefore \quad x_4 = \frac{x_3 f(x_1) - x_1 f(x_3)}{f(x_1) - f(x_3)} = 2.741$$

Also $f(x_4) = 0.0002$ so that $f(x_4) f(x_1) < 0$

$$x_5 = \frac{x_4 f(x_1) - x_1 f(x_4)}{f(x_1) - f(x_4)} = 2.7408$$

Hence the real root is 2.7408.

Example 2: Find the real root of $x^3 - 2x - 5 = 0$ lying between 2 and 3 by Regula Falsi method.

Solution: Let $f(x) = x^3 - 2x - 5$

$x = 2, f(2) = -1$ (–ve) and $x = 3, f(3) = 16 > 0$

So that $f(2) f(3) < 0$

Take $x_1 = 2, x_2 = 3$.

By Regula Falsi method

$$x_3 = \frac{x_1 f(x_2) - x_2 f(x_1)}{f(x_2) - f(x_1)} = 2.058$$

and $f(x_3) = -0.4$ so that $f(x_2) f(x_3) < 0$

$$x_4 = \frac{x_3 f(x_2) - x_2 f(x_3)}{f(x_2) - f(x_3)} = 2.081$$

and $f(x_4) = -0.15$ so that $f(x_2) f(x_4) < 0$

$$x_5 = \frac{x_4 f(x_2) - x_2 f(x_4)}{f(x_2) - f(x_4)} = 2.089$$

and $f(x_5) = -0.062$ so that $f(x_2) f(x_5) < 0$

$$x_6 = \frac{x_5 f(x_2) - x_2 f(x_5)}{f(x_2) - f(x_5)} = 2.093.$$

and $f(x_6) = $ +ve.

Hence the real root is $x = 2.09$.

Example 3: Solve $x^3 - 9x + 1 = 0$ for the root lying between 2 and 4 by Regula Falsi Method.

Solution: Let $f(x) = x^3 - 9x + 1$

$x_1 = 2, \therefore f(x_1) = -9$ and $x_2 = 4, f(x_2) = 29$.

So that $f(x_1) f(x_2) < 0$. Hence the real root lies between 2 and 4.

By Regula Falsi Method

$$x_3 = \frac{x_1 f(x_2) - x_2 f(x_1)}{f(x_2) - f(x_1)} = 2.474$$

and $f(x_3) = -6.123$ so that $fx_2) f(x_3) < 0$

$$\therefore \quad x_4 = \frac{x_3 f(x_2) - x_2 f(x_3)}{f(x_2) - f(x_3)} = 2.740$$

and $f(x_4) = -3.183$ so that $f(x_2) f(x_4) < 0$

$$\therefore \quad x_5 = \frac{x_4 f(x_2) - x_2 f(x_4)}{f(x_2) - f(x_4)} = 2.859$$

and $f(x_5) = -1.362$ so that $f(x_5) f(x_2) < 0$

$$\therefore \quad x_6 = \frac{x_5 f(x_2) - x_2 f(x_5)}{f(x_2) - f(x_5)} = 2.911$$

and $f(x_6) = -0.531$ so that $f(x_2) f(x_6) < 0$

$$\therefore \quad x_7 = \frac{x_6 f(x_2) - x_2 f(x_6)}{f(x_2) - f(x_6)} = 2.931$$

and $f(x_7) = -0.200$ so that $f(x_2) f(x_7) < 0$

$$\therefore \quad x_8 = \frac{x_7 f(x_2) - x_2 f(x_7)}{f(x_2) - f(x_7)} = 2.939$$

and $f(x_8) = -0.065$ so that $f(x_2) f(x_8) < 0$

$$\therefore \quad x_9 = \frac{x_8 f(x_2) - x_2 f(x_8)}{f(x_2) - f(x_8)} = 2.942$$

and $f(x_9) = -0.014$ so that $f(x_2) f(x_9) < 0$

$$\therefore \quad x_{10} = \frac{x_9 f(x_2) - x_2 f(x_9)}{f(x_2) - f(x_9)} = 2.943$$

and $f(x_{10}) = +\ 0.003$ so that $f(x_9) f(x_{10}) < 0$

$$\therefore \quad x_{11} = \frac{x_{10} f(x_9) - x_9 f(x_{10})}{f(x_9) - f(x_{10})} = 2.9428$$

and $f(x_{11}) = -0.0003$ so that $f(x_{10}) f(x_{11}) < 0$

$$\therefore \quad x_{12} = \frac{x_{11} f(x_{10}) - x_{10} f(x_{11})}{f(x_{10}) - f(x_{11})} = 2.9428.$$

Since $x_{11} = x_{12}$ we take the real root of the equation $f(x) = 0$ is $x = 2.9428$.

Example 4: Find the root of the equation $xe^x = \cos x$ using Regula Falsi method correct to four decimal places.

Solution: Let $f(x) = \cos x - xe^x = 0$

$x = 0, f(0) = 1$ +ve and $x = 1, f(1) = -2.17798$.

So that the real root lies between 0 and 1.

Take $x_1 = 0$ and $x_2 = 1$ $\therefore f(x_1) = 1$ and $f(x_2) = -2.17798$

By Regula Falsi method

$$x_3 = \frac{x_1 f(x_2) - x_2 f(x_1)}{f(x_2) - f(x_1)} = 0.31467$$

and $f(x_3) = 0.51987$ so that $f(x_2) f(x_3) < 0$

$$\therefore \quad x_4 = \frac{x_3 f(x_2) - x_2 f(x_3)}{f(x_2) - f(x_3)} = 0.44673$$

and $f(x_4) = 0.20356$ so that $f(x_2) f(x_4) < 0$

$$\therefore \quad x_5 = \frac{x_4 f(x_2) - x_2 f(x_4)}{f(x_2) - f(x_4)} = 0.49402$$

Repeating the process, the successive approximations are

$x_6 = 0.50995$, $x_7 = 0.51520$, $x_8 = 0.51692$ $x_9 = 0.51748$

$x_{10} = 0.51767$ and $x_{11} = 0.51775$

Hence the root is 0.5177 correct to four decimal places.

Example 5: The equation $x^6 - x^4 - x^3 - 1 = 0$ has one real root between 1.4 and 1.5. Find the root to four decimal places.

Solution: Let $f(x) = x^2 - x^4 - x^3 - 1 = 0$

Let $x_1 = 1.4$ $\therefore f(x_1) = -.056061$ and $x_2 = 1.5$ $f(x_2) = 1.953125$

So that the real root lies between 1.4 and 1.5

By Regula Falsi method

$$x_3 = \frac{x_1 f(x_2) - x_2 f(x_1)}{f(x_2) - f(x_1)} = 1.403$$

and $f(x_3) = -.0167$ so that $f(x_3)\,f(x_2) < 0$

$$\therefore \qquad x_4 = \frac{x_3 f(x_2) - x_2 f(x_3)}{f(x_2) - f(x_3)} = 1.4037$$

and $f(x_4) = .0004$ so that $f(x_4)\,f(x_3) < 0$

$$\therefore \qquad x_5 = \frac{x_4 f(x_3) - x_3 f(x_4)}{f(x_3) - f(x_4)} = 1.4036.$$

Hence the real root is $x = 1.4036$.

Example 6: Using Regula Falsi method compute the real root of the equation $xe^x - \sin x = 0$ correct to 3 decimal places.

Ans.: $x = -0.134$.

Example 7: Compute the real root of the equation $xe^x - 2 = 0$ by using Regula Falsi method.

Ans.: $x = 0.853$.

Example 8: Find by Regula Falsi method the real root of the equation

$x^3 - 3x + 4 = 0$ to 3 decimal places.

Ans.: $x = -2.198$

Example 9: Compute the real root of the equation $\tan x + \tanh x = 0$ which lies between 2 and 3 (x is in radians) by Regula Falsi method.

$$\text{Note : } \tanh x = \frac{e^x - e^{-x}}{e^x + e^{-x}} = \frac{e^{2x} - 1}{e^{2x} + 1}$$

$$\therefore \qquad f(x) = \tan x + \frac{e^{2x} - 1}{e^{2x} + 1}$$

Ans.: $x = 2.3993$ is the third approximation.

Example 10: Find the real root of the equation $\log^x = \cos x$ in (1,2) in radians by Regula Falsi method.

Ans.: $x = 1.3029$.

5.4 ITERATIVE METHODS OF SOLUTION OF A SYSTEM OF EQUATIONS

1) Gauss-Seidel iterative Method

If in each equation, the coefficient of one of the unknown is greater (numerically) than the sum of the numerical values of the other coefficients of the other unknowns and the larger coefficients in equations correspond to the different unknowns in different equations, then this type of system of simultaneous linear equation is known as a **diagonal system** and can be solved by Gauss-Seidel iterative method.

Let $a_{11} x_1 + a_{12} x_2 + \ldots\ldots + a_{1n} x_n = b_1$

$$a_{21} x_2 + a_{22} x_2 + \ldots\ldots a_{2n} x_n = b_2$$

$$\vdots \qquad\qquad \vdots$$

$$\vdots \qquad\qquad \vdots$$

$$a_{n1} x_1 + a_{n2} x_2 + \ldots\ldots a_{nn} x_n = b_n$$

be the set of n-equations in n-unknowns.

We solve each equation of the system for the unknowns with the largest coefficients in the first, second ... n^{th} equation respectively.

$$\therefore \qquad x_1 = \frac{1}{a_{11}} [b_1 - a_{12} x_2 - a_{13} x_3 \ldots\ldots - a_{1n} x_n]$$

$$x_2 = \frac{1}{a_{22}} [b_2 - a_{12} x_1 - a_{23} x_3 \ldots\ldots - a_{2n} x_n]$$

$$x_3 = \frac{1}{a_{33}} [b_3 - a_{31} x_1 - a_{32} x_2 \ldots\ldots - a_{3n} x_n]$$

$$x_n = \frac{1}{a_{nn}} [b_n - a_{n1} x_1 - a_{n2} x_2 \ldots a_{n\,n-1} x_{n-1}]$$

Now assume any initial values for $x_2, x_3 \ldots\ldots x_n$

Say $\qquad x_2 = x_3 = \ldots x_n = 0$

$$\therefore \qquad x_1^{(1)} = \frac{1}{a_{11}} [b_1 - 0 - 0 \ldots] = \frac{b_1}{a_{11}}$$

$$x_2^{(1)} = \frac{1}{a_{22}} [b_2 - a_{21} x_1^{(1)} + 0 \ldots] = \frac{1}{a_{22}} (b_2 - a_{21} x_1^{(1)})$$

$$x_3^{(1)} = \frac{1}{a_{33}} [b_3 - a_{31} x_1^{(1)} - a_{32} x_2^{(1)} - 0 \ldots]$$

$$\vdots$$

$$x_n^{(1)} = \frac{1}{a_{nn}} [b_n - a_{n1} x_1^{(1)} - a_{n2} x_2^{(1)} \ldots . a_{n\,n-1} x_{n-1}^{(1)}]$$

I iteration

$$x_1^{(2)} = \frac{1}{a_{11}} [b_1 - a_{12} x_2^{(1)} - a_{13} x_3^{(1)} \ldots . a_{1n} x_n^{(1)}]$$

$$x_2^{(2)} = \frac{1}{a_{22}} [b_2 - a_{21} x_1^{(2)} - a_{23} x_3^{(1)} \ldots . a_{2n} x_n^{(1)}]$$

$$x_3^{(2)} = \frac{1}{a_{33}} [b_3 - a_{31} x_1^{(2)} - a_{32} x_2^{(2)} \ldots a_{3n} x_n^{(1)}]$$

$$\vdots$$

$$x_n^{(2)} = \frac{1}{a_{nn}} [b_\text{n} - a_{n1} x_1^{(2)} - a_{n2} x_2^{(2)} \ldots a_{n\,n-1} x_{n-1}^{(2)}]$$

This technique is continued until $x_j^{(p)}$ and $x_j^{(p+1)}$ are equal to the desired degree of accuracy.

Note: i) This method is very useful with less work for the given system of equations whose augmented matrix have large number of zero elements.

ii) We say a matrix is diagonally dominant if the numerical value of the leading diagonal element in each row is greater than or equal to the sum of the numerical values of the other elements in that row.

iii) For Gauss-Seidel method to converge quickly, the coefficient matrix must be diagonally dominant. If it is not so we have to rearrange the equations in such a way that the coefficient matrix is diagonally dominant and then only we can apply Gauss-Seidel iterative method.

For example, consider the equations

$$8x - 3y + 2x = 7$$

$$6x + 3y + 12z = 15$$

$$4x + 11y - z = 13$$

Here the coefficient matrix is not diagonally dominant.

To make the coefficient matrix diagonally dominant interchange second and third equations as

$$8x - 3y + 2z = 7$$

$$4x + 11y - z = 13$$

$$6x + 3y + 12z = 15$$

Now, we can apply Gauss-Seidel iterative method.

Consider another example as

$$x + 7y - z = 3$$

$$5x + y + z = 9$$

$$-3x + 2y + 7z = 17$$

Here the coefficient matrix $\begin{bmatrix} 1 & 7 & 1 \\ 5 & 1 & 1 \\ -3 & 2 & 7 \end{bmatrix}$ is not diagonally dominant.

But the given equations can be rearranged as

$$7y + x + z = 3$$

$$y + 5x + z = 9$$

$$2y - 3x + 7z = 17$$

Now the coefficient matrix $\begin{bmatrix} 7 & 1 & 1 \\ 1 & 5 & 1 \\ 2 & -3 & 7 \end{bmatrix}$ is diagonally dominant.

When these equations are solved solution will be in the order y, x and z.

Example 1: Solve the system of equations

$$8x - y + z = 18$$

$$2x + 5y - 2z = 3$$

$$x + 2y - 3z = -6$$

Using Gauss-Seidel iterative method.

Solution: The system of equations is

$$8x - y + z = 18$$

$$2x + 5y - 2z = 3$$

$$x + 2y - 3z = -6$$

Clearly the coefficient matrix is diagonally dominant.

Hence we can apply Gauss-Seidel iterative method.

$$\therefore \quad x = \frac{1}{8}(18 + y - z) \quad \ldots (1)$$

$$y = \frac{1}{5}(3 - 2x + 2z) \quad \ldots (2)$$

$$z = \frac{1}{3}(6 + x + y) \quad \ldots (3)$$

Initially, we start with $x = 0$, $y = 0$, $z = 0$

$$\therefore \quad x^{(1)} = \frac{1}{8}(18 + 0 - 0) = 2.25$$

$$y^{(1)} = \frac{1}{5}(3 - 2x^{(1)} + 2 \times 0) = \frac{1}{5}[3 - 2(2.25)] = -0.3$$

$$z^{(1)} = \frac{1}{3}(6 + x^{(1)} + y^{(1)}) = \frac{1}{3}[6 + 2.25 - 0.3] = 2.65$$

II iteration

$$x^{(2)} = \frac{1}{8}(18 - y^{(1)} - z^{(1)}) = \frac{1}{8}[18 + (-0.3) - 2.65] = 1.8813$$

$$y^{(2)} = \frac{1}{5}(3 - 2x^{(2)} + 2z^{(1)}) = \frac{1}{5}[3 - 2(1.8818) + 2(2.65)] = 0.9075$$

$$z^{(2)} = \frac{1}{3}(6 + x^{(2)} + y^{(2)}) = \frac{1}{3}[6 + 1.8813 + 0.9075] = 2.9296$$

III iteration

$$x^{(3)} = \frac{1}{8}(18 + y^{(2)} - z^{(2)}) = \frac{1}{8}(18 + 0.9075 - 2.9296) = 1.9972$$

$$y^{(3)} = \frac{1}{5}(3 - 2x^{(3)} + 2z^{(2)}) = \frac{1}{5}[3 - 2(1.9972) + 2(2.9296)] = 0.9729$$

$$z^{(3)} = \frac{1}{3}(6 + x^{(3)} + y^{(3)}) = \frac{1}{3}[6 - 1.9972 + 0.9729] = 2.9900$$

IV iteration

$$x^{(4)} = \frac{1}{8}(18 + y^{(3)} - z^{(3)}) = \frac{1}{8}[18 + 0.9729 - (2.9900)] = 1.9979$$

$$y^{(4)} = \frac{1}{5}(3 - 2x^{(4)} + 2z^{(3)}) = \frac{1}{5}[3 - 2(1.9979) + 2(2.9900)] = 0.9968$$

$$z^{(4)} = \frac{1}{3}(6 + x^{(4)} + y^{(4)} = \frac{1}{3}[6 + 1.9979 + 0.9968] = 2.9982.$$

V iteration

$$x^{(5)} = \frac{1}{8}(18 + y^{(4)} - z^{(4)}) = \frac{1}{8}[18 - 0.9968 - 2.9982] = 1.9998$$

$$y^{(5)} = \frac{1}{5}(3 - 2x^{(5)} + 2z^{(4)}) = \frac{1}{5}[3 -2(1.9998) + 2(2.9982)] = 0.9994$$

$$z^{(5)} = \frac{1}{3}(6 + x^{(5)} + y^{(5)}) = \frac{1}{3}[6 + 1.9998 + 0.9994] = 2.9997.$$

VI iteration

$$x^{(6)} = \frac{1}{8}(18 + y^{(5)} - z^{(5)}) = \frac{1}{8}[18 + 0.9994 - 2.9997] = 2.000$$

$$y^{(6)} = \frac{1}{5}(3 - 2x^{(6)} + 2z^{(5)}) = \frac{1}{5}[3 - 2 \times 2 + 2(2.997)] = 0.9998$$

$$x^{(6)} = \frac{1}{3}(6 + x^{(6)} + y^{(6)}) = \frac{1}{3}[6 + 2.000 + 0.9998] = 2.999$$

After VIth iteration we get

$$x = 2, \qquad y = 0.9990, \qquad z = 2.999$$

Example 2: Using Gauss-Seidel iterative method solve the equations $8x - 3y + 2z = 20$, $6x + 3y + 12z = 35$ and $4x + 11y - z = 33$.

Solution: Rearrange the equations so that the coefficient matrix is diagonally dominant, i.e.,

$$8x - 3y + 2z = 20$$

$$4x + 11y - z = 33$$

$$6x + 3y + 12z = 35.$$

$$\therefore \quad x = \frac{1}{8}(20 + 3y - 2z) \quad \ldots (1)$$

$$y = \frac{1}{11}(33 - 4x + z) \quad \ldots (2)$$

$$z = \frac{1}{12}(35 - 6x - 3y) \quad \ldots (3)$$

We start with the initial approximation $x^{(0)} = y^{(0)} = z^{(0)} = 0$.

$$x^{(1)} = \frac{1}{8}[20 + 3y^{(0)} - 2z^{(0)}] = 2.5$$

$$y^{(1)} = \frac{1}{11}[33 - 4x^{(1)} + z^{(0)}] = 2.0909091$$

$$z^{(1)} = \frac{1}{12}[35 - 6x^{(1)} - 3y^{(1)}] = 1.1439394$$

II iteration

$$x^{(2)} = \frac{1}{8}[20 + 3y^{(1)} - 2z^{(1)}] = 2.9981061$$

$$y^{(2)} = \frac{1}{11}[33 - 4x^{(2)} + z^{(1)}] = 2.0137741$$

$$z^{(2)} = \frac{1}{12}[35 - 6x^{(2)} - 3y^{(2)}] = 0.9141701$$

III iteration

$$x^{(3)} = \frac{1}{8}[20 + 3y^{(2)} - 2z^{(2)}] = 3.0266228$$

$$y^{(3)} = \frac{1}{11}[33 - 4x^{(3)} + z^{(2)}] = 1.9825163$$

$$z^{(3)} = \frac{1}{12}[35 - 6x^{(3)} - 3y^{(3)}] = 0.9077262.$$

IV iteration

$$x^{(4)} = \frac{1}{8}[20 + 3y^{(3)} - 2z^{(3)}] = 3.0165121$$

$$y^{(4)} = \frac{1}{11}[33 - 4x^{(4)} + z] = 1.9856071$$

$$z^{(4)} = \frac{1}{12}[35 - 6x^{(4)} - 3y^{(4)}] = 0.912008$$

V iteration

$$x^{(5)} = \frac{1}{8}[20 + 3y^{(4)} - 2z^{(4)}] = 3.016605.$$

$$y^{(5)} = \frac{1}{11}[33 - 4x^{(5)} + z^{(4)}] = 1.9859643.$$

$$z^{(5)} = \frac{1}{12}[35 - 6x^{(5)} - 3y^{(5)}] = 0.9118753.$$

VI iteration

$$x^{(6)} = \frac{1}{8}[20 + 3y^{(5)} - 2z^{(5)}] = 3.0167678$$

$$y^{(6)} = \frac{1}{11}[33 - 4x^{(6)} + z^{(5)}] = 1.9858913$$

$$z^{(6)} = \frac{1}{12}[35 - 6x^{(6)} - 3y^{(6)}] = 0.9118099$$

VII iteration

$$x^{(7)} = \frac{1}{8}[20 + 3y^{(6)} - 2z^{(6)}] = 3.0167568$$

$$y^{(7)} = \frac{1}{11}[33 - 4x^{(7)} + z^{(6)}] = 1.9858894$$

$$z^{(7)} = \frac{1}{12}[35 - 6x^{(7)} - 3y^{(7)}] = 0.9118159.$$

Since at VIth and VIIth iteration the values of x, y, z are same correct to four decimal places. So we top the iteration.

$$\therefore \quad x = 3.0167, \quad y = 1.9858, \quad z = 0.9118.$$

Example 3: Using Gauss-Seidel iterative method solve the system of equations

$$10x - 2y - z - w = 3, \quad -2x + 10y - z - w = 15$$

$$-x - y + 10z - 2w = 27, \quad -x - y - 2z + 10w = -9$$

Solution: The coefficient matrix of the given system is diagonally dominant. Hence, we can apply Gauss-Seidel iteration method.

$$\therefore \quad x = \frac{1}{10}[3 + 2y + z + w] \quad \ldots (1)$$

$$y = \frac{1}{10}[15 + 2x + z + w] \quad \ldots (2)$$

$$z = \frac{1}{10}[27 + x + y + 2w] \quad \ldots (3)$$

$$w = \frac{1}{10}[-9 + x + y + 2z] \quad \ldots (4)$$

I iteration : put $y = z = w = 0$

$$x^{(1)} = \frac{1}{10}[3 + 0 + 0 + 0] = 0.3$$

$$y^{(2)} = \frac{1}{10}[15 + 2x^{(1)} + 0 + 0] = 1.56$$

$$z^{(1)} = \frac{1}{10}[27 + x^{(1)} + y^{(1)} + 0] = 2.886$$

$$w^{(1)} = \frac{1}{10}[-9 + x^{(1)} + y^{(1)} + 2z^{(1)}] = -0.1368.$$

II iteration

$$x^{(2)} = \frac{1}{10}[3 + 2y^{(1)} + z^{(1)} + w^{(1)}] = 0.88692$$

$$y^{(2)} = \frac{1}{10}[15 + 2x^{(2)} + z^{(1)} + w^{(1)}] = 2.9565$$

$$w^{(2)} = \frac{1}{10}[-9 + x^{(2)} + y^{(2)} + 2z^{(2)}] = 0.0247651$$

III iteration

$$x^{(3)} = \frac{1}{10}[3 + 2y^{(2)} + z^{(2)} + w^{(2)}] = 0.9836405$$

$$y^{(3)} = \frac{1}{10}[15 + 2x^{(3)} + z^{(2)} + w^{(2)}] = 1.9899087$$

$$z^{(3)} = \frac{1}{10}[27 + x^{(3)} + y^{(3)} + w^{(2)}] = 2.9924019$$

$$w^{(3)} = \frac{1}{10}[-9 + x^{(3)} + y^{(3)} + 2z^{(3)}] = -0.0041647.$$

IV iteration

$x^{(4)} = 0.9968054,$ $y^{(4)} = 1.9981848,$

$z^{(4)} = 2.9986661$ $w^{(4)} = -0.0007677.$

V iteration

$x^{(5)} = 0.9994268,$ $y^{(5)} = 1.99962752,$

$z^{(5)} = 2.9996752,$ $w^{(5)} = -0.0001384.$

VI iteration

$x^{(6)} = 0.9998968,$ $y^{(6)} = 1.9999412,$

$z^{(6)} = 2.9999561,$ $w^{(6)} = -0.0002498$

VII iteration

$x^{(7)} = 0.9999588,$ $y^{(7)} = 1.9999624,$

$z^{(7)} = 2.9999422,$ $w^{(7)} = -0.0001945$

∴ Up to four decimal places the values are

$x = 0.9999,\ y = 1.9999,\ z = 2.9999,\ w = -0.0002.$

Example 4: Using Gauss-Seidel iterative method solve the equations

$$20x + y - 2z = 17,\ 3x + 20y - z = -18;\ 2x - 3y + 20z = 25.$$

Solution: The coefficient matrix of the given set of equations is diagonally dominant.

Hence we apply Gauss-Seidel iterative method.

We write the given equations in the form.

$$x = \frac{1}{20}[17 - y + 2z] \qquad \ldots (1)$$

$$y = \frac{1}{20}[-18 - 3x + z] \qquad \ldots (2)$$

$$z = \frac{1}{20}[25 - 2x + 3y] \qquad \ldots (3)$$

Initially, we start with $x = y = z = 0$

$$x^{(1)} = \frac{1}{20}[17 - 0 + 0] = 0.8500$$

$$y^{(1)} = \frac{1}{20}[-18 - 3x^{(1)} + 0] = -1.0275.$$

$$z^{(1)} = \frac{1}{20}[25 - 2x^{(1)} + 3y^{(1)}] = 1.0109.$$

II iteration

$$x^{(2)} = \frac{1}{20}[17 - y^{(1)} + 2z^{(1)}] = 1.0025$$

$$y^{(2)} = \frac{1}{20}[-18 - 3x^{(2)} + z^{(1)}] = -0.998$$

$$z^{(2)} = \frac{1}{20}[25 - 2x^{(2)} + 3y^{(2)}] = 0.9998$$

III iteration

$$x^{(3)} = \frac{1}{20}[17 - y^{(2)} + 2z^{(2)}] = 1.0000$$

$$y^{(3)} = \frac{1}{20}[-18 - 3x^{(3)} + z^{(2)}] = -1.0000$$

$$z^{(3)} = \frac{1}{20}[25 - 2x^{(3)} + 3y^{(3)}] = 1.0000.$$

The values in the IInd and IIIrd iterations being practically the same. Hence the solution is

$x = 1, \quad y = -1, \quad z = 1.$

Example 5: Using Gauss-Seidel iterative method solve the following equations.

$54x + y + z = 110; \quad 2x + 15y + 6z = 72; \quad -x + 6y + 27z = 85$

Ans.: $x = 1.926, \quad y = 3.573, \quad z = 2.425.$

Example 6: Using Gauss-Seidel iterative method, solve the following equations.

$10x + y + z = 12; \quad 2x + 10y + z = 13, \quad 2x + 2y + 10z = 14.$

Ans.: $x = 1, y = 1, z = 1.$

Example 7: Solve by Gauss-Seidel iterative method

$2x + y + z = 4; \quad x + 2y + z = 4, \quad x + y + 2z = 4$

Ans.: $x = 1, y = 1, z = 1.$

Example 8: Solve by Gauss-Seidel iteration method

$10x_1 - 5x_2 - 2x_3 = 3; \quad 4x_1 - 10x + 3x_3 = -3; \quad x_1 + 6x_2 + 10x_3 = -3$

Ans.: $x_1 = 0.342, x_2 = 0.385, x_3 = -0.505$

Example 9: Solve by Gauss-Seidel iteration method

$10x_1 + 7x_2 + 8x_3 + 7x_4 = 32; \quad 7x_1 + 5x_2 + 6x_3 + 5x_4 = 24$

$8x_1 + 6x_2 + 10x_3 + 9x_4 = 33; \quad 7x_1 + 5x_2 + 9x_3 + 10x_4 = 31$

Ans.: $x_1 = 1, \quad x_2 = 1, \quad x_3 = 1, \quad x_4 = 1.$

Example 10: Solve by Gauss-Seidel iteration method

$9x + 2y + 4z = 20, \quad x + 10y + 4z = 6; \quad 2x - 4y + 10z = -15$

Ans.: $x = 2.733 \quad y = 0.986; \quad z = z = -1.652.$

5.5 RELAXATION METHOD

Consider the equations

$$a_1 x + b_1 y + c_1 z = d_1$$
$$a_2 x + b_2 y + c_2 z = d_2$$
$$a_3 x + b_3 y + c_3 z = d_3$$

Define the residuals R_1, R_2, R_3 by the relations

$$R_1 = d_1 - a_1 x - b_1 y - c_1 z$$
$$R_2 = d_2 - a_2 x - b_2 y - c_2 z \qquad ...(1)$$
$$R_3 = d_3 - a_3 x - b_3 y - c_3 z$$

To start with

i) we assume $x = y = z = 0$ and calculate the initial residuals.

ii) These residuals are reduced step by step by giving increments to the variable.

iii) Find x, y, z such that the residuals $R_1 = R_2 = R_3 = 0$. Then those values of x, y, z are the exact values.

iv) Otherwise, we liquidate the residuals smaller and smaller and finally negligible to get better approximate values of x, y, z.

For this purpose we construct an operational table as follows:

x	y	z	R_1	R_2	R_3
1	0	0	$-a_1$	$-a_2$	a_3
0	1	0	$-b_1$	$-b_2$	$-b_3$
0	0	1	$-c_1$	$-c_2$	$-c_3$

From equations (1) we can see that if x is increased by 1, keeping y and z constants, R_1, R_2, R_3 decrease by a_1, a_2, a_3 respectively.

Similarly by giving increments to y and z a unit increment, the above table shows the effect on the residuals.

Note that the operation table consists of the unit matrix I and transpose of the coefficient matrix.

At each step the numerically largest residual is reduced almost zero. To reduce a particular residual the value of corresponding variable is changed i.e., to reduce say R_2 by ky should be increased by k/b_2. When all the residuals have been reduced to almost zero, then the increments in x, y, z are added separately to give the desired solution.

Note: Check whether the residuals are negligible or not by substituting the values x, y, z in (1). If not then there is some mistake and the entire process should be rechecked.

Convergency of the relaxation method

This method can be applied successfully only if the diagonal elements of the coefficient matrix dominate the other coefficients in the corresponding row i.e., in the equations (1)

$$|a_1| \geq |b_1| + |c_1| \qquad |b_2| \geq |a_2| + |c_2| \quad \text{and} \quad |c_3| \geq |a_3| + |b_3|$$

With strict inequality for at least one row.

Example 1: Solve by relaxation method the equations

$$10x - 2y - 2z = 6; \qquad -x + 10y - 2z = 7; \qquad -x - y + 10z = 8$$

Solution: The coefficient matrix is diagonally dominant and we can apply relaxation method.

The residuals are

$R_1 = 6 - 10x + 2y + 2z$,; $R_2 = 7 + x - 10y + 2z$; $R_3 = 8 + x + y - 10z$.

The operation table is

x	y	z	R_1	R_2	R_3	
1	0	0	–10	1	1	$\leftarrow L_1$
0	1	0	2	–10	1	$\leftarrow L_2$
0	0	1	2	2	–10	$\leftarrow L_3$

Relaxation table is

x	y	z	R_1	R_2	R_3	
0	0	0	6	7	8	$\leftarrow L_4\ (x = y = z = 0)$
0	0	1	8	9	–2	$\leftarrow L_5 = L_4 + L_3\ \left(\frac{8}{10} = .8 \simeq 1\right)$
0	1	0	10	–1	–1	$\leftarrow L_6 = L_5 + L_2\ \left(\frac{9}{10} = .9 \simeq 1\right)$
1	0	0	0	0	0	$\leftarrow L_7 = L_6 + L_1\ \left(\frac{10}{10} = 1\right)$

Steps (i) In the line L_4 largest residual = 8

To reduce it we give an increament $= \dfrac{8}{c_3} = \dfrac{8}{10} = 0.8 \simeq 1$

The resulting residuals are obtained by $L_4 + (1)\ L_3$ i.e., the line L_5

Step (ii) In the line L_5 largest residual = 9

$$\therefore \quad \text{Increment} = \frac{9}{b_2} = \frac{9}{10} = 0.9 \simeq 1$$

The resulting residuals = $L_6 = L_5 + (1)\ L_2$ i.e., the line L_6.

Step (iii) In the line L_6 largest residual = 10

$$\therefore \quad \text{increment} = \frac{10}{a_1} = \frac{10}{10} = 1$$

The resulting residuals = $L_7 = L_6 + (1)\ L_1$ i.e., line L_7

In L_7, all residuals are zeros.

$\therefore$ Exact solution is arrived and it is $x = 1$, $y = 1$ and $z = 1$.

Example 2: Solve by relaxation method, the equations

$$9x - y + 2z = 9; \qquad x + 10y - 2z = 15, \qquad 2x - 2y - 13z = -17.$$

Solution: Since the coefficient matrix is diagonally dominant we apply relaxation method.

The residuals are given by

$R_1 = 9 - 9x + y - 2z$; $R_2 = 15 - x - 10y + 2z$; $R_3 = -17 - 2x + 2y + 13z$.

The operation table is

x	y	z	R_1	R_2	R_3	
1	0	0	–9	–1	–2	$\leftarrow L_1$
0	1	0	1	–10	2	$\leftarrow L_2$
0	0	1	–2	2	13	$\leftarrow L_3$

The relaxation table is.

x	y	z	R_1	R_2	R_3	
0	0	0	9	15	–17	$\leftarrow L_4$ put $x = y = z = 0$
0	0	1	7	17	–4	$L_5 = L_4 + 1L_3\left(\frac{17}{c_3} = 1.3 \simeq 1\right)$
0	1	0	8	7	–2	$L_6 = L_5 + 1L_2\left(\frac{17}{b_2} = 1.7 \simeq 1\right)$
.89	0	0	–0.01	6.11	–3.78	$L_7 = L_6 + .89L_1\left(\frac{8}{a_1} = .89\right)$
0	0.61	0	0.6	0.01	–2.56	$L_8 = L_7 + .61L_2\left(\frac{6.11}{b_2} = .61\right)$
0	0	0.19	0.22	0.39	–0.09	$L_9 = L_8 + 0.1L_3$
0	0.039	0	0.259	0	–0012	$L_{10}=L_9+0.039L_2\left(\frac{.39}{10} = .039\right)$
0.028	0	0	0.007	–0.028	–0.068	$L_{11} = L_{10} + 0.028L_1$
0	0	0.00523	–0.00346	– .01754	–0.00001	$L_{12} = L_{11} + 0.00523L_3$

Thus $\quad x = 0.89 + 0.028 = 0.918$

$$y = 1 + 0.61 + 0.039 = 1.649$$

$$z = 1 + 0.19 + 0.00523 = 1.19523.$$

Checking:

$$R_1 = 9 - 9\,(0.918) + 1.649 - 2\,(1.19523) = -\,0.00346$$

$$R_2 = 15 - 0.918 - 10\,(1.649) + 2\,(1.19523) = -\,0.01754$$

$$R_3 = -17 - 2\,(0.918) + 2\,(1.649) + 13\,(1.19523) = -\,0.00001$$

Which are in agreement with the final residuals in the table.

Example 3: Solve by relaxation method the equations

$$9x - 2y + z = 50, \qquad x + 5y - 3z = 18, \qquad -2x + 2y + 7z = 19.$$

Solution: Since the coefficient matrix is diagonally dominant we apply relaxation method.

The residuals are given by.

$R_1 = 50 - 9x + 2y - z$, $R_2 = 18 - x - 5y + 3z$, $R_3 = 19 + 2x - 2y - 7z$.

The operation table is :

x	y	z	R_1	R_2	R_3
1	0	0	–9	–1	2
0	1	0	2	–5	–2
0	0	1	–1	3	–7

Relaxation table is :

x	y	z	R_1	R_2	R_3	
0	0	0	50	18	19	L_4
5	0	0	5	13	29	$L_5 = L_4 + 5L_1 \quad \left(\frac{50}{9} = 5.5 \simeq 5\right)$
0	0	4	1	25	1	$L_6 = L_5 + 4L_3 \left(\frac{29}{9} = 4.1 \simeq 4\right)$
0	5	0	11	0	–9	$L_7 = L_6 + 5L_2 \quad \left(\frac{25}{5} = 5\right)$
1	0	0	2	–1	–7	$L_8 = L_7 + 1L_1 \quad \left(\frac{11}{9} = 1.2 \simeq 1\right)$
0	0	–1	3	–4	0	$L_9 = L_8 + (-1)L_3 \quad \left(\frac{-7}{+7} = -1\right)$
0	–0.8	0	1.4	0	1.6	$L_{10} = L_9 + (-.8)L_2 \left(+\frac{4}{+5} = -.80\right)$
0	0	.23	1.7	0.69	–.09	$L_{11} = L_{10} + .23L_3$
0.13	0	0	0	0.56	0.17	$L_{12} = L_{11} + 0.13L_1$
0	0.112	0	0.224	0	–0.054	$L_{13} = L_2 + 0.112L_2$

Thus
$$x = 5 + 1 + 0.13 = 6.13$$
$$y = 5 - 0.8 + 0.112 = 4.31$$
$$z = 4 - 1 + .23 = 3.23$$

$\therefore$ The values are $x = 6.13$, $y = 4.31$, $z = 3.23$.

Example 4: Solve by relaxation method, the equations

$$10x - 2y - 3z = 205;\ -2x + 10y - 2z = 154;\ -2x - y + 10z = 120$$

***Solution*:** The coefficient matrix is diagonally dominant and we can use relaxation method.

The residuals are given by

$R_1 = 205 - 10x + 2y + 3z$; $R_2 = 154 + 2x - 10y + 2z$; $R_3 = 120 + 2x + y - 10z$.

The operation table is

x	y	z	R_1	R_2	R_3	
1	0	0	–10	2	2	$\leftarrow L_1$
0	1	0	2	–10	–1	$\leftarrow L_2$
0	0	1	3	2	–10	$\leftarrow L_3$

Relaxation table is

x	y	z	R_1	R_2	R_3	
0	0	0	205	154	120	$\leftarrow L_4$
20	0	0	5	194	160	$\leftarrow L_5 = L_4 + 20L_1 \left(\frac{205}{+10} \simeq +20\right)$
0	11	0	43	4	179	$L_6 = L_5 + 19L_2 \left(\frac{194}{10} = 19.4 = 19\right)$
0	0	18	97	40	–1	$L_7 = L_6 + 18L_3 \quad \left(\frac{179}{10} = 18\right)$
10	0	0	–3	60	19	$L_8 = L_7 + 10L_1 \quad \left(\frac{97}{10} = 10\right)$
0	6	0	9	0	25	$L_9 = L_8 + 6L_2 \quad \left(\frac{60}{10} = 6\right)$
0	0	2	15	4	5	$L_{10} = L_9 + 2L_3 \quad \left(\frac{25}{10} = 2\right)$
2	0	0	–5	8	9	$L_{11} = L_{10} + 2L_1 \quad \left(\frac{15}{10} = 1\right)$
0	0	1	–2	10	–1	$L_{12} = L_{11} + 1L_3 \quad \left(\frac{9}{10} = 1\right)$
0	1	0	0	0	0	$L_{13} = L_{12} + 1L_2 \quad \left(\frac{10}{10} = 1\right)$

Thus $\quad x = 20 + 10 + 2 = 32$

$\quad y = 19 + 6 + 1 = 26$

and $\quad z = 18 + 2 + 1 = 21$

$\therefore$ The values $x = 32$, $y = 26$, $z = 21$.

Example 5: Solve by Relaxation method the following equations.

$$3x + 9y - 2z = 11;\ 4x + 2y + 13z = 24,\ 4x - 4y + 3z = 8$$

Ans.: $x = 1.35$, $y = 2.103$, $z = 2.845$

Example 6: $9x + 3y + 4z = -100;\ x - 7y + 3z = -80;\ 2x + 3y - 5z = -60$

Ans.: $x = 52.5$, $y = 44.5$, $z = 59.7$.

5.6 RAYLEIGH'S POWER METHOD

In some applications, it is required to compute the numerically largest eigenvalue and the corresponding eigenvector. In such cases, the iterative method is more convenient which is also well-suited for computing machines. The iterative procedure for finding the dominant eigenvalue of a matrix is known as Rayleigh's power method.

Working Procedure

Let A be a square matrix and $x^{(0)}$ be a initial eigenvector in a simple form like $[1\ 0\ 0]^1$ or $[0\ 1\ 0]^1$ or $[0\ 0\ 1]^1$. Then evaluate the matrix product $AX^{(0)}$. Take the largest component out as a common factor. (This technique is called normalization) to obtain $AX^{(0)} = \lambda^{(1)} X^{(1)}$.

Then compute $AX^{(1)}$ and again put in the form $AX^{(1)} = \lambda^{(2)} X^{(2)}$ by normalization. This iterative process is continued till two consecutive iterative values of λ and x are the same upto the desired accuracy. The values so obtained are respectively, the largest eigenvalue and the corresponding eigenvector of the given square Matrix A.

Example 1: Find the largest eigenvalue and the corresponding eigenvector of the Matrix $\begin{bmatrix} 2 & 0 & 1 \\ 0 & 2 & 0 \\ 1 & 0 & 2 \end{bmatrix}$ by power method.

Solution: Let $A = \begin{bmatrix} 2 & 0 & 1 \\ 0 & 2 & 0 \\ 1 & 0 & 2 \end{bmatrix}$ be a given matrix and let

$X^{(0)} = [1\ 0\ 0]^1$ be the initial eigenvector.

$$\therefore \quad AX^{(0)} = \begin{bmatrix} 2 & 0 & 1 \\ 0 & 2 & 0 \\ 1 & 0 & 2 \end{bmatrix} \begin{bmatrix} 1 \\ 0 \\ 0 \end{bmatrix} = \begin{bmatrix} 2 \\ 0 \\ 1 \end{bmatrix} = 2 \begin{bmatrix} 1 \\ 0 \\ .5 \end{bmatrix} = \lambda^{(1)}\ X^{(1)}$$

$$AX^{(1)} = \begin{bmatrix} 2 & 0 & 1 \\ 0 & 2 & 0 \\ 1 & 0 & 2 \end{bmatrix} \begin{bmatrix} 1 \\ 0 \\ .5 \end{bmatrix} = \begin{bmatrix} 2.5 \\ 0 \\ 2 \end{bmatrix} = 2.5 \begin{bmatrix} 1 \\ 0 \\ .8 \end{bmatrix} = \lambda^{(2)}\ X^{(2)}$$

$$AX^{(2)} = \begin{bmatrix} 2 & 0 & 1 \\ 0 & 2 & 0 \\ 1 & 0 & 2 \end{bmatrix} \begin{bmatrix} 1 \\ 0 \\ .8 \end{bmatrix} = \begin{bmatrix} 2.8 \\ 0 \\ 2.6 \end{bmatrix} = 2.8 \begin{bmatrix} 1 \\ 0 \\ .93 \end{bmatrix} = \lambda^{(3)}\ X^{(3)}$$

$$AX^{(3)} = \begin{bmatrix} 2 & 0 & 1 \\ 0 & 2 & 0 \\ 1 & 0 & 2 \end{bmatrix} \begin{bmatrix} 1 \\ 0 \\ .93 \end{bmatrix} = \begin{bmatrix} 2.93 \\ 0 \\ 2.86 \end{bmatrix} = 2.93 \begin{bmatrix} 1 \\ 0 \\ .98 \end{bmatrix} = \lambda^{(4)}\ X^{(4)}$$

$$AX^{(4)} = \begin{bmatrix} 2 & 0 & 1 \\ 0 & 2 & 0 \\ 1 & 0 & 2 \end{bmatrix} \begin{bmatrix} 1 \\ 0 \\ .98 \end{bmatrix} = \begin{bmatrix} 2.98 \\ 0 \\ 2.96 \end{bmatrix} = 2.98 \begin{bmatrix} 1 \\ 0 \\ .99 \end{bmatrix} = \lambda^{(5)} X^{(5)}$$

$$AX^{(5)} = \begin{bmatrix} 2 & 0 & 1 \\ 0 & 2 & 0 \\ 1 & 0 & 2 \end{bmatrix} \begin{bmatrix} 1 \\ 0 \\ .99 \end{bmatrix} = \begin{bmatrix} 2.99 \\ 0 \\ 2.98 \end{bmatrix} = 2.99 \begin{bmatrix} 1 \\ 0 \\ .997 \end{bmatrix} = \lambda^{(6)} X^{(6)}$$

$$AX^{(6)} = \begin{bmatrix} 2 & 0 & 1 \\ 0 & 2 & 0 \\ 1 & 0 & 2 \end{bmatrix} \begin{bmatrix} 1 \\ 0 \\ .997 \end{bmatrix} = \begin{bmatrix} 2.997 \\ 0 \\ 2.999 \end{bmatrix} = 2.99 \begin{bmatrix} 1 \\ 0 \\ .999 \end{bmatrix} = \lambda^{(7)} X^{(7)}$$

$\therefore$ after VIth iteration, we can conclude that the largest eigenvalue is approximately '3' and the corresponding eigenvector is $[1\ 0\ 1]^1$.

Example 2: Determine the largest eigenvalue and the corresponding eigenvector of the matrix $\begin{bmatrix} 2 & -1 & 0 \\ -1 & 2 & -1 \\ 0 & -1 & 2 \end{bmatrix}$ by power method.

Solution: Let $A = \begin{bmatrix} 2 & -1 & 0 \\ -1 & 2 & -1 \\ 0 & -1 & 2 \end{bmatrix}$ be a given matrix and let the initial eigenvector be $[1\ 0\ 0]^1$.

$$AX^{(0)} = \begin{bmatrix} 2 & -1 & 0 \\ -1 & 2 & -1 \\ 0 & -1 & 2 \end{bmatrix}\begin{bmatrix} 1 \\ 0 \\ 0 \end{bmatrix} = \begin{bmatrix} 2 \\ -1 \\ 0 \end{bmatrix} = 2\begin{bmatrix} 1 \\ -0.5 \\ 0 \end{bmatrix} = \lambda^{(1)} X^{(1)}$$

$$AX^{(1)} = \begin{bmatrix} 2 & -1 & 0 \\ -1 & 2 & -1 \\ 0 & -1 & 2 \end{bmatrix}\begin{bmatrix} 1 \\ -0.5 \\ 0 \end{bmatrix} = \begin{bmatrix} 2.5 \\ -2 \\ 0.5 \end{bmatrix} = 2.5\begin{bmatrix} 1 \\ -0.8 \\ 0.2 \end{bmatrix} = \lambda^{(2)} X^{(2)}$$

Similarly,

$$AX^{(2)} = 2.8\begin{bmatrix} 1 \\ -1 \\ 0.43 \end{bmatrix} = \lambda^{(3)} X^{(3)} : AX^{(3)} = 3.43\begin{bmatrix} 0.87 \\ -1 \\ 0.54 \end{bmatrix} = \lambda^{(4)} X^{(2)}$$

$$AX^{(4)} = 3.41\begin{bmatrix} 0.80 \\ -1 \\ 0.61 \end{bmatrix} = \lambda^{(5)} X^{(5)}$$

$$AX^{(5)} = 3.41\begin{bmatrix} 0.76 \\ -1 \\ 0.65 \end{bmatrix} = \lambda^{(6)} X^{(6)}$$

$$AX^{(6)} = 3.41\begin{bmatrix} 0.74 \\ -1 \\ 0.67 \end{bmatrix} = \lambda^{(7)} X^{(7)}$$

Hence the largest eigenvalue is 3.41 and the corresponding eigenvector is $[0.74, -1, 0.67]^1$

Example 3: Find the largest eigenvalue and eigenvector of the Matrix $\begin{bmatrix} 1 & 3 & -1 \\ 3 & 2 & 4 \\ -1 & 4 & 10 \end{bmatrix}$ starting with $\begin{bmatrix} 0 \\ 0 \\ 1 \end{bmatrix}$ as the initial approximation to the corresponding eigenvector by power method.

Solution: Let $A = \begin{bmatrix} 1 & 3 & -1 \\ 3 & 2 & 4 \\ -1 & 4 & 10 \end{bmatrix}$ be the given matrix and $x^{(0)} = [0\ 0\ 1]^1$ be the initial eigenvector. Then

$$AX^{(0)} = \begin{bmatrix} 1 & 3 & -1 \\ 3 & 2 & 4 \\ -1 & 4 & 10 \end{bmatrix} \begin{bmatrix} 0 \\ 0 \\ 1 \end{bmatrix} = \begin{bmatrix} -1 \\ 4 \\ 10 \end{bmatrix} = 10 \begin{bmatrix} -0.1 \\ 0.4 \\ 1 \end{bmatrix} = \lambda^{(1)} X^{(1)}$$

$$AX^{(1)} = \begin{bmatrix} 1 & 3 & -1 \\ 3 & 2 & 4 \\ -1 & 4 & 10 \end{bmatrix} \begin{bmatrix} -0.1 \\ 0.4 \\ 1 \end{bmatrix} = \begin{bmatrix} 0.1 \\ 4.5 \\ 11.7 \end{bmatrix} = 11.7 \begin{bmatrix} 0.008547 \\ 0.384615 \\ 1 \end{bmatrix} = \lambda^{(2)} X^{(2)}$$

Continuing the process, we get

$$AX^{(2)} = (11.5299) \begin{bmatrix} 0.01408 \\ 0.41586 \\ 1 \end{bmatrix} = \lambda^{(3)} X^{(3)}$$

$$AX^{(3)} = (11.64936) \begin{bmatrix} 0.02246 \\ 0.41839 \\ 1 \end{bmatrix} = \lambda^{(4)} X^{(4)}$$

$$AX^{(4)} = (11.6511) \begin{bmatrix} 0.02383 \\ 0.42092 \\ 1 \end{bmatrix} = \lambda^{(5)} X^{(5)}$$

$$AX^{(5)} = (11.6599) \begin{bmatrix} 0.02458 \\ 0.42139 \\ 1 \end{bmatrix} = \lambda^{(6)} X^{(6)}$$

$$AX^{(6)} = (11.66098)\begin{bmatrix} 0.02476 \\ 0.42162 \\ 1 \end{bmatrix} = \lambda^{(7)} X^{(7)}$$

Thus, we observe that $\lambda^{(6)}$ and $\lambda^{(7)}$ are equal upto an accuracy of 0.01. Hence we stop the iteration and take the largest eigenvalue as $\lambda = 11.66$ and the corresponding eigenvector as $X = [0.02476, 0.42162, 1]^1$.

Example 4: Using power method find the largest eigenvalue and the corresponding eigenvector of the matrix $\begin{bmatrix} 2 & -1 & 0 \\ -1 & 2 & -1 \\ 0 & -1 & 2 \end{bmatrix}$ by taking the initial approximation to the corresponding eigenvector as $[1, 1, 1]^1$

Solution: Let $A = \begin{bmatrix} 2 & -1 & 0 \\ -1 & 2 & -1 \\ 0 & -1 & 2 \end{bmatrix}$ be the given matrix and $X^{(0)} = [1, 1, 1]^1$ be the corresponding initial eigenvector.

$$\therefore \quad AX^{(0)} = \begin{bmatrix} 2 & -1 & 0 \\ -1 & 2 & -1 \\ 0 & -1 & 2 \end{bmatrix}\begin{bmatrix} 1 \\ 1 \\ 1 \end{bmatrix} = \begin{bmatrix} 1 \\ 0 \\ 1 \end{bmatrix} = 1\begin{bmatrix} 1 \\ 0 \\ 1 \end{bmatrix} = \lambda^{(1)} X^{(1)}$$

$$AX^{(1)} = \begin{bmatrix} 2 & -1 & 0 \\ -1 & 2 & -1 \\ 0 & -1 & 2 \end{bmatrix}\begin{bmatrix} 1 \\ 0 \\ 1 \end{bmatrix} = \begin{bmatrix} 2 \\ -2 \\ 2 \end{bmatrix} = 2\begin{bmatrix} 1 \\ -1 \\ 1 \end{bmatrix} = \lambda^{(2)} X^{(2)}$$

$$AX^{(2)} = \begin{bmatrix} 2 & -1 & 0 \\ -1 & 2 & -1 \\ 0 & -1 & 2 \end{bmatrix}\begin{bmatrix} 1 \\ -1 \\ 1 \end{bmatrix} = \begin{bmatrix} 3 \\ -4 \\ 3 \end{bmatrix} = 4\begin{bmatrix} 0.75 \\ -1 \\ 0.75 \end{bmatrix} = \lambda^{(3)} X^{(3)}$$

Continuing the process, we get

$$AX^{(3)} = (3.5)\begin{bmatrix} 0.7143 \\ -1.0000 \\ 0.7143 \end{bmatrix} = \lambda^{(4)} X^{(4)}$$

$$AX^{(4)} = (3.4286)\begin{bmatrix} 0.7083 \\ -1.0000 \\ 0.7083 \end{bmatrix} = \lambda^{(5)} X^{(5)}$$

$$AX^{(5)} = (3.4166)\begin{bmatrix} 0.7073 \\ -1.0000 \\ 0.7073 \end{bmatrix} = \lambda^{(6)} X^{(6)}$$

$$AX^{(6)} = (3.4166)\begin{bmatrix} 0.7071 \\ -1.0000 \\ 0.7071 \end{bmatrix} = \lambda^{(7)} X^{(7)}$$

$$AX^{(7)} = (3.4142)\begin{bmatrix} 0.7071 \\ -1.0000 \\ 0.7071 \end{bmatrix} = \lambda^{(8)} X^{(8)}$$

Since $\lambda^{(7)}$ and $\lambda^{(8)}$ are the same, we take the eigenvalue $\lambda = 3.4142$ and the corresponding eigenvector

$$X = [0.7071, \ -1, \ 0.7071]^1.$$

Example 5: Using the power method find the dominant eigenvalue and the corresponding eigenvector of the matrix $\begin{bmatrix} 5 & 4 \\ 1 & 2 \end{bmatrix}$.

Solution: Let $A = \begin{bmatrix} 5 & 4 \\ 1 & 2 \end{bmatrix}$ be the given matrix and let $x^{(0)} = [1 \ 0]^1$ be the corresponding eigenvector.

Then

$$AX^{(0)} = \begin{bmatrix} 5 & 4 \\ 1 & 2 \end{bmatrix}\begin{bmatrix} 1 \\ 0 \end{bmatrix} = \begin{bmatrix} 5 \\ 1 \end{bmatrix} = 5\begin{bmatrix} 1 \\ 0.2 \end{bmatrix} = \lambda^{(1)} X^{(1)}$$

$$AX^{(1)} = \begin{bmatrix} 5 & 4 \\ 1 & 2 \end{bmatrix}\begin{bmatrix} 1 \\ 0.2 \end{bmatrix} = \begin{bmatrix} 5.8 \\ 1.4 \end{bmatrix} = 5.8\begin{bmatrix} 1 \\ 0.2414 \end{bmatrix} = \lambda^{(2)} X^{(2)}$$

$$AX^{(2)} = \begin{bmatrix} 5 & 4 \\ 1 & 2 \end{bmatrix}\begin{bmatrix} 1 \\ 0.2414 \end{bmatrix} = \begin{bmatrix} 5.9656 \\ 1.4828 \end{bmatrix} = (5.9656)\begin{bmatrix} 1 \\ 0.2486 \end{bmatrix} = \lambda^{(3)} X^{(3)}$$

$$AX^{(3)} = \begin{bmatrix} 5 & 4 \\ 1 & 2 \end{bmatrix}\begin{bmatrix} 1 \\ 0.2486 \end{bmatrix} = \begin{bmatrix} 5.9944 \\ 1.4972 \end{bmatrix} = (5.9944)\begin{bmatrix} 1 \\ 0.2498 \end{bmatrix} = \lambda^{(4)} X^{(4)}$$

$$AX^{(4)} = \begin{bmatrix} 5 & 4 \\ 1 & 2 \end{bmatrix}\begin{bmatrix} 1 \\ 0.2498 \end{bmatrix} = \begin{bmatrix} 5.9992 \\ 1.4996 \end{bmatrix} = (5.9992)\begin{bmatrix} 1 \\ 0.25 \end{bmatrix} = \lambda^{(5)} X^{(5)}$$

$$AX^{(5)} = \begin{bmatrix} 5 & 4 \\ 1 & 2 \end{bmatrix} \begin{bmatrix} 1 \\ 0.25 \end{bmatrix} = \begin{bmatrix} 6 \\ 1.5 \end{bmatrix} = 6 \begin{bmatrix} 1 \\ 0.25 \end{bmatrix} = \lambda^{(6)} X^{(6)}$$

We observe that 5th and 6th iterations are the same upto three decimal places and we stop the iteration process we take the largest eigenvalue $\lambda = 6$ and the corresponding eigenvector as $[1\ 0.25]^1$.

Example 6: Find the numerically largest eigenvalue and the corresponding eigenvector of the matrix.

$$A = \begin{bmatrix} 4 & 1 & -1 \\ 2 & 3 & -1 \\ -2 & 1 & 5 \end{bmatrix}$$

Ans.: $\lambda = 6$ and $X = [1,\ 1\ -1]^1$.

Example 7: By using power method find the largest eigenvalue and the corresponding eigenvector of the matrix

$$A = \begin{bmatrix} 7 & 3 & -2 \\ 3 & 4 & -1 \\ -2 & -1 & 3 \end{bmatrix}$$

Ans.: $\lambda = 9.6055\ X = [1,\ 0.6056,\ -0.3944]^1$.

Example 8: Find the dominant eigenvalue and the corresponding

eigenvector of the matrix $A = \begin{bmatrix} 1 & 6 & 1 \\ 1 & 2 & 0 \\ 0 & 0 & 3 \end{bmatrix}$

Ans.: $\lambda = 4,\ X = [2,\ 1,\ 0]^1$.

Example 9: Using Rayleigh's power method find the largest eigenvalue and the corresponding eigenvector of the matrix

$$A = \begin{bmatrix} 25 & 1 & 2 \\ 1 & 3 & 0 \\ 2 & 0 & -4 \end{bmatrix}$$

Ans.: $\lambda = 25.1821,\ X = [1,0.0451,\ 0.0685]^1$.

5.7 QUESTION BANK

1) Find the real root of the equation $3x = \cos x + 1$ correct to four decimal place by Newton's Method.

2) Find the smallest and the largest roots of $e^x - 4x = 0$ correct to 4 decimal places by Newton-Raphson Method.

3) Compute the real root of the equation $x \log_{10}^{x} - 1.2 = 0$ correct to five decimal places using Regula Falsi method.

4) Show that a real root of the equation $\tan x + \tan hx = 0$ lies between 2 and 3. Then apply the Regula Falsi method to find third approximation.

5) Using Regula Falsi method, find the real positive root of the equation $x \tan x = -1$.

6) Explain Newton-Raphson method for solving the equation $f(x) = 0$. Use it to find the real root of the equation $x\, e^x = 2$ correct to the decimal places.

7) Using Newton-Raphson method find the root lies near $x = 4.5$ of the equation $\tan x = x$. Carry out three iterations (here, x is in radians)

8) Apply Gauss-Seidel iterative method to solve

$$6x + 15y + 2z = 72;\ x + y + 54z = 110;\ 27x + 6y - z = 85$$

Take $x = 1,\ y = 0,\ z = 1$ as an initial approximation. Carry out three iterations.

9) Apply Gauss-Seidel iterative method to find the third approximate solution of the system of equations

$5x + 2y + z = 12;\ x + 4y + 2z = 15\ ;\ x + 2y + 5z = 20$ with initial approximate solution (1, 0, 3).

10) Solve the system of equations using Gauss-Seidel method :

$10x + y + z = 12;\ 2x + 10y + z = 13,\ 2x + 2y + 10z = 14$ using

$x_0 = 0,\ y_0 = 0,\ z_0 = 0.$

11) Solve by Gauss-Seidel method (carry out three iterations)

$5x + 2y + z = 12;\ x + 4y + 2z = 15;\ x + 2y + 5z = 0.$

12) Using Relaxation method solve the following equations

$3x + 9y - 2z = 11\ ;\ 4x + 2y + 13z = 24\ ;\ 4x - 4y + 3zi = -8.$

13) Solve by Relaxation method the equations

$10x - 2y - 2z = 6\ ;\ x + 10y - 2z = 7\ ;\ -x - y + 10z = 8.$

14) Solve by Relaxation method the following equations

$-9x + 3y + 4z = -100;\ x - 7y + 3z = -80;\ 2x + 3y - 5z = -60.$

15) Find the largest eigenvalue and the corresponding eigenvector of the matrix by using power method $A=\begin{bmatrix} 6 & -2 & 2 \\ -2 & 3 & -1 \\ 2 & -1 & 3 \end{bmatrix}$ taking $(1, 1, 1)^T$ as the initial eigenvector perform five iterations.

16) Find the largest eigenvalue and the corresponding eigenvector of the matrix $\begin{bmatrix} 2 & -3 & 1 \\ 3 & 1 & 3 \\ -5 & 2 & -4 \end{bmatrix}$ by power method. Start with the initial vector (1, 0, 0). (Carry out 4 iterations.)

17) Use Rayleigh's power method to find the largest eigenvalue and the corresponding eigenvector of the matrix $\begin{bmatrix} 2 & -1 & 0 \\ -1 & 2 & -1 \\ 0 & -1 & 2 \end{bmatrix}$ by taking initial vector a $[1, 1, 1]^1$

18) Using power method to determine the numerically largest eigenvalue and corresponding eigenvector of the matrix $\begin{bmatrix} 1 & -3 & 2 \\ 4 & 4 & -1 \\ 6 & 3 & 5 \end{bmatrix}$

19) Find the largest eigenvalue and the corresponding eigenvector of the matrix $\begin{bmatrix} 25 & 1 & 2 \\ 1 & 3 & 0 \\ 2 & 0 & -4 \end{bmatrix}$ by power method.

Take $(1, 0\ 0)^T$ as the initial vector.

20) Find the largest eigenvalue and the corresponding eigenvector of the matrix $\begin{bmatrix} 2 & 0 & 1 \\ 0 & 2 & 0 \\ 1 & 0 & 2 \end{bmatrix}$ by power method. Taking $[1, 0, 0]^T$ as the initial eigenvector (carry out four iterations).

NUMERICAL METHODS – 2

UNIT 6

6.1 CALCULUS OF FINITE DIFFERENCES

Calculus of finite differences are very important tools to provide practical methods for calculating the solutions of problems in applied mathematics to a desired degree of accuracy. Let $y = f(x)$ for $a \le x \le b$, this statement from numerical point of view means that one or more values of $f(x)$ exist corresponding to every value of x in the given range. In case $f(x)$ is specified by a given explicit formula, we can compute the value or values of $f(x)$ corresponding to some given value of x. If it is not the case even then we can find an approximate representative of the function. This is done with the help of calculus of finite differences. Thus, "The calculus of finite differences deals with the changes in the values of the function (dependent variable) due to changes in the independent variable." Through this we study the relations of the values which the function assumes, whenever the independent variable changes by finite jumps whether equal or unequal. Here we shall discuss the cases when independent variable changes by equal or unequal intervals.

Forward differences

Let $y = f(x)$ be a function of x and the independent variable x, takes the consecutive values, $a, a + h, a + 2h, a + 3h \,\ldots\ldots a + nh$. Here '$h$' is called the interval of differencing. The value of the independent variable is called **Argument** and the value of the dependent variable is called the **Entry.**

The first differences of $y = f(x)$ are defined by

$$\begin{aligned}
\Delta f(a) &= f(a+h) - f(a) \\
\Delta f(a+h) &= f(a+2h) - f(a+h) \\
\Delta f(a+2h) &= f(a+3h) - f(a+2h) \text{ etc.}
\end{aligned}$$

Where Δ is the basic differencing operator and is called the first forward difference.

The differences of the first forward differences are called second forward differences and are given by

$$\Delta^2 f(a) = \Delta f(a+h) - \Delta f(a)$$
$$\Delta^2 f(a+h) = \Delta f(a+2h) - \Delta f(a+h)$$
$$\Delta^3 f(a+2h) = \Delta f(a+3h) - \Delta f(a+2h) \text{ etc.}$$

Also $\Delta^2 f(a) = \Delta f(a+b) - \Delta f(a)$

$$= f(a+2h) - f(a+h) - f(a+h) + f(a)$$
$$= f(a+2h) - 2f(a+h) + f(a).$$

Similarly

$$\Delta^2 f(a+2h) = f(a+3h) - 2f(a+2h) + f(a+h) \text{ etc.}$$

In general, we have

$$\Delta^2 f(x) = \Delta f(x+h) - \Delta f(x)$$
$$= f(a+2h) - 2f(x+h) + f(x).$$

Again, the differences of the second differences are called third differences and are given by

$$\Delta^3 f(a) = \Delta^2 f(a+h) - \Delta^2 f(a)$$
$$= f(a+3h) - 3f(a+2h) + 3f(a+h) - f(a).$$

and so on,

In general the nth difference is given by

$$\Delta^n f(x) = \Delta^{n-1} f(x+h) - \Delta^{n-1} f(x).$$

Note: (i) $f(a)$, the first entry is termed the leading term and the differences $\Delta f(a)$, $\Delta^2 f(a)$ ..... $\Delta^n f(a)$ are known as the leading differences.

(ii) It should be kept in mind that Δ is an operator, Δ^2 is not the square of the operator Δ but it is the repetition of the operator Δ.

(iii) The differences of a constant term are zero i.e., $\Delta C = 0$ where C is constant.

(iv) The operator Δ is commutative in its operations as regards constants

$$\Delta \{ c f(x)\} = c \Delta f(x).$$

(v) If a and b are constants then

$$\Delta \{a f(x) + bg(x)\} = a \Delta f(x) + b \Delta g(x)$$

Forward difference table

While using higher differences we form the forward difference table as follows.

Argument x	Entry $y = f(x)$	I^{st} diff $\Delta f(x)$	II diff $\Delta^2 f(x)$	III diff $\Delta^3 f(x)$	IV diff $\Delta^4 f(x)$
a	$f(a)$				
		$\Delta f(a)$			
$a + h$	$f(a + h)$		$\Delta^2 f(a)$		
		$\Delta f(a + h)$		$\Delta^3 f(a)$	
$a + 2h$	$f(a + 2h)$		$\Delta^2 f(a + h)$		$\Delta^4 f(a)$
		$\Delta f(a + 2h)$		$\Delta^3 f(a + h)$	
$a + 3h$	$f(a + 3h)$		$\Delta^2 f(a + 2h)$		
		$\Delta f(a + 3h)$			
$a + 4h$	$f(a + 4h)$				

Backward Differences

The forward differences are defined by points to the right of x, in this case, i.e., backward differences are defined by points to the left of x. Let $y = f(x)$ and $a, a + h, a + 2h \ldots, a + nh$ be the values of x and the corresponding to these values of x, the values of $y = f(x)$ be, $f(a), f(a + h), f(a + 2h)\ldots.. f(a + nh)$ respectively. Then the first backward differences are defined by

$$\nabla f(a + h) = f(a + h) - f(a)$$

$$\nabla f(a + 2h) = f(a + 2h) - f(a + h)$$

$$\nabla f(a + 3h) = f(a + 3h) - f(a + 2h) \text{ etc.}$$

The differences of first backward differences are said to be second backward differences and are

$$\begin{aligned}\nabla f(a + h) &= \nabla f(a + 2h) - \nabla f(a + h)\\ &= f(a + 2h) - f(a + h) - f(a + h) + f(a)\\ &= f(a + 2h) - 2 f(a + h) + f(a).\end{aligned}$$

In general, first backward difference is given by

$$\nabla f(x) = f(x) - f(x - h)$$

and second backward difference is given by

$$\nabla^2 f(x) = \nabla f(x) - \nabla f(x-h)$$
$$= f(x) - 2f(x-h) + f(x-2h)$$

Thus we can show that the nth backward difference of $y = f(x)$ is

$\nabla^n f(x) = \nabla^{n-1} f(x) - \nabla^{n-1} f(x-h).$

Backward difference table

Argument x	Entry $y = f(x)$	$\nabla f(x)$	$\nabla^2 f(x)$	$\nabla^3 f(x)$	$\nabla^4 f(x)$
a	$f(a)$				
		$\nabla f(a+b)$			
$a+h$	$f(a+h)$		$\nabla^2 f(a+2h)$		
		$\nabla f(a+2h)$		$\nabla^3 f(a+3h)$	
$a+2h$	$f(a+2h)$		$\nabla^2 f(a+3h)$		$\nabla^4 f(a+4h)$.
		$\nabla f(a+3h)$		$\nabla^3 f(a+4h)$	
$a+3h$	$f(a+3h)$		$\nabla^2 f(a+4h)$		
		$\nabla f(a+4h)$			
$a+4h$	$f(a+4h)$				

Fundamental theorem of the difference calculus

If $f(x)$ be a polynominal of nth degree in x, then the nth difference of $f(x)$ is constant and $\Delta^{n+1} f(x) = 0$ i.e., $\Delta^n f(x) = n!\ h^n\ an$

and $\Delta^{n+1} f(x) = 0.$

The operator *E*

The shift operator E is given by the rule

$$E\,[f(x)] = f(x+h)$$

We know that

$$\Delta f(x) = f(x+h) - f(x) = E f(x) - f(x)$$
$$\therefore \quad E f(x) = \Delta f(x) + f(x)$$
$$\therefore \quad E f(x) = (\Delta + 1) f(x)$$

Since $f(x)$ is arbitrary, we have

$$E \equiv 1 + \Delta \text{ or } \Delta \equiv E - 1$$

Again $E^n f(x) = \underbrace{E\,.\,E\,.\,E \ldots\ldots\ldots E\,.\,E}_{n \text{ times}}\ f(x) = f(x+nh).$

6.2 INTERPOLATION WITH EQUAL INTERVALS

Interpolation means insertion or filling up intermediate terms of the series. It is the technique of obtaining the value of a function for any intermediate value of the dependent variable. The process of computing the value of the function outside the range of given values of the variable is called extrapolation.

6.3 NEWTON-GREGORY FORMULA FOR FORWARD INTERPOLATION

Let $y = f(x)$ represent a function which assumes the values $f(a)$, $f(a + h)$ $f(a + nh)$ for $(n + 1)$ equidistant values $a,\ a + h,\ a + 2h$ .... $a + nh$ of the independent variable x and let $P_n(x)$ be a polynominal in x of degree n.

i.e., $P_n(x) = A_0 + A_1\,(x-a) + A_2\,(x-a)\,(x-a-h)...(x-a-\overline{n-1}\,h)$ ....(1)

Where A_0, A_1, A_2 .... A_n are constants and are chosen such that

$$p_n(a) = f(a),\ p_n(a+h) = f(a+h).....\ p_n(a+nh) = f(a+nh)$$

Now, putting successively the values $a,\ a + h,\ a + 2h$ ...... in (1) we get

$$p_n(a) = f(a) = A_0 \quad \text{or} \quad A_0 = f(a)$$

$$p_n(a+h) = f(a+h) = A_0 + hA_1 = f(a) + hA_1$$

$$\therefore \quad A_1 = \frac{f(a+h)-f(a)}{h} = \frac{\Delta f(a)}{h}$$

$$f(a+2h) = A_0 + 2hA_1 + 2h\,.\,hA_2$$

And

$$A_2 = \frac{1}{2h^2}\,[f(a+2h) - 2f(a+h) + f(a)]$$

$$= \frac{1}{2h^2}\Delta^2 f(a) = \frac{1}{2!\,h^2}\Delta^2 f(a)$$

Similarly, we get

$$A_3 = \frac{1}{3!\,h^3}\Delta^3 f(a),\ A_4 = \frac{1}{4!h^4}\Delta^4 f(a)\ ....\ A_n = \frac{1}{n!\,h^n}\Delta^n f(a)$$

Substituting the values of A_0, A_1, A_2 .... A_n in (1) we get

$$p_n(x) = f(a) + \frac{\Delta f(a)}{h}(x-a) + \frac{\Delta^2 f(a)}{2!\,h^2}(x-a)\,(x-a-h) + \ ....$$

$$... + \frac{\Delta^n f(a)}{n!\,h^n}(x-a)\,(x-a-h)\ ......\ (x-a-\overline{n-1}\,h) \quad ....(2)$$

putting $\dfrac{x-a}{h} = u$ or $x = a + hu$ we get

$$p_n(x) = p_n(a + hu) = f(a) + u\ \Delta f(a) + \frac{u(u-1)}{2!}\Delta^2 f(a)$$

$$+ \frac{u(u-1)(u-2)}{3!}\Delta^2 f(a) + \ldots\ldots + \frac{1}{n!}\ u(u-1)(u-2)\ldots$$

$$\ldots\ldots.(u - n + 1)\ \Delta^n f(a) \qquad \ldots.(3)$$

This is Newton Gregory formula for forward interpolation.

Corollary: Taking $u^{(n)} = u(u-1)(u-2) \ldots\ldots (u-n+1)$ we write the above formula as

$$p_n(x) = f(a) + u^{(1)}\Delta f(a) + \frac{u^{(2)}}{2!}\Delta^2 f(a) + \ldots\ldots\ldots\ldots + \frac{1}{n!}u^{(n)}\Delta^n f(a)$$

Note: This formula is used mainly for interpolating the values of y near the beginning ($x > a$) of a set of tabulated values.

6.4 NEWTON-GREGORY FORMULA FOR BACKWARD INTERPOLATION

To find the formula consider the polynominal $p_n(x)$ as

$$p_n(x) = A_0 + A_1(x - a - nh) + A_2(x - a - nh)(x - a - nh + h)$$

$$+ A_3(x - a - nh)(x - a - nh + h)(x - a - nh + 2h) + \ldots$$

$$+ \ldots\ldots.. + A_n(x - a - nh) \ldots\ldots\ldots. (x - a - h) \qquad \ldots.(1)$$

Where $A_0, A_1, A_2\ldots\ldots\ldots A_n$ are constants and are chosen such that

$$p_n(a + nh) = f(a + nh),\ p_n(a + (n-1)h)$$

$$= f(a + \overline{(n-1)}\,h) \ldots\ldots.p_n(a) = f(a)$$

Putting in (1) $x = a + nh,\ a + nh - h \ldots\ldots a$ we get

$$f(a + nh) = A_0, \text{ or } A_0 = f(a + nh)$$

$$f(a + nh - h) = A_0 + A_1(-h)$$

or $$A_1 = \frac{f(a+nh) - f(a+nh-h)}{h}$$

$\therefore$ $$A_1 = \frac{\nabla f(a+nh)}{h}$$

Next,

$$f(a + nh - 2h) = A_0 + A_1(-2h) + A_2(-2h)(-h)$$

or $$A_2 = \frac{f(a+nh) - 2f(a+\overline{n-1}\,h) + f(a+\overline{n-2h})}{2h^2}$$

$$= \frac{1}{2!\,h^2}\nabla^2 f(a+nh)$$

Similarly, we have

$$A_3 = \frac{1}{3!\,h^3}\nabla^3 f(a+nh) \,..........\, A_n = \frac{1}{n!\,h^n}\nabla^n f(a+nh)$$

Substituting these values of $A_1, A_2, A_3 \,......A_n$ in (1) we get

$$p_n(x) = f(a+nh) + \frac{\nabla f(a+nh)}{h}(x-\overline{a+nh})$$

$$+\frac{\nabla^2 f(a+nh)}{2!\,h^2}(x-\overline{a+nh})\,(x-\overline{a+nh-h}) + ...$$

$$....+\frac{\nabla^n f(a+nh)}{2!\,h^n}(x-\overline{a+nh})\,........\,(x-a-h) \qquad ...(2)$$

put $$u = \frac{x-(a+nh)}{h}$$ or $x = a + nh + hu$ in (2) we get

$$p_n(x) = f(a + nh) + u\,\nabla f(a + nh) + \frac{u(u+1)}{2!}\nabla^2 f(a+nh)$$

$$+\frac{u(u+1)\,(u+2)}{3!}\nabla^3 f(a+nh) + ...$$

$$+\frac{u(u+1)\,(u+2)\,............(u+n-1)}{n!}\nabla^n f(a+nh)$$

This is Newton-Gregory formula for backward interpolation.

Example 1: Using Newton's formula for interpolation, estimate the population for the year 1905.

Year	Population
1891	98,752
1901	132,285
1911	168,076

1921	195,690
1931	246,050

Solution: The difference table is as under

Year (x)	Population $f(x)$	$\Delta f(x)$	$\Delta^2 f(x)$	$\Delta^3 f(x)$	$\Delta^4 f(x)$
1891	98,752				
		33533			
1901	132,285		2258		
		35791		–10435	
1911	168,076		–8177		41358
		27614		30923	
1921	195,690		22746		
		50360			
1931	246,050				

$$\text{Here } a = 1891, \quad u = \frac{x-a}{h} = \frac{1905-1891}{10} = 1.4$$

By Newton's interpolation formula, we get

$$f(1905) = 98752 + (1.4)\,(33533) + \frac{(1.4)\,(0.4)}{2!}(2258) + \frac{(1.4)\,(0.4)(-0.6)}{3!} \times(-10435) + \frac{(1.4)(0.4)(-0.6)(-1.6)(41358)}{3!}$$

$$= 147841 \text{ approx.}$$

Example 2: Using Newton's formula for interpolation, estimate the increase in the population during the period from 1946 to 1948.

Year :	1911	1921	1931	1941	1951	1961
Population : (in thousands)	12	15	20	27	39	52

Solution: The difference table is as under.

Year x	Population $y = f(x)$	$\Delta f(x)$	$\Delta^2 f(x)$	$\Delta^3 f(x)$	$\Delta^4 f(x)$	$\Delta^5 f(x)$
1911	12					
		3				
1921	15		2			
		5		0		
1931	20		2		3	
		7		3		–10
1941	27		5		–7	
		12		–4		
1951	39		1			
		13				
1961	52					

i) When $x = 1946$, $a = 1911$, $h = 10$

$$u = \frac{x-a}{h} = \frac{1946-1911}{10} = 3.5$$

Newton's formula for forward interpolation is

$$f(x) = f(a) + u\Delta f(a) + \frac{u(u-1)}{2!}\Delta^2 f(a) + \frac{u(u-1)(u-2)}{3!}\Delta^3 f(a) + \ldots$$

$$= 12 + (3.5)(3) + \frac{(3.5)(2.5)}{2!}(2) + \frac{(3.5)(2.5)(1.5)}{3!}(0)$$

$$+ \frac{(3.5)(2.5)(1.5)(.5)}{4!}(3) + \frac{(3.5)(2.5)(1.5)(.5)(-.5)}{5!}(-10)$$

$$= 12 + 10.5 + 8.75 + 0.8203 + .2734$$

$$= 32.3437$$

ii) When $x = 1948$, $u = \dfrac{1948-1911}{10} = 3.7$

$$f(x) = 12 + (3.7)(3) + \frac{(3.7)(2.7)}{2!}(2) + \frac{(3.7)(2.7)(1.7)}{3!}(0)$$

$$+ \frac{(3.7)(2.7)(1.7)(.7)}{4!}(3) + \frac{(3.7)(2.7)(1.7)(.7)(-.3)}{5!}(-10)$$

$= 12 + 11.1 + 9.19 + 1.4860 + 0.2972$

$= 34.8732$

$\therefore$ increase in the population during the period from 1946 to 1948 $= 34.8732 - 32.3437 = 2.5295$ thousands

$= 2.53$ thousands app.

Example 3: From the following table find the number of students who obtained less than 45 marks.

Marks	No of students
30 – 40	31
40 – 50	42
50 – 60	51
60 – 70	35
70 – 80	31

Solution: The difference table is as under.

Marks x	No of Students $f(x)$	$\Delta f(x)$	$\Delta^2 f(x)$	$\Delta^3 f(x)$	$\Delta^4 f(x)$
Below 40	31				
		42			
Below 50	73		9		
		51		–25	
Below 60	124		– 16		37
		35		12	
Below 70	159		– 4		
		31			
Below 80	190				

Here $a = 40,\ h = 10,\ x = 45 \quad \therefore\ u = \dfrac{x-a}{h} = \dfrac{45-40}{10} = .5$

By using Newton's forward formula we have

$$f(45) = f(40) + .5\ \Delta f(40) + \frac{(.5)\ (.5-1)}{2!}\Delta^2 f(40)$$

$$+\frac{(.5)(.5-1)(.5-2)}{3!}\Delta^3 f(40)+\frac{(.5)(.5-1)(.5-2)(.5-3)}{4!}\Delta^4 f(40)$$

$$= 31 + .5\,(42) + \frac{(.5)(.5-1)}{2!}(9)+\frac{(.5)(.5-1)(.5-2)}{3!}(-25)$$

$$+\frac{.5\,(.5-1)(.5-2)(.5-3)}{4!}(37)$$

$$= 47.868$$

Hence the number of students getting marks not more than 45 is 48.

Example 4: Find the cubic polynomial which takes the following values

x :	0	1	2	3
$f(x)$	1	2	1	10

Hence or otherwise evaluate f(4).

Solution: The difference table is

X	$f(x)$	$\Delta f(x)$	$\Delta^2 f(x)$	$\Delta^3 f(x)$
0	1			
		1		
1	2		–2	
		–1		12
2	1		10	
		9		
3	10			

i) We take a = 0 and $u - \frac{x-a}{h} = \frac{x-0}{1} = x$

∴ Using Newton's forward interpolation formula, we get

$$f(x) = f(0) + x\Delta f(0) + \frac{x(x-1)}{2!}\Delta^2 f(0)+\frac{x(x-1)(x-2)}{3!}\Delta^2 f(0)$$

$$= 1 + x(1) + \frac{x(x-1)}{2}(-2)+\frac{x(x-1)(x-2)}{6}(12)$$

$= 2x^3 - 7x^2 + 6x + 1$ which is the required polynomial.

ii) Using Newton's backward interpolation formula

$$f(x) = f(a+nh) + u\nabla f(a+nh) + \frac{u(u+1)}{2!}\nabla^2 f(a+nh) + ...$$

$$f(x) = 10 + 1\times 9 + \frac{1(1+1)}{2!}(10) + \frac{1(1+1)(1+2)}{3!}(12)$$

$$= 10 + 9 + 10 + 12 = 41.$$

Example 5: The following table gives the distances in nautical miles of the visible horizon for the given heights in feet above the earths surface.

Height(x):	100	150	200	250	300	350	400
Distance(y):	10.63	13.03	15.04	16.81	18.42	19.90	21.27

Find the value of y when x = 218ft and 410ft.

Solution:

i) To find y when x = 218ft we use Newton's forward formula for interpolation and

ii) To find y when x = 410ft we use Newton's backward formula for interpolation.

The difference table is as under.

x	$f(x)$	Δ	Δ^2	Δ^3	Δ^4	Δ^5	Δ^6
100	10.63						
		2.40					
150	13.03		–0.39				
		2.01		0.15			
200	**15.04**		–0.24		–0.07		
		1.77		0.08		0.02	
250	16.81		**–0.16**		–0.05		0.02
		1.61		**0.03**		0.04	
300	18.42		–0.13		**–0.01**		
		1.48		0.02			
350	19.90		–0.11				
		1.37					
400	21.27						

i) When $x = 218$ we take $a = 200$, $h = 50$ and

$$u = \frac{x-a}{h} = \frac{218-200}{50} = 0.36$$

∴ Using Newton's forward interpolation formula.

$$f(x) = f(a) + u\Delta f(a) + \frac{u(u-1)}{2!}\Delta^2 f(a) + \ldots\ldots\ldots\ldots\ldots$$

$$= 15.04 + (0.36)\,(1.77) + \frac{(0.36)\,(0.36-1)}{2!}\,(-0.16)$$

$$+\frac{(0.36)(0.36-1)\,(0.36-2)}{3!}(0.03)$$

$$+\frac{(0.36)(0.36-1)(0.36-2)(0.36-3)}{4!}\,(-.01)$$

$$= 15.04 + 0.637 + 0.018 + 0.001 + \ldots\ldots$$

$= 15.696$ i.e., 15.7 nautical miles.

ii) When $x = 410$, $u = \frac{x-(a+nh)}{h} = \frac{410-400}{50} = 0.2$

∴ Using Newton's backward interpolation formula

$$f(x) = f(a+nh) + u\nabla f(a+nh) + \frac{u(u+1)}{2!}\nabla^2 f(a+nh) + \ldots\ldots$$

$$= 21.27 + (0.2)\,(1.37) + \frac{(0.2)\,(0.2+1)}{2!}\,(-0.11)$$

$$+\frac{(0.2)\,(0.2+1)\,(.2+2)}{3!}(0.02) + \ldots\ldots\ldots\ldots$$

$= 21.53$ nautical miles.

Example 6: Given sin 45° = 0.7071. sin 50° = 0.7660, sin 55° = 0.8192 sin 60° = 0.8660. Find sin 52° by using any method of interpolation.

Solution: Here, we take $a = 45°$ $x = 52°$ and $h = 5$

$$\therefore \quad u = \frac{x-a}{h} = \frac{52-45}{5} = 1.4$$

We use Newton's forward formula

$$f(x) = f(a) + u\Delta f(a)\ \frac{u(u-1)}{2!}\Delta^2 f(a) + \frac{u(u-1)(u-2)}{3!}\Delta^3 f(a) + \ldots \quad \ldots(1)$$

The difference table is as under

x	$f(x)$	$\Delta f(x)$	$\Delta^2 f(x$	$\Delta^3 f(x)$
45	0.7071			
		0.0589		
50	0.8192		–0.0057	
		0.0532		–0.0007
55	0.8192		–0.0064	
		0.0468		
60	0.8660			

Substituting in (1)

$$f(x) = 0.7071 + 1.4(0.0589) + \frac{(1.4)\,(0.4)}{2!}\,(-0.0057)$$

$$+ \frac{(1.4)\,(0.4)\,(-0.6)}{3!}\,(-0.0007)$$

$$= .7880032.$$

$$\therefore \quad \sin 45° = 0.7880032.$$

Example 7 : Find the number of men getting wages between Rs 10 and Rs. 15 from the following table.

Wages in Rs :	0–10	10 – 20	20 – 30	30 – 40
Frequency :	9	30	35	45

Solution: The difference table is as under.

x	$f(x)$	$\Delta f(x)$	$\Delta^2 f(x)$	$\Delta^3 f(x)$
Under 10	9			
		30		
Under 20	39		5	
		35		2
Under 30	74		7	
		42		
Under 40	116			

Here a = 10, $u = \frac{x-a}{h} = \frac{15-10}{10} = 0.5$

By using Newton's forward formula

$$f(15) = 9 + 0.5\ (30) + \frac{(0.5)(-0.5)(5)}{2!} + \frac{(0.5)(-0.5)(-1.5)(2)}{3!}$$

$$= 24 \text{ approximate}$$

Number of persons getting wages between Rs 10 and Rs 15 is 24–9=15.

6.5 INTERPOLATION WITH UNEQUAL INTERVALS

In case the values of the arguments are given at unequal intervals then the various differences will be affected with the change in values of the argument and consequently definition of various differences for equal intervals will not hold good. The differences defined taking into consideration the changes in the values of the argument are called divided differences.

Divided differences

Let $y_{o,}\ y_1, y_2\ \ldots\ldots\ y_n$ be the values of the function $y = f(x)$ corresponding to the values $x_0, x_1, x_2\ \ldots.\ x_n$ of the arguments x, where $x_1 - x_0,\ x_2 - x_1,\ x_3 - x_2\ \ldots\ldots\ ,\ x_n - x_{n-1}$ are not necessarily equal, i.e., when there is a case unequal intervals.

The first divided difference of $y = f(x)$ for the arguments x_0, x_1 is defined as

$\frac{y_1 - y_0}{x_1 - x_0}$ and is denoted by $[x_0, x_1]$ or by $\underset{x_1}{\Delta} y_0$

Similarly the other divided differences of $y = f(x)$ for the arguments $x_1, x_2\ ;\ x_2, x_3\ ;\ \ldots\ldots;x_{n-1}, x_n$ are

$$[x_1, x_2] = \frac{y_2 - y_1}{x_2 - x_1} = \underset{x_2}{\Delta} y_1$$

$$[x_2, x_3] = \frac{y_3 - y_2}{x_3 - x_2} = \underset{x_3}{\Delta} y_2$$

$$\vdots$$

$$[x_{n-1}, x_n] = \frac{y_n - y_{n-1}}{x_n - x_{n-1}} = \underset{x_n}{\Delta} y_{n-1}$$

Again the second and higher divided differences are defined in terms of lower divided differences e.g. the second divided difference of $y = f(x)$ for the three arguments x_0 , x_1, x_2 is given by

$$\frac{[x_1, x_2]-[x_0, x_1]}{x_2 - x_0} \text{ and is denoted by } [x_0, x_1, x_2]$$

Similarly, the nth divided difference is given by

$$\frac{[x_1, x_2, x_3 \ldots\ldots\ldots x_n]-[x_0, x_1, x_2 \ldots\ldots\ldots x_{n-1}]}{x_n - x_0} \text{ and is denoted by }$$

$$[x_0, x_1, x_2 \ldots\ldots x_n]$$

6.6 NEWTON'S FORMULA FOR UNEQUAL INTERVALS

Let $y_0, y_1, y_2 \ldots\ldots y_n$ be the values of $y = f(x)$ corresponding to the arguments $x_0, x_1, x_2 \ldots\ldots x_n$. Then by the definition of divided differences, we have

$$[x, x_0] = \frac{y - y_0}{x - x_0} \text{ so that}$$

$$\therefore \quad y = y_0 + (x - x_0)\,[x\ x_0] \qquad \ldots\ldots (1)$$

$$\text{Again , } [x, x_0, x_1] = \frac{[x, x_0]-[x_0, x_1]}{x - x_1}$$

$$\therefore \quad [x, x_0] = [x_0, x_1] + (x - x_1)\,[x, x_0, x_1]$$

Substituting this value of $[x, x_0]$ in (1) we get

$$y = y_0 + (x - x_0)\,[x_0\, x_1] + (x - x_0)\,(x - x_1)\,[x, x_0, x_1] \quad \ldots. (2)$$

$$\text{Also } [x, x_0, x_1, x_2] = \frac{[x, x_0, x_1]-[x_0, x_1, x_2]}{x - x_2}$$

$$\therefore \quad [x\ x_0\ x_1] = [x_0, x_1, x_2] + (x - x_2)\,[x, x_0, x_1, x_2]$$

Substituting in 2 we get

$$y = y_0 + (x - x_0)\,[x_0, x_1] + (x - x_0)\,(x - x_1)\,[x_0, x_1, x_2] + (x - x_0)\,(x - x_1)(x - x_2)\,[\,x, x_0, x_1, x_2]$$

Continuing in this manner we get

$$y = y_0 + (x - x_0)\,[x_0, x_1] + (x - x_0)(x - x_1)\,[x_0, x_1, x_2]$$
$$+ (x - x_0)(x - x_1)(x - x_2)\,[x_{0,}\, x_1, x_2, x_3] + \ldots\ldots.$$
$$+ \ldots. + (x - x_0)(x - x_1)\ldots(x - x_{n-1})\,[x, x_{0,}\, x_1 \ldots. x_n] \quad ..(3)$$

Which is called Newton's general interpolation formula with divided differences.

Example 1: By means of Newton's divided difference formula find the value of $f(8)$ and $f(15)$ from the following table.

x :	4	5	7	10	11	13
$f(x)$:	48	100	294	900	1210	2028.

Solution: The divided difference table is as under

x	$y = f(x)$	Δ	Δ_y^2	Δ_y^3	Δ_y^4
4	48				
		$\frac{100-48}{5-4} = 52$			
5	100		$\frac{97-52}{7-4} = 15$		
		$\frac{294-100}{7-5} = 97$		$\frac{21-15}{10-4} = 1$	
7	294		$\frac{202-97}{10-5} = 21$		0
		$\frac{900-294}{10-7} = 202$		$\frac{27-21}{11-5} = 1$	
10	900		$\frac{310-202}{11-7} = 27$		0
		$\frac{1210-900}{11-10} = 310$		$\frac{33-27}{13-7} = 1$	
11	1210		$\frac{409-310}{13-10} = 33$		
		$\frac{2023-1210}{13-11} = 409$			
13	2023				

Newton's interpolation formula for unequal intervals is given by

$$f(x) = f(x_0) + (x - x_0)[x_0\, x_1] + (x - x_0)(x - x_1)[x_0\, x_1\, x_2] + (x - x_0)(x - x_1)(x - x_2)[x_0\, x_1\, x_2\, x_3] + \ldots\ldots$$

When $x = 8$

$$f(8) = 48 + (8 - 4)(52) + (8 - 4)(8 - 5)(15) + (8 - 4)(8 - 5)(8 - 7)(1) + 0$$

$$\therefore \quad f(8) = 448$$

When $x = 15$

$$f(15) = 48 + (15 - 4)(52) + (15 - 4)(15 - 5)(15) + (15 - 4)(15 - 5)(15 - 7)(1) + 0 = 3150.$$

Example 2: Given the values

x :	5	7	11	13	17
$f(x)$:	150	392	1452	2366	5202.

Evaluate $f(9)$ using Newton's divided difference formula.

Solution: The divided difference table is

x	$y = f(x)$	Δ	Δ_y^2	Δ_y^3	Δ_y^4
5	150				
		$\frac{392-150}{7-5}=121$			
7	392		$\frac{265-121}{11-5}=24$		
		$\frac{1452-392}{11-7}=265$		$\frac{32-24}{13-5}=1$	
11	1452		$\frac{457-265}{13-7}=32$		0
		$\frac{2366-1452}{13-11}=457$		$\frac{42-32}{17-7}=1$	
13	2366		$\frac{705-457}{17-11}=42$		
		$\frac{5202-2366}{17-13}=709$			
17	5202				

Taking $x = 9$ and using Newton's divided difference formula

$$f(a) = 150 + (9-5)(121) + (9-5)(9-7)(24) + (9-5)(9-7)(9-11)(1) + 0$$
$$= 810.$$

Example 3: The observed values of a function are respectively 168, 120, 72 and 63 at the four positions 3,7,9 and 10 of the independent variable, what is the best estimate you can give for the value of the function at the position 6 of the independent variable.

Solution: The divided difference table is as under.

x	$y = f(x)$	Δ	Δ_y^2	Δ_y^3
3	168			
		$\dfrac{120-168}{7-3} = -12$		
7	120		$\dfrac{-24-(-12)}{9-3} = -2$	
		$\dfrac{72-120}{9-7} = -24$		$\dfrac{5-(-2)}{10-3} = 1$
9	72		$\dfrac{-9-(-24)}{10-7} = 5$	
		$\dfrac{63-72}{10-9} = -9$		
10	63			

$$\therefore \quad f(x) = 168 + (x-3)(-12) + (x-3)(x-7)(-2) + (x-7)(x-9)(1)$$
$$= x^3 - 21x^2 + 119x - 27$$

At $x = 6$,

$$f(6) = 6^3 - 21 \times 6^2 + 119 \times 6 - 27$$
$$= 147.$$

Example 4: Find the polynominal of the lowest possible degree of x by using Newton's divided difference formula and also find the value of the polynominal at $x = 1$ for the following table.

x :	-4	-1	0	2	5
$f(x)$:	1245	33	5	9	1335

Solution: The divided difference table is as under.

x	$y = f(x)$	Δ	Δ_y^2	Δ_y^3	Δ_y^4
–4	1245				
		$\frac{33-1245}{-1-(-4)} = -404$			
–1	33		$\frac{-28-(-404)}{0-(-4)} = 94$		
		$\frac{5-33}{0-(-1)} = -28$		$\frac{10-94}{2-(-4)} = -14$	
0	5		$\frac{2-(-28)}{2-(-1)} = 10$		$\frac{13-(-14)}{5-(-4)} = 3$
		$\frac{9-5}{2-0} = 2$		$\frac{88-10}{5-(-1)} = 13$	
2	9		$\frac{442-2}{5-0} = 88$		
		$\frac{1335-9}{5-2} = 442$			
5	1335				

Newton's divided difference formula is

$$f(x) = f(x_0) + (x - x_0)\,[x_0\, x_1] + (x - x_0)\,(x - x_1)\,[x_0\, x_1\, x_2] + \dots\dots\dots$$

$$= 1245 + (x + 4)\,(-404) + (x + 4)\,(x + 1)\,(94) + (x + 4)$$

$$(x + 1)\,(x - 0)\,(-14) + (x + 4)\,(x + 1)\,(x - 0)\,(x - 2)\,(3)$$

$$\therefore \quad f(x) = 5 - 14x + 6x^2 - 5x^3 + 3x^4$$

At $x = 1$, $f(1) = 5 - 14 + 6 - 5 + 3 = -5$.

Example 5: If log 2 = 0.30103, log 3 = 0.47712, log 5 = 0.69897 and $\log^7$ = 0.84510 find the value of log 4.7 to four places of decimals, by using divided difference formula.

Ans.: log 4.7 = 0.69110 approx.

Example 6 : Use Newton's divided difference formula to obtain $f(x)$ from the following table

x :	0	1	2	4	5	6
$f(x)$	1	14	15	5	6	19

Ans.: $f(x) = x^3 - 9x^2 + 21x + 1.$

Example 7 : Given $f(0) = -18, f(1) = 0, f(3) = 0, f(5) = -248, f(6) = 0, f(9) = 13104$ find $f(x)$.

Ans.: $f(x) = x^5 - 9x^4 + 18x^3 - x^2 + 9x - 19.$

Example 8: The following table gives the normal weights of babies during the first 12 months of life.

Age in Months :	0	2	5	8	10	12
Weights in lbs :	7.5	10.25	15	16	18	21

Estimate the weight of the baby at the age of seven months.

Ans.: 15,67 lbs.

6.7 LAGRANGE'S INTERPOLATION FORMULA FOR UNEQUAL INTERVALS

Let $y_0, y_1, y_2 \ldots y_n$ be the (n + 1) values of the function $y = f(x)$. Corresponding to the arguments $x_0, x_1, x_2 \ldots\ldots x_n$, not necessarily equally spaced. It is assumed that function $f(x)$ is a polynomial in x and since $n + 1$) values of $f(x)$ are given so $(n + 1)^{th}$ differences are zero. Thus $f(x)$ is a polynominal in x of degree n.

Let the polynominal be of the form

$$f(x) = A_0 (x - x_1)(x - x_2) \ldots (x - x_n) + A_1(x - x_0)(x - x_2) \ldots (x - x_n) + \ldots + A_n (x - x_0)(x - x_1) \ldots (x - x_n) \quad \ldots (1)$$

Where $A_0, A_1, A_2 \ldots\ldots A_n$ are constants.

put $x = x_0$

$$y_0 = f(x_0) = A_0 (x_0 - x_1)(x_0 - x_2) \ldots (x_0 - x_n)$$

$$\therefore \quad A_0 = \frac{y_0}{(x_0 - x_1)(x_0 - x_2) \ldots (x_0 - x_n)}$$

Similarly

$$A_1 = \frac{y_1}{(x_1 - x_0)(x_1 - x_2) \ldots\ldots\ldots (x_1 - x_n)}$$

$$\vdots$$

$$A_n = \frac{y_n}{(x_n - x_0)(x_n - x_1) \ldots\ldots\ldots (x_n - x_{n-1})}$$

Substituting the values of $A_0, A_1, A_2 \ldots\ldots A_n$ in (1) we get

$$f(x) = \frac{(x - x_1)(x - x_2)\ldots\ldots\ldots(x - x_n)}{(x_0 - x_1)(x_0 - x_1)\ldots\ldots\ldots(x_0 - x_n)} y_0$$

$$+ \frac{(x - x_0)(x - x_2)\ldots\ldots\ldots(x - x_n)}{(x_1 - x_0)(x_1 - x_2)\ldots\ldots\ldots(x_1 - x_n)} y_1$$

$$+ \ldots\ldots\ldots\ldots\ldots\ldots\ldots\ldots\ldots\ldots\ldots\ldots\ldots\ldots$$

$$+ \frac{(x - x_0)(x - x_1)\ldots\ldots\ldots(x - x_{n-1})}{(x_n - x_0)(x_n - x_1)\ldots\ldots\ldots(x_n - x_{n-1})} y_n \qquad \ldots.(2)$$

Which is Lagrange's interpolation formula for unequal intervals.

Example 1: Using Lagrange's interpolation formula find the value of y when $x = 10$ for the following table

X:	5	6	9	11
Y:	12	13	14	16.

Solution: Here $x_0 = 5, x_1 = 6, x_2 = 9, x_3 = 11$ and $x_4 = 10$

$y_0 = 12, y_1 = 13, y_2 = 14, y_3 = 16.$

By using Lagrange's formula, we get

$$y = \frac{(10-6)(10-9)(10-11)}{(5-6)(5-9)(5-11)} \times 12 + \frac{(10-5)(10-9)(10-11)}{(6-5)(6-9)(6-11)} \times 13$$

$$+ \frac{(10-5)(10-6)(10-11)}{(9-5)(9-6)(9-11)} \times 14 + \frac{(10-5)(10-6)(10-9)}{(11-5)(11-6)(11-9)} \times 16$$

$$= 2 - 4.33 + 11.66 + 5.3.$$

$$y = 14.63.$$

Example 2: Using Lagrange's formula, evaluate $f(9)$ for the following table

x :	5	7	11	13	17
y :	150	392	1452	2366	5202

Solution: Here, $x_0 = 5$, $x_1 = 7$, $x_2 = 11$, $x_3 = 13$, $x_4 = 17$ and $x=9$

$y_0 = 150$, $y_1 = 392$, $y_2 = 1452$ $y_3 = 2366$, $y_4 = 5202$.

By using Lagrange's interpolation formula we get

$$f(9) = \frac{(9-7)(9-11)(9-13)(9-17)}{(5-7)(5-11)(5-13)(5-17)} \times 150$$

$$+ \frac{(9-5)(9-11)(9-13)(9-17)}{(7-5)(7-11)(7-13)(7-17)} \times 392$$

$$+ \frac{(9-5)(9-7)(9-13)(9-17)}{(11-5)(11-7)(11-13)(11-17)} \times 1452$$

$$+ \frac{(9-5)(9-7)(9-11)(9-17)}{(13-5)(13-7)(13-11)(13-17)} \times 2366$$

$$+ \frac{(9-5)(9-7)(9-11)(9-13)}{(17-5)(17-7)(17-11)(17-13)} \times 5202$$

$$= \frac{-50}{3} + \frac{3136}{15} + \frac{3872}{3} - \frac{2366}{3} + \frac{578}{5}$$

$$= 810$$

$\therefore$ When $x = 9$, $y = 810$.

Example 3: The following table gives the normal weights of babies during the first 12 months of life.

Age in months :	0	2	5	8	10	12
Weight in lbs :	$7\frac{1}{2}$	$10\frac{1}{4}$	15	16	18	21

Estimate the weight of the baby at the age of 7 months.

Solution: Here

$x_0 = 0$, $x_1 = 2$, $x_2 = 5$, $x_3 = 8$, $x_4 = 10$, $x_5 = 12$

$y_0 = 7\frac{1}{2}$, $y_1 = 10\frac{1}{4}$, $y_2 = 15$, $y_3 = 16$, $y_4 = 18$, $y_5 = 21$

By using Lagrange's interpolation formula we get

$$f(7) = \frac{(7-2)(7-5)(7-8)(7-10)(7-12)}{(0-2)(0-5)(0-8)(0-10)(0-12)} \times \frac{15}{2}$$

$$+ \frac{(7-0)(7-5)(7-8)(7-10)(7-12)}{(2-0)(2-5)(2-8)(2-10)(2-12)} \times \frac{41}{4}$$

$$+ \frac{(7-0)(7-2)(7-8)(7-10)(7-12)}{(5-0)(5-2)(5-8)(5-10)(5-12)} \times 15$$

$$+ \frac{(7-0)(7-2)(7-5)(7-10)(7-12)}{(8-0)(8-2)(8-5)(8-10)(8-12)} \times 16$$

$$+ \frac{(7-0)(7-2)(7-5)(7-8)(7-12)}{(10-0)(10-2)(10-5)(10-8)(10-12)} \times 18$$

$$+ \frac{(7-0)(7-2)(7-5)(7-8)(7-10)}{(12-0)(12-2)(12-5)(12-8)(12-10)} \times 21$$

$$= \frac{15}{128} - \frac{287}{984} + 5 + \frac{175}{12} - \frac{63}{16} + \frac{21}{32} = 15.67 \text{ lbs}$$

$\therefore$ When $x = 7$, $y = 15.67$ lbs.

Example 4: Find the form of the function given by

x :	3	2	1	– 1
$f(x)$:	3	12	15	– 21

Solution: Here $x_0 = 3$, $x_1 = 2$, $x_2 = 1$, $x_3 = -1$

$y_0 = 3$, $y_1 = 12$, $y_2 = 15$, $y_3 = -21$

$\therefore$ By Lagrange's formula we have.

$$f(x) = \frac{(x-2)(x-1)(x+1)}{(3-2)(3-1)(3+1)} \times 3 + \frac{(x-3)(x-1)(x+1)}{(2-3)(2-1)(2+1)} \times 12$$

$$+ \frac{(x-3)(x-2)(x+1)}{(1-3)(1-2)(1+1)} \times 15 + \frac{(x-3)(x-2)(x-1)}{(-1-3)(-1-2)(-1-1)} \times (-21)$$

$\therefore$ $f(x) = x^3 - 9x^2 + 17x + 6.$

Example 5: Find the form of the function, given

x :	0	1	2	3	4
$f(x)$:	3	6	11	18	27.

Solution: Here $x_0 = 0,\quad x_1 = 1,\quad x_2 = 2,\quad x_3 = 3,\quad x_4 = 4$

$y_0 = 3,\quad y_1 = 6,\quad y_2 = 11,\quad y_3 = 18,\quad y_4 = 27$

$\therefore$ By Lagrange's formula we have

$$f(x) = \frac{(x-1)(x-2)(x-3)(x-4)}{(0-1)(0-2)(0-3)(0-4)}\times 3$$

$$+\frac{(x-0)(x-2)(x-3)(x-4)}{(1-0)(1-2)(1-3)(1-4)}\times 6$$

$$+\frac{(x-0)(x-1)(x-3)(x-4)}{(2-0)(2-1)(2-3)(2-4)}\times 11$$

$$+\frac{(x-0)(x-1)(x-2)(x-4)}{(3-0)(3-1)(3-2)(3-4)}\times 18$$

$$+\frac{(x-0)(x-1)(x-2)(x-3)}{(4-0)(4-1)(4-2)(4-3)}\times 27$$

$$\therefore \quad f(x) = x^2 + 2x + 3.$$

Example 6: Given $\log_{10}^{654} = 2.8156$, $\log_{10}^{658} = 2.8182$, $\log_{10}^{659} = 2.8189$

$\log_{10}^{661} = 2.8202$, find by using Lagrange's formula the value of $\log_{10}^{656}$.

Ans.: $\log_{10}^{656} = 2.8168$

Example 7 : Use Lagrange's formula to find the form of $f(x)$, given

x :	0	2	3	6
$f(x)$:	648	704	729	792

Ans.: $f(x) = 648 + 30x - x^2$.

Example 8: The following values of the function $f(x)$ for values of x are given.

$f(1) = 4,\qquad f(2) = 5,\qquad f(7) = 5,\qquad f(8) = 4.$

Find the value of $f(6)$ and also the value of x for which $f(x)$ is maximum or minimum.

Solution: Here $x_0 = 1,\quad x_1 = 2,\quad x_2 = 7,\quad x_3 = 8$

$y_0 = 4,\quad y_1 = 5,\quad y_2 = 5\quad y_3 = 4$

$\therefore$ By Lagrange's formula, we have

$$f(x) = \frac{(x-2)(x-7)(x-8)}{(1-2)(1-7)(1-8)}\times 4 + \frac{(x-1)(x-7)(x-8)}{(2-1)(2-7)(2-8)}\times 5$$

$$+\frac{(x-1)(x-2)(x-8)}{(7-1)(7-2)(7-8)}\times 5 + \frac{(x-1)(x-2)(x-7)}{(8-1)(8-2)(8-7)}\times 4$$

$$\therefore \qquad f(x) = \frac{1}{6}[-x^2 + 9x + 16]$$

At $x = 6$, $f(6) = \dfrac{1}{6}[-36 + 54 + 16] = 6.66.$

For maximum or minimum of $f(x)$, we have

$f^1(x) = 0 \qquad$ i.e., $-2x + 9 = 0$ or $x = 4.5$

Also $f^{11}(x) = -2 < 0$

Hence $f(x)$ is maximum at the point $x = 4.5$.

Example 9: From the following data estimate the number of persons having incomes between Rs 1000 and Rs 1500.

Income :	Below 500	500–1000	1000–2000	2000–3000	3000–4000
No of persons :	6000	4250	3600	1500	650.

Solution: We put all the frequencies in the form of cumulative frequencies and form the table as follows.

Income less than:	500	1000	2000	3000	4000
No of persons:	6000	10250	13850	15350	16000

Now shift the origin to 500, and divide the scale by 500 in this way we have known $f(x)$ for $x = 0$, 1,3,5 and 7 we shall now find the number of persons whose income is less than 1500 i.e., $f(x)$ for $x = \dfrac{1500-500}{500} = 2$.

By Lagrange's formula,

$$f(z) = \frac{(2-1)(2-3)(2-5)(2-7)}{(0-1)(0-3)(0-5)(0-7)} \times 6000$$

$$+ \frac{(2-0)(2-3)(2-5)(2-7)}{(1-0)(1-3)(1-5)(1-7)} \times 10250$$

$$+ \frac{(2-0)(2-1)(2-5)(2-7)}{(3-0)(3-1)(3-5)(3-7)} \times 13850$$

$$+ \frac{(2-0)(2-1)(2-3)(2-7)}{(5-0)(5-1)(5-3)(5-7)} \times 15350$$

$$+ \frac{(2-0)(2-1)(2-3)(2-5)}{(7-0)(7-1)(7-3)(7-5)} \times 16000$$

$$f(z) = 12570 \text{ approx.}$$

i.e., number of persons having income less than Rs 1500 = 12570. But number of persons having income less than Rs 1000 = 10250. Therefore persons having income between Rs 1000 and Rs 1500 = 12570 – 10250 = 2320.

Example 10: Find the missing term from the following table

x :	0	1	2	3	4
y:	1	3	9	–	81

Solution: Take 'a' as the missing term and form the difference table as follows:

x	y	Δy	$\Delta^2 y$	$\Delta^3 y$	$\Delta^4 y$
0	1				
		2			
1	3		4		
		6		$a-19$	
2	9		$a-15$		$-4a+124$
		$a-9$		$-3a+105$	
3	a		$-2a+90$		
		$81-a$			
4	81				

Since there is only one unknown the forward differences are carried upto the stage of getting a single value of a.

i.e., assuming $y = f(x)$ to be a polynominal of degree 3.

Now put $\Delta^4 y = 0$

$\therefore \quad -4a + 124 = 0 \qquad \therefore \; a = 31$

Thus $y = 31$ at $x = 3$.

Example 11: Find the missing terms in the following table

x :	-0.2	0	0.2	0.4	0.6	0.8	1.0
$f(x)$:	2.6	–	3.4	4.28	–	14.2	29

Solution: Let $f(0) = a$ and $f(0.6) = b$.

Let us assume that $f(x)$ be a polynominal of degree 4 so that $\Delta^5 f(x) = 0$.

x	$f(x)$	$\Delta f(x)$	$\Delta^2 f(x)$	$\Delta^3 f(x)$	$\Delta^4 f(x)$	$\Delta^5 f(x)$
-0.2	2.6					
		$a-2.6$				
0	a		$-2a+6$			
		$3.4-a$		$3a-8.53$		
0.2	3.4		$a-2.52$		$b-4a+5.88$	
		.88		$b-a-2.64$		$-5b+5a+20.4$
0.4	4.28		$b-5.16$		$-4b+a+26.28$	
		$b-4.28$		$-3b+23.64$		$10b-a-67.8$
0.6	b		$-2b+18.48$		$6b-41.52$	
		$14.2-b$		$3b-17.88$		
0.8	14.2		$b+0.6$			
		14.8				
1.0	29					

$\therefore \quad -5b + 5a + 20.4 = 0$ Solving $a = 3$, $b = 7.08$

$10b - a - 67.8 = 0$

Thus $f(0) = 3$ and $f(0.6) = 7.08$.

Example 12: Under suitable assumption find u_2 and u_4 given that

$$u_1 = 8,\ u_3 = 64,\ u_5 = 216$$

Ans.: $u_2 = 24$ and $u_4 = 128$.

13) Given the values of x and y

x :	1.2	2.1	2.8	4.1	4.9	6.2
y :	4.2	6.8	9.8	13.4	15.5	19.6

Find the values of x corresponding to $y = 12$ using Lagrange's technique.

14) Use Lagrange's interpolation formula to fit a polynomial for the data

x :	0	1	3	4
y :	–12	0	6	12

Hence estimate y at $x = 2$.

6.8 INVERSE INTERPOLATION FORMULA

Consider the function $y = f(x)$, where y is the dependent variable and x is the independent variable. The technique of determining y for x lying between two tabulated values of argument, is known as interpolation whereas the technique of determining the value of x corresponding to the value of y lying between two tabulated values of y is known as inverse interpolation.

"The technique of determining i.e., the value of the argument corresponding to the given value of the function when the function lies between two tabulated values, is known as inverse interpolation."

For example, if $y = \tan x$ then $x = \tan^{-1} y$; if $y = \sin x$ then $x = \sin^{-1} y$, which shows that the values of x can be calculated by substituting the values of y.

Lagrange's method resembles to the method of direct interpolation as discussed already. Here the main difference is that we assume x to be a polynomial in y and the formula is

$$\therefore \qquad x = \frac{(y-y_1)(y-y_2)\,............(y-y_n)}{(y_0-y_1)(y_0-y_2)\,............(y_0-y_n)}x_0$$

$$+ \frac{(y-y_0)(y-y_2)\,............(y-y_n)}{(y_1-y_0)(y_1-y_2)\,............(y_1-y_n)}x_1$$

$$+ \ ....$$

$$+\frac{(y-y_0)(y-y_1)\,..........(y-y_{n-1})}{(y_n-y_0)(y_n-y_1)\,..........(y_n-y_{n-1})}x_n.$$

This formula gives the value of x when y is known.

Example 1 : Apply Lagrange's inverse formula to find the value of x when $y = 13.6$ from the following table.

x :	30	35	40	45	50
y :	15.9	14.9	14.1	13.3	12.5

Solution: Here $x_0 = 30$, $x_1 = 35$, $x_2 = 40$, $x_3 = 45$, $x_4 = 50$,

$y_0 = 15.9$, $y_1 = 14.9$, $y_2 = 14.1$, $y_3 = 13.3$, $y_4 = 12.5$.

When $y = 13.6$

$$x = \frac{(13.6-14.9)(13.6-14.1)(13.6-13.3)(13.6-12.5)}{(15.9-14.9)(15.9-14.1)(15.91-13.3)(15.9-12.5)}\times 30$$

$$+\frac{(13.6-15.9)(13.6-14.1)(13.6-13.3)(13.6-12.5)}{(14.9-15.9)(15.9-14.1)(15.9-13.3)(15.9-12.5)}\times 35$$

$$+\frac{(13.6-15.9)(13.6-14.9)(13.6-13.3)(13.6-12.5)}{(14.1-15.9)(14.1-14.9)(14.1-13.3)(14.1-12.5)}\times 40$$

$$+\frac{(13.6-15.6)(13.6-14.9)(13.6-14.1)(13.6-12.5)}{(13.3-15.6)(13.3-14.9)(13.3-14.1)(13.3-12.5)}\times 45$$

$$+\frac{(13.6-15.6)(13.6-14.9)(13.6-14.1)(13.6-13.3)}{(12.5-15.6)(12.5-14.9)(12.5-14.1)(12.5-13.3)}\times 50$$

$$= 40.1$$

Thus when $y = 13.6$ $x = 40.1$.

Example 2: The following table gives the values of x and y

x :	1.2	2.1	2.8	4.1	4.9	6.2
y :	4.2	6.8	9.8	13.4	15.5	19.6

Find the value of x corresponding to $y = 12$ using Lagrange's technique.

Solution: Here $x_0 = 1.2$, $x_1 = 2.1$, $x_2 = 2.8$, $x_3 = 4.1$, $x_4 = 4.9$, $x_5 = 6.2$

$y_0 = 4.2$, $y_1 = 6.8$, $y_2 = 9.8$, $y_3 = 13.4$, $y_4 = 15.5$, $y_5 = 19.6$

Taking $y = 12$ and using Lagrange's inverse formula we get

$$x = \frac{(12-6.8)(12-9.8)(12-13.4)(12-15.5)(12-19.6)}{(4.2-6.8)(4.2-9.8)(4.2-13.4)(4.2-15.5)(4.2-19.6)} \times 1.2$$

$$+ \frac{(12-4.2)(12-9.8)(12-13.4)(12-15.5)(12-19.6)}{(6.8-4.2)(6.8-9.8)(6.8-13.4)(6.8-15.5)(6.8-19.6)} \times 2.1$$

$$+ \frac{(12-4.2)(12-6.8)(12-13.4)(12-15.5)(12-19.6)}{(9.8-4.2)(9.8-6.8)(9.8-13.4)(9.8-15.5)(9.8-19.6)} \times 2.8$$

$$+ \frac{(12-4.2)(12-6.8)(12-9.8)(12-15.5)(12-19.6)}{(13.4-4.2)(13.4-6.8)(13.4-9.8)(13.4-15.5)(13.4-19.6)} \times 4.1$$

$$+ \frac{(12-4.2)(12-6.8)(12-9.8)(12-13.4)(12-19.6)}{(15.5-4.2)(15.5-6.8)(15.5-9.8)(15.5-13.4)(15.5-19.6)} \times 4.9$$

$$+ \frac{(12-4.2)(12-6.8)(12-9.8)(12-13.4)(12-15.5)}{(19.6-4.2)(19.6-4.2)(19.6-9.8)(19.6-13.4)(19.6-15.5)} \times 19.6$$

$$= 0.022 - 0.234 + 1.252 + 3.419 - 0.964 + 0.055 = 3.55$$

$\therefore x = 3.55$ when $y = 12$.

Example 3: Apply Lagrange's formula inversely to obtain the root of the equation $f(x) = 0$ given that

$$f(30) = -30, f(34) = -13, f(38) = 3 \text{ and } f(42) = 18.$$

Ans.: Here

$x_0 = 30,\quad x_1 = 34,\quad x_2 = 38,\quad x_3 = 42$

$y_0 = -30,\quad y_1 = -13,\quad y_2 = 3,\quad y_3 = 18.$

Using inverse Lagrange's formula we get

$x = 37.23.$

Example 4: Obtain the value of t when $A = 85$ from the following table, using Lagrange's method

t :	2	5	8	14
A :	94.8	87.9	81.3	68.7

Ans.: 6.5928.

Example 5: Apply Lagrange's method to find the value of x when $f(x) = 15$ from the given data.

x :	5	6	9	11
$f(x)$:	12	13	14	16.

Ans.: $x = 10.5$.

6.9 NUMERICAL INTEGRATION

It is the process of finding or evaluating a definite integral $I = \int_a^b f(x)dx$ from a set of numerical values of the integrand. If it is applied to the integration of a function of a single variable the process is known as quadrature. The problem of numerical integration is solved by first approximating the integrand by a polynomial with the help of an interpolation formula and then integrating the formula between the desired limits.

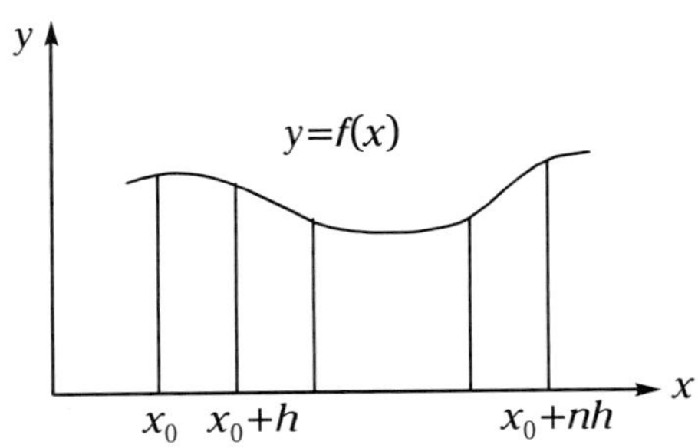

Let $I = \int_a^b f(x)dx$ be the integral, where $f(x)$ takes the values

$y_0, \; y_1, \; y_2 \ldots\ldots y_n$ for

$x_0, \; x_1, \; x_2 \ldots\ldots x_n.$

Let us divide the integral (a, b) so that $x_0 = a, x_1 = a + h$

$x_2 = a + 2h \ldots\ldots\ldots\ldots x_n = x_0 + nh$

put $\qquad x = x_0 + rh \qquad \therefore \; dx = hdr$

$$= h\int_0^n f(x_0 + rh)dr$$

$$= h\int_0^n \left[y_0 + r\Delta y_0 + \frac{r(r-1)}{2!}\Delta^2 y_0 + \frac{r(r-1)(r-2)}{3!}\Delta^3 y_0 + ... \right] dr$$

[By Newton's forward interpolation formula]

Integrating term by term we obtain

$$\int_{x_0}^{x_0+nh} f(x)dx = \left[y_0 + \frac{n}{2}\Delta y_0 + \frac{n(2n-3)}{12}\Delta^2 y + \frac{n(n-2)^2}{24}\Delta^3 y_0 + ... \right] \qquad ....(1)$$

Which is known as the Newton-Cotes quadrature formula. This is the general quadrature formula, we can obtain a number of quadrature formulae from this by putting $n = 1,2......$

1) The Trapezoidal Rule

Put $n = 1$ in (1) and neglect second and higher differences, we get

$$\int_{x_0}^{x_0+nh} ydx = \frac{h}{2} [(y_0 + y_n) + 2 (y_1 + y_2 + y_3 + ... + y_{n-1})]$$

$$= \frac{h}{2} [(\text{Sum of first and last ordinates}) + 2(\text{sum of the remaining ordinates})]$$

2) Simpson's $\frac{1}{3}$ rule

Put $n = 2$ in (1) we get

$$\int_{x_0}^{x_0+nh} ydx = \frac{h}{3} [(y_0 + y_n) + 4 (y_1 + y_3 + ... + y_{n-1}) + 2 (y_2 + y_4 + ... y_{n-2})]$$

This formula is known as Simpson's $\frac{1}{3}$ rule.

3) Simpson's $\frac{3}{8}$ rule

Put $n = 3$ in (1) we get

$$\int_{x_0}^{x_0+nh} y\,dx = \frac{3h}{8}\left[(y_0 + y_n) + 3(y_1 + y_2 + y_4 + y_5 + \ldots y_{n-1}) + 2\,(y_3 + y_6 + \ldots\ldots + y_{n-3})\right]$$

This formula is known as Simpson's $\frac{3}{8}$ rule.

Note : While applying this rule the number of sub-intervals should be taken as multiple of 3.

4) Weddle's rule

Put $n = 6$ in (1), we get

$$\int_{x_0}^{x_0+nh} y\,dx = \frac{3h}{10}\left[y_0 + 5y_1 + y_2 + 6y_3 + y_4 + 5y_5 + 2y_6 + 5y_7 + y_8 + 6y_9 + y_{10} + y_{11} + 2y_{12} + 5y_{13} + y_{14} + 6y_{15} + y_{16} + 5y_{17} + \ldots\ldots\right]$$

This formula is known as Weddle's rule.

Note : While applying this formula the number of sub-intervals i.e., n should be taken as multiple of 6.

Grouping,

1 5	1 6	1 5
2 5	1 6	1 5
2 5	1 6	1 5
2 etc.		

Example 1: A curve is drawn to pass through the points given by the following table.

x :	1	1.5	2	2.5	3	3.5	4
y :	2	2.4	2.7	2.8	3	2.6	2.1

Estimate the area bounded by the curve, x-axis and the lines $x = 1$, $x = 4$, by Simpson's $\frac{1}{3}$ rule. ii) by Weddle's rule

Solution: Here $h = 0.5$

x :	1	1.5	2	2.5	3	3.5	4
y :	2	2.4	2.7	2.8	3	2.6	2.1
	y_0	y_1	y_2	y_3	y_4	y_5	y_6

i) By Simpson's $\frac{1}{3}$ rule we have.

$$\int_1^4 y\,dx = \frac{h}{3}[(y_0+y_6)+4(y_1+y_3+y_5)+2(y_2+y_4)]$$

$$= \frac{0.5}{3}[(2+2.1)+4(2.4+2.8+2.6)+2(2.7+3)]$$

$$= \frac{5}{30}\times 46.7 = 7.783 \text{ units of area.}$$

ii) By Weddle's rule, we have

$$\int_1^4 y\,dx = \frac{3h}{10}[y_0+5y_1+y_2+6y_3+y_4+5y_5+y_6]$$

$$= \frac{3\times 0.5}{10}[2+5\times 2.4+2.7+6\times 2.8+3+5\times 2.6+2.1]$$

$$= \frac{3\times 0.5}{10}[2+12.00+2.7+16.8+3+13.00+2.1]$$

$$= 7.74 \text{ units of area.}$$

Example 2: Evaluate $\int_0^1 \frac{dx}{1+x^2}$ by using Simpson's '$\frac{1}{3}$' and '$\frac{3}{8}$' rule.

Obtain the approximate value of π in each case.

Solution: Divide the range (0,1) into 6 equal parts so that

$$h = \frac{b-a}{n} = \frac{1-0}{6} = \frac{1}{6}$$

Let $y = \frac{1}{1+x^2}$ then form the table as follows,

$x:$	0	$\frac{1}{6}$	$\frac{2}{6}$	$\frac{3}{6}$	$\frac{4}{6}$	$\frac{5}{6}$	1
$y:$	1.0000	0.97297	0.90000	0.80000	0.69231	.59016	0.50000
	y_0	y_1	y_2	y_3	y_4	y_5	$y_6.$

i) By Simpson's $\frac{1}{3}$ rule, we have

$$\int_0^1 \frac{dx}{1+x^2} = \frac{h}{3}\left[(y_0+y_6)+4\,(y_1+y_3+y_5)+2\,(y_2+y_4)\right]$$

$$= \frac{1}{18}\left[(0+.5000)+4\,(0.97292+0.80000+0.59016)\right.$$
$$\left.+\,2\,(0.90000+0.69231)\right]$$

$$= .785395 \quad \ldots (1)$$

ii) By Simpson's $\frac{3}{8}$ rule

$$\int_0^1 \frac{dx}{1+x^2} = \frac{3h}{8}\left[(y_0+y_6)+3(y_1+y_2+y_4+y_5)+2(y_3)\right]$$

$$= \frac{1}{16}\left[(1.0000+0.50000)+3\,(0.97297+0.90000\right.$$
$$\left.+\,0.69231+0.59016)+2\,(0.8000)\right]$$

$$= .785395 \quad \ldots. (2)$$

Also,

$$\int_0^1 \frac{dx}{1+x^2} = \frac{dx}{1+x^2} = \left[\tan^{-1} x\right]_0^1 = \tan^{-1} 1 - \tan^{-1} 0 = \frac{\pi}{4} \quad \ldots(3)$$

We have from (1), (2) and (3)

$$\frac{\pi}{4} = 0.785395$$

$$\therefore \quad \pi = 4 \times 0.785395$$

$$\pi = 3.141580.$$

Example 3: Evaluate $\int_0^6 \frac{dx}{1+x^2}$ by using (i) Simpson's $\frac{1}{3}$ rule .

(ii) Simpson's $\frac{3}{8}$ rule and (iii) Weddle's rule.

Solution: Divide the range (0.6) into six equal parts

So that $h = \frac{b-a}{n} = \frac{6-0}{6} = 1.$

Let $y = \frac{1}{1+x^2}$ and form the table as follows

x :	0	1	2	3	4	5	6
y :	1.00	.500	.200	.100	.058824	.038462	.027027
	y_0	y_1	y_2	y_3	y_4	y_5	y_6.

i) By Simpson's $\frac{1}{3}$ rule, we have

$$\int_0^6 \frac{dx}{1+x^2} = \frac{h}{3} [(y_0 + y_6) + 4 (y_1 + y_3 + y_5) + 2 (y_2 + y_4)]$$

$$= \frac{1}{3} [(1 + .027027 + 4 (.500 + .100 + .038462)$$

$$+ 2 (.200 + 0.58824)]$$

$$= 1.366174.$$

ii) By Simpson's $\frac{3}{8}$ rule

$$\int_0^6 \frac{dx}{1+x^2} = \frac{3}{8} [(y_0 + y_6) + 3 (y_1 + y_2 + y_4 + y_5) + 2 (y_3)]$$

$$= \frac{3}{10} [(1.00 + .027027 + 3 (.500 + .200 + .058828$$

$$+ .038462) + 2 (.100)]$$

$$= 1.357082$$

iii) By Weddle's rule

$$\int_0^6 \frac{dx}{1+x^2} = \frac{3h}{10}[y_0 + 5y_1 + y_2 + 6y_3 + y_4 + 5y_5 + y_6]$$

$$= \frac{3}{10}[1.000 + 5 \times .500 + .200 + 6 \times .100 + 0.58824 + 5 \times .038462 + 0.27027]$$

$$= 1.3734483.$$

Also $\int_0^6 \frac{dx}{1+x^2} = \tan^{-1} x \Big]_0^6 \tan^{-1} 6 = 1.4056.$

This shows that the value of the integral found by Weddle's rule is nearest to the actual value followed by its value given by Simpson's $\frac{1}{3}$ rule.

Example 4: Evaluate $\int_4^{5.2} \log_e^x \, dx$ by using (i) Simpson's $\frac{1}{3}$ rule (ii) Simpson's $\frac{3}{8}$ rule and (iii) Weddle's rule. After finding the true value of the integral compare the errors in the three cases.

Solution: Divide the range (4, 5.2) into 6 equal parts

So that $h = \frac{b-a}{n} = \frac{5.2-4}{6} = .2$

Let $y = \log_e^x$ and form the table as follows.

x :	4	4.2	4.4	4.6
y :	1.38629436	1.43508453	1.48160454	1.52605630
	y_0	y_1	y_2	y_3
	4.8	5.0	5.2	
	1.56861592	1.60943791	1.64865863	
	y_4	y_5	y_6.	

i) Simpson's $\frac{1}{3}$ rule, we have

$$\int_{4}^{5.2} \log_e^x = \frac{h}{3}[(y_0 + y_6) + 4(y_1 + y_3 + y_5) + 2(y_2 + y_4)]$$

$$= 1.82784726 \quad \text{.... (1)}$$

ii) Simpson's $\frac{3}{8}$ rule

$$\int_{4}^{5.2} leg_e^x dx = \frac{3h}{8}[(y_0 + y_6) + 3(y_1 + y_2 + y_4 + y_5) + 2(y_3)]$$

$$= 1.82784707 \quad \text{.... (2)}$$

iii) Weddle's rule

$$\int_{4}^{5.2} \log_e^x dx = \frac{3h}{10}[y_0 + 5y_1 + y_2 + 6y_3 + y_4 + 5y_5 + y_6]$$

$$= 1.82784741 \quad \text{..... (3)}$$

Also $\int_{4}^{5.2} \log_e^x dx = \left[x \log_e^x - 1\right]_4^{5.2}$

$$= 1.82784741.$$

∴ By Weddle's rule error is zero and we conclude that in this case Weddle's rule is more accurate than Simpson's '$\frac{1}{3}$' and '$\frac{3}{8}$' rules.

Example 5: The velocity of a train which starts from rest is given by the following table. The time being recorded in minutes from the start and the speed in km/hr.

Time :	0	2	4	6	8	10	12	14	16	18	20
Speed :	0	10	18	25	29	32	20	11	5	2	0

Estimate approximately the total distance travelled in 20 minutes.

Solution: If S be the distance travelled in time t, then we know that $\frac{dS}{dt} = v$

i.e., the velocity,

$$\therefore \quad S = \int_0^t v\, dt$$

Here $h = 2$ minutes $= \frac{1}{30}$ hrs.

$\therefore$

t :	0	2	4	6	8	10	12	14	16	18	20
v :	0	10	18	25	29	32	20	11	5	2	0
	v_0	v_1	v_2	v_3	v_4	v_5	v_6	v_7	v_8	v_9	v_{10}

By using Simpson's $\frac{1}{3}$ rule we have

$$\int_0^{20} v\, dt = \frac{h}{3}\,[(v_0 + v_{10}) + 4\,(v_1 + v_3 + v_5 + v_7 + v_9) + 2\,(v_2 + v_4 + v_6 + v_8)]$$

$$= \frac{1}{90}\,[(0 + 0) + 4\,(10 + 25 + 32 + 11 + 2) + 2\,(18 + 29 + 20 + 5)$$

$$= 5.16 \text{ km approx.}$$

Example 6: Use Simpson's $\frac{1}{3}$ rule to find the approximate area of the cross-section of a river 80 ft wide. The depth y (in feet) at a distance x from one bank being given by the following table.

x :	0	10	20	30	40	50	60	70	80
y :	0	4	7	9	12	15	14	8	3.

Solution: Here $h = 10$ and we use Simpson's $\frac{1}{3}$ rule

$$\therefore \quad \int_0^{80} y\, dx = \frac{h}{3}\,[(y_0 + y_8) + 4\,(y_1 + y_3 + y_5 + y_7) + 2\,(y_2 + y_4 + y_6)]$$

$$= \frac{10}{3}\ [(0 + 3) + 4\ (4 + 9 + 15 + 8) + 2\ (7 + 12 + 14)]$$

$$= 710 \text{ square ft.}$$

Example 7: A rocket is launched from the ground. Its acceleration is registered during first 80 seconds and is given as follows. Using Simpson's $\frac{1}{3}$ rule find the velocity of the rocket at 80 seconds.

Time :	0	10	20	30	40	50	60	70	80
A :	30	31.63	33.34	35.47	37.75	40.75	43.35	46.69	50.67

Solution: If S be the distance travelled with velocity then $\frac{d^2S}{dt^2} = \frac{dv}{dt} = A$ is the acceleration.

$$\therefore \qquad v = \int A\,dt$$

Time t :	0	10	20	30	40	50	60	70	80
Acc^n :	30	31.63	33.34	35.47	37.75	40.75	43.35	46.69	50.67
	A_0	A_1	A_2	A_3	A_4	A_5	A_6	A_7	A_8

By Simpson's $\frac{1}{3}$ rule.

$$v = \frac{h}{3}\ [(A_0 + A_8) + 4\ (A_1 + A_3 + A_5 + A_7) + 2\ (A_2 + A_4 + A_6)]$$

Where $h = 10$

$$= \frac{10}{3}\ [(30 + 50.67) + 4\ (31.63 + 35.47 + 40.75 + 46.69) + 2\ (33.34 + 37.75 + 43.35)]$$

$$= 30.87 \text{ m / see.}$$

Example 8: Find the value of $\int_0^{\pi/2} e^{\sin x} dx$ correct to four decimal places by Simpson's $\frac{3}{8}$ rule.

Solution: Divide the range $\left(0, \frac{\pi}{2}\right)$ into 3 equal parts

So that $$h = \frac{b-a}{n} = \frac{\frac{\pi}{2} - 0}{3} = \frac{\pi}{6}$$

$$y = e^{\sin x}.$$

$\therefore$ x :	0	$\frac{\pi}{6}$	$\frac{\pi}{3}$	$\frac{\pi}{2}$
$y = e^{\sin x}$:	1	1.64872	2.45960	2.71828
	y_0	y_1	y_2	y_3

By Simpson's $\frac{3}{8}$ rule

$$\int_0^{\pi/2} e^{\sin x} dx = \frac{3h}{8} [(y_0 + y_3) + 3 (y_1 + y_2)]$$

$$= \frac{3}{8} \times \frac{\pi}{6} [(1 + 2.71828) + 3 (1.64872 + 2.45960)]$$

$$= 3.1500\,722930.$$

$$= 3.1501, \text{ approx.}$$

Example 9: Evaluate $\int_0^4 e^x \, dx$ by using Simpson's rule to the data given $e = 2.72$, $e^2 = 7.39$, $e^3 = 20.09$, $e^4 = 54.60$ and compare it with the actual value.

Ans.: $\int_0^4 e^x \, dx = 53.873$ and actual value = 53.60.

Example 10: Find the value of $\log^2$ from $\int_0^1 \frac{x^2}{1+x^3} dx$ using Simpson's $\frac{1}{3}$ rule by dividing the range into four equal parts

Ans.: $\log^2 = .69315$.

Example 11: Compute the value of the definite integral

$\int_{.2}^{1.4} \left(\sin x - \log_e^x + e^x\right)dx$ dividing the range into six equal parts by using Weddle's rule and compare with the exact value.

Ans.: $I = 4.05098$ and actual value = 4.05095.

Example 12: Calculate an approximate value of $\int_0^{\pi/2} \sin x \, dx$ by Simpson's rule, using 11 ordinates.

Ans.: $I = 1.0006$.

Example 13 : Using Simpson's $\frac{3}{8}$ rule find the value of $\int_0^{0.3} (1-8x^3)^{1/2} dx$

Ans.: 0.2916.

Example 14: Evaluate $\int_0^1 \frac{x}{1+x^2} dx$ by using Weddle's rule, taking six equal intervals. Hence evaluate $\log^2$.

Ans.: $\log^2 = 0.693134$.

Example 15: Evaluate $\int_0^1 \frac{x}{1+x^2} dx$ by Simpson's $\frac{3}{8}$ rule taking six equal parts.

Ans.: 1.1873.

6.10 QUESTION BANK

1) Given $f(a) = 1. f(1) = 3, f(2) = 7, f(3)$ 13. Find $f(0.1)$ and $f(2.9)$ using Newton Interpolation formula.
2) Given $f(40) = 184$, $f(50) = 204$, $f(60) = 226$, $f(70) = 250$, $f(80) = 276$, $f(90)$ 304, find $f(85)$ using Newton's backward interpolation formula.

3) A function $y = f(x)$ is given by the following table.

x:	1	1.2	1.4	1.6	1.8	2.0
y:	0.0	0.128	0.544	1.296	2.432	4.00

Find the approximate value of y when $x = 1.1$

By using suitable interpolation formula.

4) Under the suitable assumptions find the missing terms in the following table.

x :	– 0.2	0.0	0.2	0.4	0.6	0.8	1.0
$f(x)$:	2.6	–	3.4	4.28	–	14.2	29

5) From the data given in the following table, find the number of students who obtained marks between 40 and 45.

Marks :	30 – 40	40 – 50	50 – 60	60 – 70	70 – 80
No of Students:	31	42	51	35	31.

6) The population of a town is given by the table

Year :	1951	1961	1971	1981	1991
Population: in '000'	19.96	39.65	58.81	77.21	94.61.

Estimate the population of the town in the year 1955.

7) Using Newton's divided difference formula evaluate $f(8)$ and $f(15)$ given that

x :	4	5	7	10	11	13
$f(x)$:	48	100	294	900	1210	2028.

8) Given the values of x and y

x :	1.2	2.1	2.8	4.1	4.9	6.2
y :	4.2	6.8	9.8	13.4	15.5	19.6.

Find the value of x corresponding to $y = 12$ using Lagrange's technique.

9) Use Lagrange's interpolation formula to fit a polynomial for the data.

x :	0	1	3	4
y :	– 12	0	6	12

Hence evaluate y when $x = 2$.s

10) Compute the value of x when $y = 8$ using Lagrange's interpolation formula given that

x :	-2	-1	1	2
y:	-7	2	0	11

11) Determine $f(x)$ as a polynomial in x for the following using Newton's divided difference formula

x :	-4	-1	0	2	5
y :	1245	33	5	9	1335

12) Construct an interpolating polynomial for the data given below using Newton's divided difference formula

x :	2	4	5	6	8	10
$f(x)$:	10	96	196	350	868	1746.

13) Evaluate $\int_0^1 \frac{1}{1+x^2}$ using Simpson's $\frac{3}{8}$ in rule by considering seven ordinates and hence find the approximate value of π.

14) Given

x :	0	1	2	3	4	5	6
y :	0	2	2.5	2.3	2	1.7	1.5.

Evaluate $\int_0^6 y\,dx$, by Weddle's rule.

15) Evaluate $\int_0^{\pi/2} \sqrt{\cos\theta}\,d\theta$ do by Simpson's $\frac{1}{3}$ rule, dividing the interval $\left(0, \pi/2\right)$ into eight equal parts.

16) Using Simpson's $\frac{3}{8}$ rule, evaluate $\int_0^{0.3} \sqrt{1-8x^3}\, dx$ by taking seven ordinates.

17) Evaluate $\int_0^5 \frac{dx}{4x+5}$ by using the Simpson's $\frac{1}{3}$ rule taking 10 equal parts. Hence find $\log^5$.

18) Evaluate $\int_0^6 \frac{1}{1+x^2} dx$ using Weddle's rule taking seven ordinates.

19) Evaluate $\int_4^{5.2} \log_e^x \, dx$ by using Weddle's rule taking seven ordinates.

UNIT 7

NUMERICAL SOLUTIONS OF PARTIAL DIFFERENTIAL EQUATIONS

7.1 NUMERICAL SOLUTIONS OF PARTIAL DIFFERENTIAL EQUATIONS

In Applied Mathematics, we often come across partial differential equations. But only a few of them can be solved by analytical methods. So in most cases, we go in for numerical methods to approximate solution. Of all the numerical methods available for the solution of partial differential equations, the method of finite differences is most commonly used. In this method, the derivatives appearing in the equation and the boundary conditions are replaced by their finite difference approximations. Then the given equation is changed into a system of linear equations which are solved by iterative procedures.

Classification of second order equations

The general second order linear partial differential equation is of the form

$$A\frac{\partial^2 u}{\partial x^2}+B\frac{\partial^2 u}{\partial y\,\partial x}+C\frac{\partial^2 u}{\partial y^2}+D\frac{\partial u}{\partial x}+E\frac{\partial u}{\partial y}+Fu=G$$

Which can be written as

$$Au_{xx}+Bu_{xy}+Cu_{yy}+Du_x+Eu_y+Fu=G \qquad \text{.... (1)}$$

Where A, B, C, D, E, F and G are functions of x and y Equation (1) can be classified with respect to the sign of the discriminant $\Delta = B^2 - 4AC$ as follows.

1) If $B^2 - 4AC < 0$ Elliptic
2) If $B^2 - 4AC = 0$ Parabolic and
3) If $B^2 - 4AC > 0$ Hyperbolic.

Example 1: Classify the following equation.

$$\frac{\partial^2 u}{\partial x^2} + 4\frac{\partial^2 u}{\partial x\,\partial y} + 4\frac{\partial^2 u}{\partial y^2} - \frac{\partial u}{\partial x} + 2\frac{\partial u}{\partial y} = 0$$

Solution: Comparing this equation with equation (1)

We get $A = 1, B = 4, C = 4$.

$\therefore$ $\Delta = B^2 - 4AC = 4^2 - 4 \times 1 \times 4 = 0$

$\therefore$ The given equation is parabolic.

Example 2: Classify the equation

$$x^2 \frac{\partial^2 u}{\partial x^2} + (1 - y^2)\frac{\partial^2 u}{\partial y^2} = 0 \qquad -\infty < x < \infty, \quad -1 < y < 1$$

Solution: Comparing with the general equation (1)

We have $A = x^2, B = 0, C = 1 - y^2$

$\therefore$ $\Delta = B^2 - 4AC = 0 - 4x^2(1 - y^2)$

$= 4x^2(y^2 - 1)$

For all x between $-\infty$ and ∞ x^2 is positive and for all y between -1 and $+1$, $y^2 < 1$.

$\therefore$ $B^2 - 4AC < 0$

Hence the equation is elliptic.

Example 3: Classify the equation

$$(1 + x^2)\frac{\partial^2 u}{\partial x^2} + (5 + 2x^2)\frac{\partial^2 u}{\partial x\,\partial t} + (4 + x^2)\frac{\partial^2 u}{\partial t^2} = 0$$

Solution: Comparing with the general equation (1) we have $A = 1 + x^2$ $B = 5 + 2x^2$, $C = 4 + x^2$

$\therefore$ $\Delta = B^2 - 4AC = (5 + 2x^2)^2 - 4(1 + x^2)(4 + x^2)$

$= 9 > 0.$

Since $\Delta = 9 > 0$, the equation is hyperbolic.

Example 4: Determine whether the following equation is elliptic or hyperbolic.

$$(x + 1)\, u_{xx} - 2(x + 2)\, u_{xy} + (x + 3)\, u_{yy} = 0$$

Ans.: Hyperbolic.

Example 5: Classify the equation

$$y^2 \frac{\partial^2 u}{\partial x^2} - 2y \frac{\partial^2 u}{\partial x\, \partial y} + \frac{\partial^2 u}{\partial y^2} - \frac{\partial u}{\partial y} = 8y$$

Ans.: Parabolic.

7.2 FINITE DIFFERENCE APPROXIMATIONS TO DERIVATIVES

Let the (x, y) plane be divided into a network of rectangles of sides $\Delta\, x = h$ and $\Delta\, y = k$ by drawing the sets of lines. The points of intersection of the dividing lines are called **mess points** or **nodal points** or **grid points** or **lattice points.**

Now, we have the finite difference approximations for the partial derivatives in x – direction.

$$\frac{\partial u}{\partial x} = \frac{u(x+h,\, y) - u(x,\, y)}{h} + 0(h)$$

$$= \frac{u(x, y) - y(x-h,\, y)}{h} + 0(h)$$

$$= \frac{u(x+h,\, y) - u(x-h, y)}{2h} + 0(h^2)$$

and $$\frac{\partial^2 u}{\partial x^2} = \frac{u(x-h, y) - 2u(x,\, y) + u(x+h,\, y)}{h^2} + 0(h^2)$$

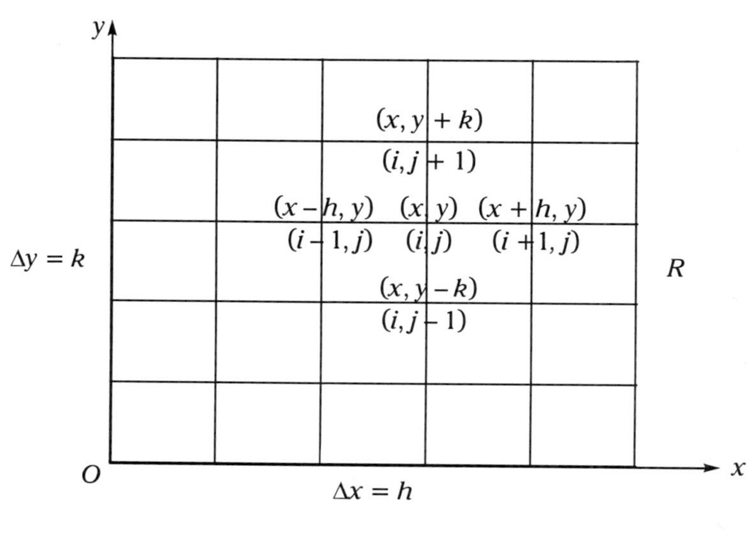

Writing $u(x, y) = u\ (ih, jk)$ as simply u_{ij} the above approximations become.

$$U_x = \frac{u_{i+1,j} - u_{i,j}}{h} + 0(h)$$

$$= \frac{u_{i,j} - u_{i-1,j}}{h} + 0(h)$$

$$= \frac{u_{i+1,j} - u_{i-1,j}}{2h} + 0(h^2)$$

and $$U_{nn} = \frac{u_{i-1,\ j} - 2u_{i,j} + u_{i+1,\ j}}{h^2} + 0(h^2)$$

Similarly, we have the approximations for the derivatives with respect to y.

$$U_y = \frac{u_{i,j+1} - u_{ij}}{k} + 0(k)$$

$$= \frac{u_{ij} - u_{i,\ j-1}}{k} + 0(k)$$

$$= \frac{u_{i,\ j+1} - u_{i,\ j-1}}{2k} + 0(k^2)$$

and $$U_{yy} = \frac{u_{i,\ j-1} - 2u_{ij} + u_{ij+1}}{k^2} + 0(k^2)$$

Replacing the derivatives in any partial differential equation by their corresponding difference approximations we obtain the finite difference analogues of the given equations.

7.3 ELLIPTIC EQUATIONS

The Laplace's equation

$$\nabla^2 u = \frac{\partial^2 U}{\partial x^2} + \frac{\partial^2 U}{\partial y^2} = 0 \qquad ....(1)$$

and the Poisson's equation

$$\frac{\partial^2 U}{\partial x^2} + \frac{\partial^2 U}{\partial y^2} = f(x, y) \qquad ....(2)$$

are examples of elliptic partial differential equations, Laplace's equation arises in steady - state flow and potential problems. Poisson's equation arises in fluid mechanics, electricity and magnetism and torsion problems.

The solution of these equations is a function $U(x,y)$ which is satisfied at every point of a region R subject to certain boundary conditions specified on the closed curve. In general, problems concerning steady viscous flow, equilibrium stresses in classic structures etc lead to elliptic type of equations.

7.4 SOLUTION OF LAPLACE'S EQUATION

$$\frac{\partial^2 U}{\partial x^2}+\frac{\partial^2 U}{\partial y^2}=0 \qquad \text{....(1)}$$

Consider a rectangular region R for which $U(x, y)$ is known at boundary. Divide this region into a network of square mesh of side 'h'. Replacing the derivatives in (1) by their difference approximations, we have.

$$\frac{1}{h^2}\left[u_{i-1,\,j}-2u_{ij}+u_{i+1,\,j}\right]+\frac{1}{h^2}\left[u_{i,\,j-1}-2u_{ij}+u_{i,\,j+1}\right]=0$$

$$\therefore \quad u_{ij}=\frac{1}{4}\left[u_{i-1,j}+u_{i+1,j}+u_{i,\,j+1}+u_{i,\,j-1}\right] \qquad \text{....(2)}$$

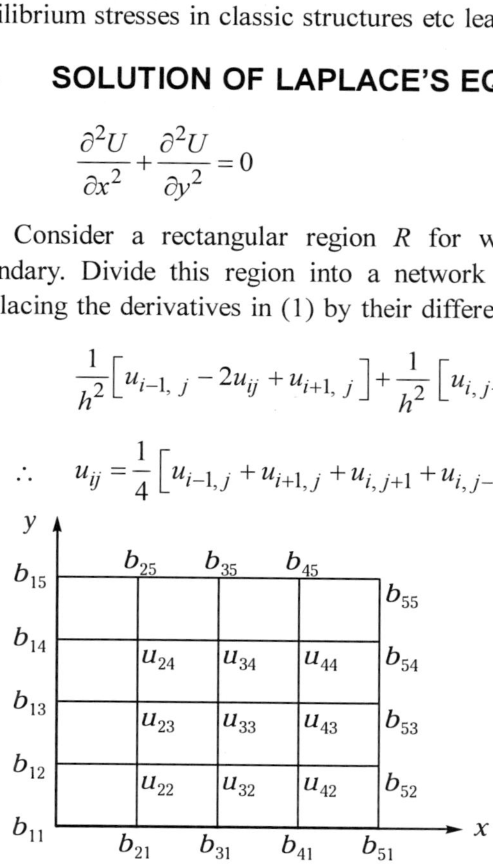

....(1)

This shows that the value of u_{ij} at any interior mesh point is the average of its values at four neighbouring points to the left, right above and below (2) is called the **standard 5-point formula.**

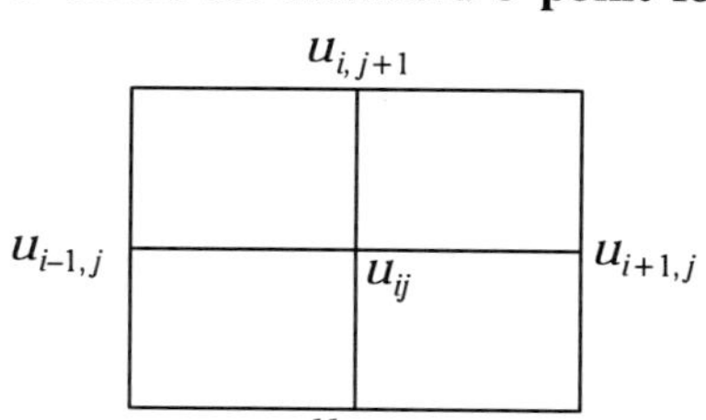

...(2)

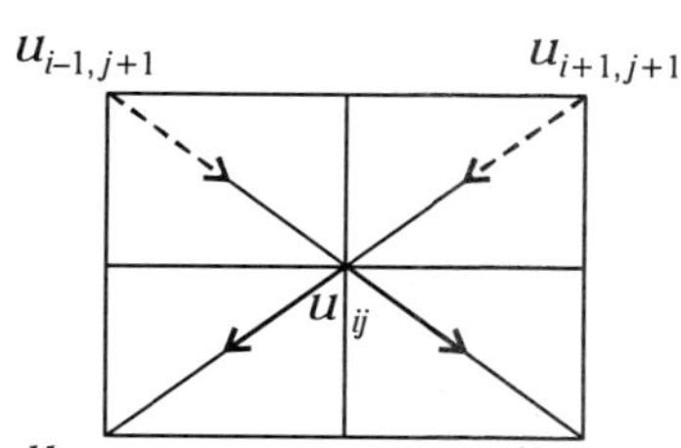

...(3)

Sometimes a formula similar to (2) is used which is given by

$$u_{ij} = \frac{1}{4}\left[u_{i-1,\, j+1} + u_{i+1,\, j-1} + u_{i+1, j+1} + u_{i-1, j-1}\right] \qquad ...(3)$$

This shows that the value of u_{ij} is the average of its values at the four neighbouring diagonal mesh points (3) is called the **diagonal 5-point formula.**

Although (3) is less accurate than (2), yet it serves as a reasonably good approximation for obtaining the starting values at the mesh points.

Now to find the initial values of U at the interior mesh points, we first use diagonal five point formula (3) and compute $u_{3,3,}$ $u_{2,4}$, $u_{4,4}$ $u_{4,2}$, and $u_{2,2}$.

$$u_{3,3} = \frac{1}{4}\ [b_{1,5}, + b_{5.1} + b_{55} + b_{11}]$$

$$u_{2,4} = \frac{1}{4}\ [b_{15} + u_{33} + b_{3,5} + b_{13}]$$

$$u_{4,4} = \frac{1}{4}\ [b_{3,5} + b_{5,3} + b_{5,5} + u_{33}]$$

$$u_{4,2} = \frac{1}{4}\ [u_{33} + b_{5,1} + b_{3,1} + b_{5,3}]$$

$$u_{2,2} = \frac{1}{4}\ [b_{1,3} + b_{31} + u_{33} + b_{1,1}]$$

The values at the remaining interior points i.e., $u_{2,3}$, $u_{3,4}$, $u_{4,3}$ and $u_{3,2}$ are computed by the standard 5-point formula (2) thus

$$u_{2,3} = \frac{1}{4}\ [b_{1,3} + u_{3,3} + u_{2,4} + u_{22}]$$

$$u_{3,4} = \frac{1}{4}\ [u_{2,4} + u_{44} + b_{3,5} + u_{33}]$$

$$u_{4,3} = \frac{1}{4}\ [u_{3,3} + b_{5,3} + u_{4,4} + u_{4,2}]$$

$$u_{3,2} = \frac{1}{4}\ [u_{2,2} + u_{4,2} + u_{3,3} + u_{3,1}]$$

Having found all the nine values of u_{ij}, once their accuracy is improved by repeated application of (2) in the form

$$U_{i,j}^{(n+1)} = \frac{1}{4}\left[u_{i-1,j}^{(n+1)} + u_{i+1,j}^{(n)} + u_{i,j+1}^{(n+1)} + u_{i,j-1}^{(n)}\right]$$

This formula utilises the latest iterative value available and scans the mesh points symmetrically from left to right along successive rows. This process is repeated till the difference of values in one round and the next becomes negligible. This is known as **Liemann's iteration process.**

Example 1: Solve the Laplace's equation $U_{xx} + U_{yy} = 0$ for the square mesh given in the following figure with the boundary values.

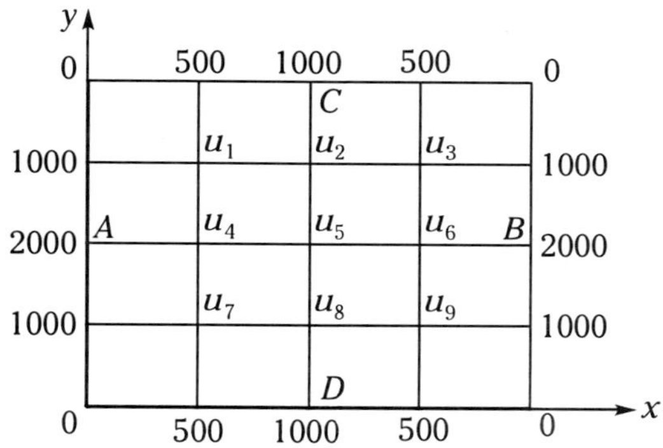

Let $u_1, u_2, u_3 \ldots. u_9$ be the values of U at the interior mesh points, since the boundary values of U are symmetrical about AB.

$$\therefore \quad u_1 = u_7, \quad u_8 = u_2, \quad u_9 = u_3$$

Also the values of U being symmetrical about CD

$$u_3 = u_1, \quad u_6 = u_4, \quad u_9 = u_7$$

Thus it is sufficient to find the values of u_1, u_2, u_4 and u_5.

Now we find their initial values as follows.

$$u_5^{(0)} = \frac{1}{4}[2000 + 2000 + 1000 + 1000] = 1500 \text{ S.F.}$$

$$u_1^{(0)} = \frac{1}{4}[0 + 1500 + 1000 + 2000] = 1125 \text{ D.F.}$$

$$u_2^{(0)} = \frac{1}{4}[1125 + 1125 + 1000 + 1500] = 1188 \text{ S.F.}$$

$$u_4^{(0)} = \frac{1}{4}[2000 + 1500 + 1125 + 1125] = 1438 \text{ S.F.}$$

We carry out the iteration process using the formulae

$$u_1^{(n+1)} = \frac{1}{4}\left[1000 + u_2^{(n)} + 500 + u_4^{(n)}\right]$$

$$u_2^{(n+1)} = \frac{1}{4}\left[u_1^{(n+1)} + u_3^{(n)} + 1000 + u_5^{(n)}\right]$$

$$u_4^{(n+1)} = \frac{1}{4}\left[2000 + u_5^{(n)} + u_1^{(n+1)} + u_7^{(n)}\right]$$

$$u_5^{(n+1)} = \frac{1}{4}\left[u_4^{(n+1)} + u_6^{(n)} + u_2^{(n+1)} + u_8^{(n)}\right]$$

I iteration (put $n = 0$)

$$u_1^{(1)} = \frac{1}{4}\,[1000 + 1188 + 500 + 1438] = 1032$$

$$u_2^{(1)} = \frac{1}{4}\,[1032 + 1125 + 1000 + 1500] = 1141$$

$$u_4^{(1)} = \frac{1}{4}\,[2000 + 1500 + 1032 + 1032] = 1391$$

$$u_5^{(1)} = \frac{1}{4}\,[1391 + 1391 + 1141 + 1141] = 1266.$$

II iteration (put $n = 1$)

$$u_1^{(2)} = \frac{1}{4}\,[1000 + 1141 + 500 + 1391] = 1008$$

$$u_2^{(2)} = \frac{1}{4}\,[1008 + 1008 + 1000 + 1266] = 1069$$

$$u_4^{(2)} = \frac{1}{4}\,[2000 + 1266 + 1008 + 1008] = 1321$$

$$u_5^{(2)} = \frac{1}{4}\,[1321 + 1321 + 1069 + 1069] = 1195$$

III iteration (put $n = 2$)

$$u_1^{(3)} = \frac{1}{4}\,[1000 + 1069 + 500 + 1321] = 973$$

$$u_2^{(3)} = \frac{1}{4}[973 + 973 + 1000 + 1195] = 1035$$

$$u_4^{(3)} = \frac{1}{4}[2000 + 1195 + 973 + 973] = 1288$$

$$u_5^{(3)} = \frac{1}{4}[1288 + 1288 + 1035 + 1035] = 1162.$$

IV iteration (put $n = 3$)

$$u_1^{(4)} = \frac{1}{4}[1000 + 1135 + 500 + 1288] = 956$$

$$u_2^{(4)} = \frac{1}{4}[956 + 956 + 1000 + 1162] = 1091$$

$$u_4^{(4)} = \frac{1}{4}[2000 + 1162 + 956 + 956] = 1269$$

$$u_5^{(4)} = \frac{1}{4}[1269 + 1269 + 1091 + 1091] = 1144.$$

V iteration (put $n = 4$)

$$u_1^{(5)} = \frac{1}{4}[1000 + 1091 + 500 + 1269] = 947$$

$$u_2^{(5)} = \frac{1}{4}[947 + 947 + 1000 + 1144] = 1010$$

$$u_4^{(5)} = \frac{1}{4}[2000 + 1144 + 947 + 947] = 1260$$

$$u_5^{(5)} = \frac{1}{4}[1260 + 1260 + 1010 + 1010] = 1135$$

Similarly we can proceed, as follows,

$u_1^{(6)} = 942$ $\quad u_2^{(6)} = 1005$ $\quad u_4^{(6)} = 1255$ $\quad u_5^{(6)} = 1130$

$u_1^{(7)} = 940$ $\quad u_2^{(7)} = 1003$ $\quad u_4^{(7)} = 1253$ $\quad u_5^{(7)} = 1128$

$u_1^{(8)} = 939$ $\quad u_2^{(8)} = 1002$ $\quad u_4^{(8)} = 1252$ $\quad u_5^{(8)} = 1128$

$u_1^{(9)} = 939$ $\quad u_2^{(6)} = 1001$ $\quad u_4^{(9)} = 1251$ $\quad u_5^{(9)} = 1126$

Thus there is a negligible difference between the valued obtained in 8th and 9th iterations.

Hence $u_1 = 939$, $u_2 = 1001$, $u_4 = 1251$ and $u_5 = 1126$.

Example 2: Given the values of $U(x, y)$ on the boundary of the square in the following table, evaluate the function $U(x, y)$ satisfying the Laplace equation $\nabla^2 U = 0$ at the pivotal points of this figure.

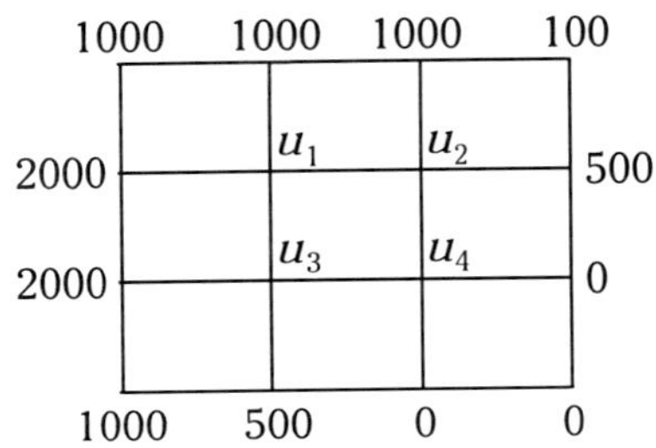

To get the initial values of u_1, u_2, u_3, u_4 we assume $u_4 = 0$. Then

$$u_1 = \frac{1}{4}[1000 + 0 + 1000 + 2000] = 1000 \text{ D.F.}$$

$$u_2 = \frac{1}{4}[1000 + 500 + 1000 + 0] = 625 \text{ S.F.}$$

$$u_3 = \frac{1}{4}[2000 + 0 + 1000 + 500] = 875 \text{ S.F.}$$

$$u_4 = \frac{1}{4}[875 + 0 + 625 + 0] = 375 \text{ S.F.}$$

We carry out the successive iterations using the formula *e*.

$$u_1^{(n+1)} = \frac{1}{4}\left[2000 + u_2^{(n)} + 1000 + u_3^{(n)}\right]$$

$$u_2^{(n+1)} = \frac{1}{4}\left[u_1^{(n+1)} + 500 + 1000 + u_4^{(n)}\right]$$

$$u_3^{(n+1)} = \frac{1}{4}\left[2000 + u_4^{(n)} + u_1^{(n+1)} + 500\right]$$

$$u_4^{(n+1)} = \frac{1}{4}\left[u_3^{(n+1)} + 0 + u_2^{(n+1)} + 0\right]$$

I iteration (put $n = 0$)

$$u_1^{(1)} = \frac{1}{4}[2000 + 625 + 1000 + 875] = 1125$$

$$u_2^{(1)} = \frac{1}{4}[1125 + 500 + 1000 + 375] = 750$$

$$u_3^{(1)} = \frac{1}{4}[2000 + 375 + 1125 + 500] = 1000$$

$$u_4^{(1)} = \frac{1}{4}[1000 + 0 + 750 + 0] = 438.$$

II iteration (put $n = 1$)

$$u_1^{(2)} = \frac{1}{4}[2000 + 750 + 1000 + 1000] = 1188$$

$$u_2^{(2)} = \frac{1}{4}[1188 + 500 + 1000 + 438] = 782$$

$$u_3^{(2)} = \frac{1}{4}[2000 + 438 + 1188 + 500] = 1032$$

$$u_4^{(2)} = \frac{1}{4}[1032 + 0 + 782 + 0] = 454.$$

III iteration (put $n = 2$)

$$u_1^{(3)} = \frac{1}{4}[2000 + 782 + 1000 + 1032] = 1204$$

$$u_2^{(3)} = \frac{1}{4}[1204 + 500 + 1000 + 454] = 789$$

$$u_3^{(3)} = \frac{1}{4}[2000 + 454 + 1204 + 500] = 1040$$

$$u_4^{(3)} = \frac{1}{4}[1040 + 0 + 789 + 0] = 458.$$

Similarly,

$$u_1^{(4)} = 1207 \qquad u_2^{(4)} = 791 \qquad u_3^{(4)} = 1041 \qquad u_4^{(4)} = 458$$

$$u_1^{(5)} = 1208 \qquad u_2^{(5)} = 791.5 \qquad u_3^{(5)} = 1041.5 \qquad u_4^{(5)} = 458.5$$

Thus there is no significant difference between IVth and Vth iterations, Hence we take,

$$u_1 = 1208, \quad u_2 = 792, \quad u_3 = 1042, \quad u_4 = 458.$$

Example 3: Solve $U_{xx} + U_{yy} = 0$ in $0 \le x \le 4$, $0 \le y \le 4$ given that

$U(0,y) = 0$, $y(4,y) = 8 + 2y$, $U(x,0) = \dfrac{x^2}{2}$ and $U(x,4) = x^2$ take $h = k = 1$

and obtain the result correct to one decimal.

Solution: Let us divide the given region R i.e., $0 \le x \le 4$ $0 \le y \le 4$ into 16 square meshes. The numerical values of the boundary, using $U(0,y) = 0$,

$U(4,y) = 8 + 2y$, $U(x, 0\ ) = \dfrac{x^2}{2}$ and $U(x, 4) = x^2$ are calculated.

Let $u_1, u_2, u_3, u_4 \ldots u_9$ be the values of u at the interior mesh points.

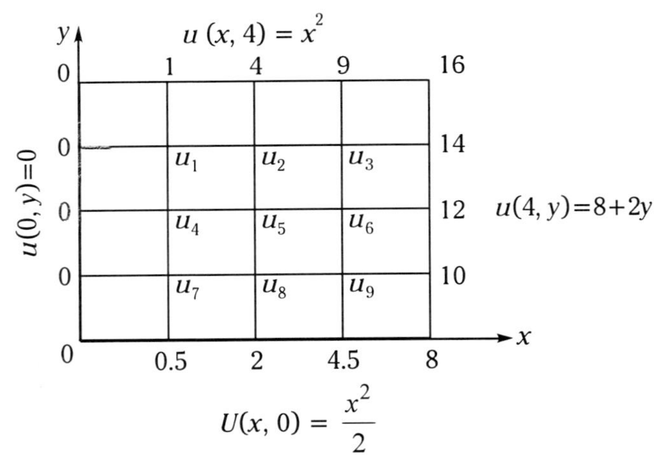

Now, we calculate the initial values of U's by using D. F. or S.F.

$$u_5^{(0)} = \frac{0+12+4+2}{4} = 4.5 \qquad \text{(S.F)}$$

$$u_1^{(0)} = \frac{0+4+0+4.5}{4} = 2.125 \qquad \text{(D.F.)}$$

$$u_3^{(0)} = \frac{4.5+16+4+12}{4} = 9.125 \qquad \text{(D.F.)}$$

$$u_7^{(0)} = \frac{0+2+0+4.5}{4} = 1.625 \qquad \text{(D.F.)}$$

$$u_9^{(0)} = \frac{4.5+8+2+12}{4} = 6.625 \qquad \text{(D.F.)}$$

$$u_2^{(0)} = \frac{4+4.5+2.125+9.125}{4} = 4.9375 \qquad \text{(S.F.)}$$

$$u_4^{(0)} = \frac{0+4.5+2.125+1.625}{4} = 2.0625 \qquad \text{(S.F.)}$$

$$u_6^{(0)} = \frac{4.5+12+9.125+6.625}{4} = 0.0625 \qquad \text{(S.F.)}$$

$$u_8^{(0)} = \frac{1.625+6.625+4.5+2}{4} = 3.6875 \qquad \text{(S.F.)}$$

We use Leibmann iteration formula

$$u_{i,j}^{(n+1)} = \frac{1}{4}\left[u_{i-1,j}^{(n+1} + u_{i+1,j}^{(n)} + u_{i,j-1}^{(n)} + u_{i,j+1}^{(n+1)}\right]$$

To improve the above results

$$\therefore \quad u_1^{(n+1)} = \frac{1}{4}\left[0 + u_2^{(n)} + u_4^{(n)} + 1\right]$$

$$u_2^{(n+1)} = \frac{1}{4}\left[u_2^{(n+1)} + u_3^{(n)} + u_5^{(n)} + 4\right]$$

$$u_3^{(n+1)} = \frac{1}{4}\left[u_2^{(n+1)} + 14 + u_6^{(n)} + 9\right]$$

$$u_4^{(n+1)} = \frac{1}{4}\left[0 + u_5^{(n)} + u_7^{(n)} + u_1^{(n+1)}\right]$$

$$u_5 = \frac{1}{4}\left[u_4^{(n+1)} + u_6^{(n)} + u_2^{(n+2)} + u_8^{(n)}\right]$$

$$u_6 = \frac{1}{4}\left[u_5^{(n+1)} + 12 + u_9^{(n)} + u_3^{(n+1)}\right]$$

$$u_7 = \frac{1}{4}\left[0 + u_8^{(n)} + 0.5 + u_4^{(n+1)}\right]$$

$$u_8 = \frac{1}{4}\left[u_7^{(n+1)} + u_9^{(n)} + 2 + u_5^{(n+1)}\right]$$

$$u_9 = \frac{1}{4}\left[u_8^{(n+1)} + 10 + 4.5 + u_6^{(n+1)}\right]$$

I iteration ($n = 0$)

$$u_1^{(1)} = \frac{1}{4}\left[0 + u_2^{(0)} + u_4^{(0)} + 1\right] = \frac{1}{4}\ [0 + 4.9375 + 2.0625 + 1] = 2$$

$$u_2^{(1)} = \frac{1}{4}\left[u_1^{(1)} + u_3^{(0)} + u_5^{(0)} + 4\right] = \frac{1}{4}\ [2 + 9.125 + 4.5 + 4] = 4.90625$$

$$u_3^{(1)} = \frac{1}{4}\left[u_2^{(1)} + 14 + u_6^{(0)} + 9\right] = \frac{1}{4}\ [4.9375 + 14 + 8.0625 + 9]$$
$$= 8.9921875$$

$$u_4^{(1)} = \frac{1}{4}\left[0 + u_5^{(0)} + u_7^{(0)} + u_1^{(1)}\right] = \frac{1}{4}\ [0 + 4.5 + 1.625 + 2] = 2.03125$$

$$u_5^{(1)} = \frac{1}{4}\left[u_4^{(1)} + u_6^{(0)} + u_8^{(0)} + u_2^{(1)}\right] = \frac{1}{4}\ [2.03125 + 8.0625 + 3.6875$$
$$+ 4.90625] = 4.671875$$

$$u_6^{(1)} = \frac{1}{4}\left[u_5^{(1)} + 12 + u_9^{(0)} + u_3^{(1)}\right] = \frac{1}{4}\ [4.671875 + 12 + 6.625$$
$$+ 8.9921875] = 8.0722656.$$

$$u_7^{(1)} = \frac{1}{4}\left[0 + u_8^{(1)} + 0.5 + u_4^{(0)}\right] = \frac{1}{4}\ [0 + 3.6875 + 0.5 + 2.03125]$$
$$= 1.554685.$$

$$u_8^{(1)} = \frac{1}{4}\left[u_7^{(1)} + u_9^{(0)} + 2 + u_5^{(0)}\right] = \frac{1}{4}\ [1.5546875 + 6.625 + 2$$
$$+ 4.671875] = 3.71289.6.$$

$$u_9^{(1)} = \frac{1}{4}\left[u_8^{(1)} + 10 + 4.5 + u_6^{(0)}\right] = \frac{1}{4}\ [3.7128906 + 10 + 4.5$$
$$+ 8.0722656] = 6.571289.$$

Similarly, we can proceed as follows.

II Iteration ($n = 1$)

$u_1^{(2)} = 1.98437,$ $\quad u_2^{(2)} = 4.9121094 \quad u_3^{(2)} = 8.9960937$

$u_4^{(2)} = 2.0527344 \quad u_5^{(2)} = 4.6875 \quad u_6^{(2)} = 8.0637207$

$u_7^{(2)} = 1.5664063 \quad u_8^{(2)} = 3.7062988 \quad u_9^{(2)} = 6.5675049$

III iteration (n = 2)

$u_1^{(3)} = 1.991211$ $\quad u_2^{(3)} = 4.9187012$ $\quad u_3^{(3)} = 8.9956055$

$u_4^{(3)} = 2.0612793$ $\quad u_5^{(3)} = 4.6875$ $\quad u_6^{(3)} = 8.0626526$

$u_7^{(3)} = 1.5668945$ $\quad u_8^{(3)} = 3.7054749$ $\quad u_9^{(3)} = 6.56707319.$

$\therefore$ we take

$u_1 = 1.99$ $\quad u_2 = 4.91$ $\quad u_3 = 8.99$ $\quad u_4 = 2.06$ $\quad u_5 = 4.69$

$u_6 = 8.06$ $\quad u_7 = 1.57,$ $\quad u_8 = 3.71$ $\quad u_9 = 6.57.$

7.5 SOLUTION OF POISSON'S EQUATION

Let $\frac{\partial^2 U}{\partial x^2} + \frac{\partial^2 U}{\partial y^2} = f(x, y)$ ... (1) be the Poisson's equation.

This is an elliptic equation which can be solved numerically at the interior mesh points of the square network when the boundary values are known. The standard 5 - point formula for (1) takes the form.

$$u_{i-1,j} + u_{i+1,j} + u_{i,j+1} + u_{i,j-1} - 4u_{ij} = h^2 f(ih, jh) \quad \text{... (2)}$$

By applying (2) at each mesh-points, we arrive at linear equations in the pivotal values i, j. These can be solved by Gauss-Seidal iteration method.

Example 1: Solve the partial differential equation.

$$\frac{\partial^2 U}{\partial x^2} + \frac{\partial^2 U}{\partial y^2} = -10(x^2 + y^2 + 10)$$

over the square with side $x = 0, y = 0; x = 3, y = 3$ with u_0 on the boundary and mesh length $h = 1$.

Solution: Here $h = 1$.

$\therefore$ The standard 5-point formula for the given equation is,

$$u_{i-1,j} + u_{i+1,j} + u_{i,j+1} + u_{i,j-1} - 4u_{ij} = 10(i^2 + j^2 + 10) \quad \text{...(1)}$$

For u_1 ($i = 1, j = 2$) (1) gives

$$0 + u_2 + 0 + u_3 + 4u_1 = -10(1 + 4 + 10)$$

$$\therefore \quad u_1 = \frac{1}{4}[u_2 + u_3 + 150] \quad \text{..... (2)}$$

For u_2 $(i = 2, j = 2)$ (1) gives

$$u_2 = \frac{1}{4}[u_1 + u_4 + 180] \quad \text{.. (3)}$$

For u_3 $(i = 1, j = 1)$

$$u_3 = \frac{1}{4}[u_1 + u_4 + 120] \quad \text{..... (4)}$$

For u_4 $(i = 2, j = 1)$

$$u_4 = \frac{1}{4}[u_2 + u_3 + 150] \quad \text{... (5)}$$

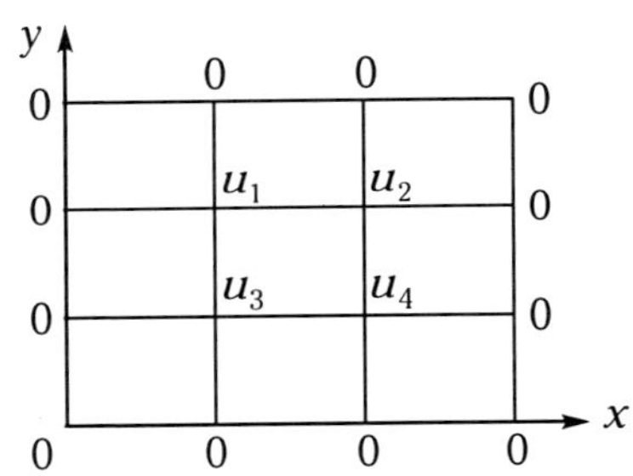

The equations (2) and (4) show that $u_4 = u_1$. Then the above equations reduce to

$$u_1 = \frac{1}{4}[u_2 + u_3 + 150] \quad \text{.... (6)}$$

$$u_2 = \frac{1}{4}[u_1 + 90] \quad \text{..(7)}$$

$$u_3 = \frac{1}{2}[u_1 + 60] \quad \text{.. (8)}$$

Now, let us solve these equations by Gauss-Seidel iteration method.

I iteration

Starting from the approximations $u_1 = 0$, $u_3 = 0$ we obtain

$u_1^{(1)} = 37.5$ Then

$$u_2^{(1)} = \frac{1}{2}(37.5 + 90) = 64$$

$$u_3^{(1)} = \frac{1}{2}\ (37.5 + 60) = 49.$$

II iteration

$$u_1^{(2)} = \frac{1}{4}\ (64 + 49 + 150) = 60$$

$$u_2^{(2)} = \frac{1}{2}\ (66 + 90) = 78$$

$$u_3^{(2)} = \frac{1}{2}\ (66 + 60) = 63.$$

III iteration

$$u_1^{(3)} = \frac{1}{4}\ (78 + 63 + 150) = 73.$$

$$u_2^{(3)} = \frac{1}{2}\ (73 + 90) = 82$$

$$u_3^{(3)} = \frac{1}{2}\ (73 + 60) = 67$$

IV iteration

$$u_1^{(4)} = \frac{1}{4}\ (82 + 67 + 150) = 75$$

$$u_2^{(4)} = \frac{1}{2}\ (75 + 90) = 82.5$$

$$u_3^{(4)} = \frac{1}{2}\ (73 + 60) = 67.5.$$

V iteration

$$u_1^{(5)} = \frac{1}{4}\ (82.5 + 67.5 + 150) = 75.$$

$$u_2^{(5)} = \frac{1}{2}\ (75 + 90) = 82.5$$

$$u_3^{(5)} = \frac{1}{2}\ (75 + 90) = 67.5.$$

Since Vth and IVth iterations are the same we take

$$u_1 = 75, \quad u_2 = 82.5, \quad u_3 = 67.5 \quad \text{and} \quad u_4 = 75.$$

7.6 PARABOLIC EQUATION

The one-dimensional heat equation

$\dfrac{\partial U}{\partial t} = c^2 \dfrac{\partial^2 U}{\partial x^2}$ ... (1) where $c^2 = \dfrac{k}{s\rho}$ (s = specific heat of the material, ρ is the density and K is thermal conductivity) is a well-known example of a parabolic equation. The solution of it is a function of x and t i.e., $U(x, t)$. It is defined for values of x from $x = 0$ to $x = l$ and for values of time t from $t = 0$ to $t = \infty$.

The solution is not defined in a closed form but propagates in an open ended region from initial values satisfying the prescribed boundary conditions as shown in the figure.

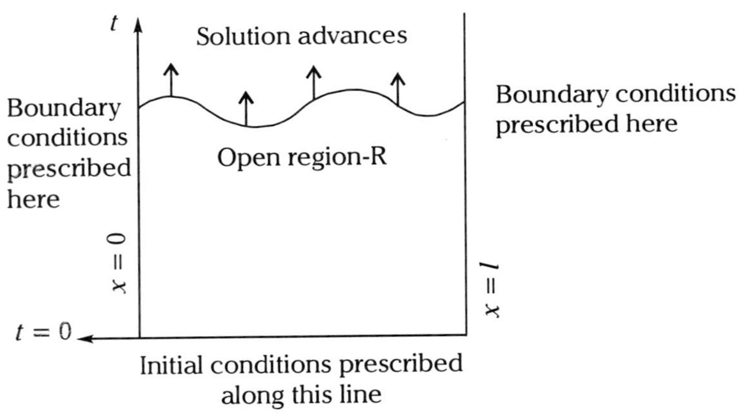

In general, the steady pressure of waves in a fluid, propagation of heat and unsteady state problems lead to parabolic type of equations.

7.7 BENDER-SCHMIDT METHOD

Consider the heat flow equation of one dimension

$\dfrac{\partial U}{\partial t} = c^2 \dfrac{\partial^2 U}{\partial x^2}$ .....(1) where $c^2 = \dfrac{K}{S\rho}$ is the diffusivity of the substance (cm^2/sec).

Consider a rectangular mesh in the $x - t$ plane with spacing h along x direction and k along time t – direction. Denoting a mesh point $(x,t) = (ih, jk)$ as (i, j) we have,

$$\frac{\partial U}{\partial t} = \frac{u_{i\ j+1} - u_{i,\ j}}{k}$$

and
$$\frac{\partial^2 U}{\partial x^2} = \frac{u_{i-1,j} - 2u_{i,\ j} + u_{i-1,\ j}}{h^2}$$

Substituting in (1) we get

$$u_{i,\ j+1} - u_{i,\ j} = \frac{kc^2}{h^2}\left[u_{i-1,j} - 2u_{i,j} + u_{i+1,j}\right]$$

or
$$u_{i,\ j+1} = \alpha u_{i-1,j} + (1 - 2\alpha)\ u_{i,\ j} + \alpha u_{i+1,j} \qquad \text{...(2)}$$

Where $\alpha = \dfrac{kc^2}{h^2}$ is the mesh ratio parameter.

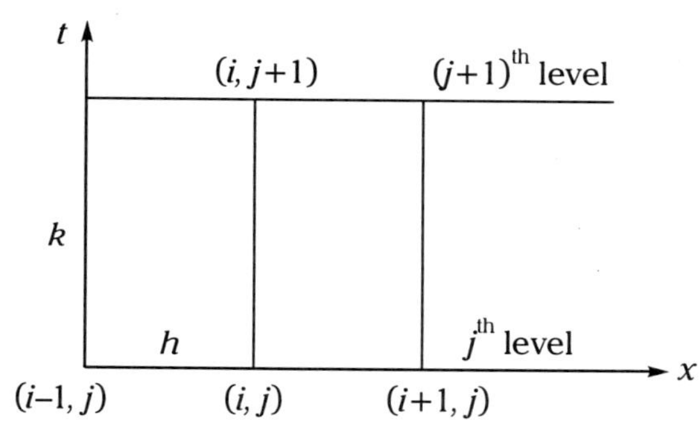

This formula enables us to determine the value of u at the $(i, j + 1)^{th}$ mesh point in terms of the known function values at the points x_{i-1}, x_i and x_{i+1} at the instant t_j. It is a relation between the function values at the two time levels $j + 1$ and j and is therefore, called a 2-level formula. In schematic form (2) is shown in the above figure which is called the **Schmidt explicit formula** which is valid only for $0 < \alpha < \dfrac{1}{2}$.

Obs : When $\alpha = \dfrac{1}{2}$ Relation (2) reduces to

$$u_{i,j+1} = \frac{1}{2}[u_{i-1,}j + u_{i+1,}j] \qquad \dots (3)$$

Which shows that the value of x_i at time t_{j+1}, is the mean of the u - values at x_{i-1}, and x_{i+1} at time t_j. This relation known as **Bender Schmidt recurrence relation**, gives the values of u at the interval mesh points with the help of boundary conditions.

Example 1: Solve $\frac{\partial u}{\partial t} = \frac{1}{2}\frac{\partial^2 u}{\partial x^2}$ with the conditions $u(0,t) = 0$ $u(4,t) = 0$

$u(x,0) = x(4 - x)$ taking $h = 1$ and employing the Bender-Schmidt recurrence equation. Continue the solution through ten time steps.

Solution: The general equation is

$$U_{xx} = aUt$$

Here $a = 2$, $h = 1$, $\therefore \quad \lambda = \frac{k}{h^2 a} = \frac{k}{2}$

$\therefore \quad \lambda = \frac{1}{2}$, and k = 1. Now using the Bender-Schmidt recurrence relation,

$$u_{i,j+1} = \frac{1}{2}[u_{i-1,j} + u_{i+1,j}] \qquad \dots.(1)$$ We have

$$x = ih \quad \Rightarrow \quad x_i = ih = i \qquad (\because h = 1 \text{ and } k = 1)$$

$$t = jk \quad \Rightarrow \quad t_j = jk = j$$

Given $u(x,0) = x(4 - x)$ or $u(i,0) = i(4 - i)$

Now for $0 \le x \le 4$ i.e,. $0 \le i \le 4$ we have

$$u(i,0) = 0, 3, 4, 3, 0 \qquad (\because i = 0, 1, 2, 3, 4)$$

Given $u(0, t) = 0 \ \forall\ t$ i.e., $u(0,j) = 0 \ \forall\ j$. Hence the entries in the first column are zero.

Also $u(4,t) = 0 \ \forall\ t$ i.e., $u(0, 4) = 0 \ \forall\ j$. Hence the entries in the last column are zero.

From (1) putting $j = 0$ in (1) $u_{i,1} = \frac{1}{2}\left[u_{i+1,0} + u_{i-1,0}\right] \qquad \dots(2)$

putting $i = 1$ in (2) we get, $u_{11} = \frac{1}{2}\left[u_{2,0} + u_{0,0}\right] = \frac{1}{2}[4+0] = 2$

$i = 2$ in (2) we get, $u_{21} = \frac{1}{2}\left[u_{3,0} + u_{1,0}\right] = \frac{1}{2}[3+3] = 3$

$i = 3$ in (2) we get, $u_{31} = \frac{1}{2}\left[u_{4,0} + u_{2,0}\right] = \frac{1}{2}[0+4] = 2$

are the values in the second row. Similarly putting $j = 1, 2, 3 \ldots\ldots 9$ the other values can be found as given in the table below.

j \ i	0	1	2	3	4
0	0	3	4	3	0
1	0	2	3	2	0
2	0	1.5	2	1.5	0
3	0	1	1.5	1	0
4	0	0.75	1	0.75	0
5	0	0.5	0.75	0.5	0
6	0	0.375	0.5	0.375	0
7	0	0.25	0.375	0.25	0
8	0	0.1875	0.25	0.1875	0
9	0	0.125	0.1875	0.125	0
10	0	0.094	0.125	0.094	0

Example 2: Find the value of $u(x,t)$ satisfying the parabolic equation $\frac{\partial U}{\partial t} = 4\frac{\partial^2 U}{\partial x^2}$ and the boundary conditions $u(0,t) = 0$ $u(8, t) = 0$ and

$u(x, 0) = 4x - \frac{1}{2}x^2$ at the points $x = i$

$i = 0, 1, 2 \ldots\ldots 8$ and $t = \frac{1}{8}j$; $j = 0\ 1, 2, 3, 4, 5.$

Solution: Let $U_t = 4\ U_{xx} \ldots\ldots (1)$ be the given parabolic equation.

Here $c^2 = 4$, $h = 1$ and $k = \frac{1}{8}$. Then $\alpha = \frac{c^2 k}{h^2} = \frac{1}{2}$

$\therefore$ We have Bender-Schmidt's recurrence relation is

$$u_{i,\,j+1} = \frac{1}{2}\left[u_{i-1,\,j} + u_{i+1,\,j}\right] \qquad ...(2)$$

Now since $u(0,t) = 0$ and $u(8,t) = 0$ we have

$u_{0,j} = 0$ and $u_{8,j} = 0$ and for all values of j i.e., the entries in the first and last columns are zero.

Since $u(x, 0) = 4x - \frac{1}{2}x^2$

$\therefore$ $u_{i\,0} = 0, 3.5, 6, 7.5, 8, 7.5, 6, 3.5$

For $i = 0, 1, 2, 3, 4, 5, 6, 7$ at $t = 0$.

Putting $j = 0$ in (2)

$$u_{i,\,1} = \frac{1}{2}\left[u_{i-1,0} + u_{i+1,\,0}\right]$$

Taking $i = 1, 2, 3, 4, 5, 6, 7$ successively we get

$$u_{11} = \frac{1}{2}(u_{0,0} + u_{2,1}) = \frac{1}{2}(0+6) = 3\,;$$

$$u_{2,\,1} = \frac{1}{2}(u_{1,0} + u_{3,0}) = \frac{1}{2}(3.5+7.5) = 5.5\,;$$

$$u_{3,\,1} = \frac{1}{2}(u_{2,0} + u_{4,0}) = \frac{1}{2}(6+8) = 7\,; \quad u_{4,\,1} = 7.5$$

$$u_{5,\,1} = 7, \quad u_{6,\,1} = 5.5 \quad \text{and} \quad u_{7,\,1} = 3.$$

These are the entries in the second row.

Similarly putting $j = 2, 3, 4$ successively in (2) we get the next entries which are shown in the following table.

j \ i	0	1	2	3	4	5	6	7	8
0	0	3.5	6	7.5	8	7.5	6	3.5	0
1	0	3	5.5	7	7.5	7	5.5	3	0
2	0	2.75	5	6.5	7	6.5	5	2.75	0
3	0	2.5	4.625	6	6.5	6	4.625	2.5	0
4	0	2.3125	4.25	5.5625	6	5.5625	4.25	2.3125	0
5	0	2.125	3.9375	5.125	5.5625	5.125	3.9375	2.125	0

Example 3: Solve the equation $\frac{\partial u}{\partial t} = \frac{\partial^2 u}{\partial x^2}$ subject to the conditions $u(x,0) = \sin \pi x.\ 0 \le x \le 1$, $u(0,t) = u(1,t) = 0$ carry out computations for two levels, taking $h = \frac{1}{3}$, $k = \frac{1}{36}$.

Solution: Here $c^2 = 1$, $h = \frac{1}{3}$, $k = \frac{1}{36}$ so that $\alpha = \frac{kc^2}{h^2} = \frac{1}{4}$

Also $u_{1,0} = \sin\frac{\pi}{3} = \frac{\sqrt{3}}{2}$, $u_{2,0} = \sin\frac{2\pi}{3} = \frac{\sqrt{3}}{2}$ and all other boundary values are zero as in the figure.

Schemidt's formula is

$$u_{i,\,j+1} = \alpha u_{i-1,\,j} + (1-2\alpha)\,u_{i,\,j} + \alpha u_{i+1,\,j}$$

which reduces to

$$u_{i,\,j+1} = \frac{1}{4}\left[u_{i-1,\,j} + 2u_{ij} + u_{i+1,\,j}\right]$$

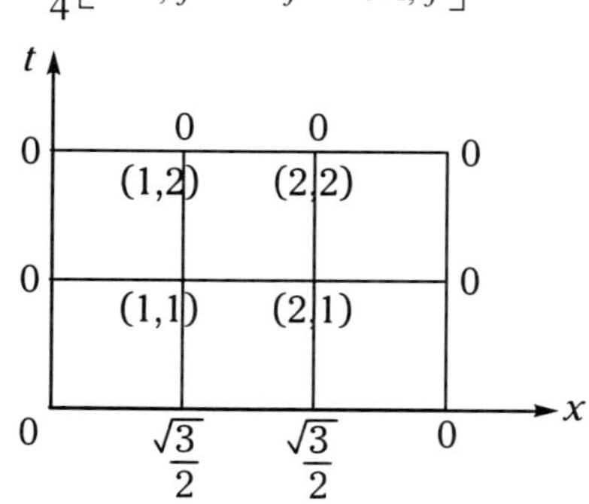

For $i = 1, 2$: $j = 0$

$$u_{1,1} = \frac{1}{4}\left[u_{0,0} + 2u_{1,0} + 4_{2,0}\right] = \frac{1}{4}\left[0 + 2\times\frac{\sqrt{3}}{2} + \frac{\sqrt{3}}{2}\right] = 0.65$$

$$u_{2,1} = \frac{1}{4}\left[u_{1,0} + 2u_{2,0} + 4_{3,0}\right] = \frac{1}{4}\left[\frac{\sqrt{3}}{2} + 2\times\frac{\sqrt{3}}{2} + 0\right] = 0.65$$

For $i = 1, 2, \quad j = 1$

$$u_{1,2} = \frac{1}{4}\left[u_{0,1} + 2u_{1,1} + 4_{2,1}\right] = \frac{1}{4}[0 + 2\times 0.65 + 0.65] = 0.49$$

$$u_{2,2} = \frac{1}{4}\left[u_{1,1} + 2u_{2,1} + 4_{3,1}\right] = \frac{1}{4}[0.65 + 2(.65) + 0] = 0.49$$

7.8 CRANK-NICHOLSON DIFFERENCE SCHEME FOR SOLVING PARABOLIC EQUATIONS

Let $U_{xx} = \alpha \; U_t$ .... (1) be the partial differential equation to be solved subject to the conditions.

$$u(0, t) = T_0,\; u(l\, t) = T_l \text{ and } u(x,0) = f(x).$$

We have the finite difference approximation for U_{xx} at the point U_{ij} is

$$U_{xx} = \frac{1}{h^2}\left[u_{i+1,j} - 2u_{i,j} + u_{i-1,j}\right] \quad ...(2)$$

At the point $u_{i,j+1}$, the finite difference approximation for U_{xx} is

$$U_{xx} = \frac{1}{h^2}\left[u_{i+1,j+1} - 2u_{i,j+1} + u_{i-1,j+1}\right] \quad ...(3)$$

Average of (2) and (3) is

$$U_{xx} = \frac{1}{2h^2}\left[u_{i+1,j+1} - 2u_{i,j+1} + u_{i-1,j+1} + u_{i+1,j} - 2u_{ij} + u_{i-1,j}\right] \quad ..(4)$$

For U_t the difference approximation is

$$U_t = \frac{1}{k}\left[u_{i,j+1} - u_{ij}\right] \quad ...(5)$$

Substituting (4) and (5) in (1) we get

$$\frac{\lambda}{2}\;[u_{i+1,j+1} - 2u_{i,j+1} + u_{i-1,j+1} + u_{i+1,j} - 2u_{i,j} + u_{i-1,j}]$$

$$= u_{i,\,j+1} - u_{i,\,j} \quad \text{where} \quad \lambda = \frac{k}{h^2 \alpha}$$

or $$\frac{\lambda}{2} u_{i+1,j+1} - (\lambda + 1) u_{i,\,j+1} + \frac{\lambda}{2} u_{i-1,\,j+1}$$

$$= -\frac{\lambda}{2} u_{i+1,j} + (\lambda - 1) u_{i,j} - \frac{\lambda}{2} u_{i-1,\,j}$$

or $$\lambda\,(u_{i+1,\,j+1} + u_{i-1,\,j+1}) + 2\,(\lambda + 1)\,u_{i,\,j+1}$$

$$= 2\,(\lambda - 1)\,u_{i,\,j} - \lambda\,(u_{i+1,\,j} - u_{i-1,\,j}) \qquad ....(6)$$

Equation (6) is called Crank-Nicholson difference method.

Note: Choosing λ is very important very often a good choice of λ is $\lambda = 1$. In such case the Crank - Nicholson scheme becomes,

$$u_{i,\,j+1} = \frac{1}{4}\left[u_{i-1,\,j+1} + u_{i+1,\,j+1} + u_{i-1,\,j} + u_{i+1,\,j}\right]$$

Subject to $k = ah^2$.

Example: Using Crank - Nicholson's method solve $U_{xx} = 16Ut$, $0 < x < 1$, $t > 0$ given $u(x,0) = 0$, $u(0,t) = 0$, $u(l, t) = 50\,t$. Compute U for two steps in t direction taking $h = \frac{1}{4}$

Solution: Here $\alpha = 16$, $h = \frac{1}{4}$

$\therefore \quad k = \alpha h^2 = 16\left(\frac{1}{16}\right) = 1$. The Crank-Nicholson's scheme is given by

$$u_{i,\,j+1} = \frac{1}{4}\left[u_{i+1,\,j+1} + u_{i-1,\,j+1} + u_{i+1,j} + u_{i-1,j}\right] \qquad ...(1)$$

$\rightarrow$ x increasing

j \ i	0	0.25	0.5	0.75	1
0	0	0	0	0	0
1	0	u_1	u_2	u_3	50
2	0	u_3	u_4	u_5	100

Applying (1) at the mesh points u_1, u_2, u_3 we get

$$u_1 = \frac{1}{4}(0+0+0+u_2) = \frac{1}{4}u_2 \qquad \text{... (i)}$$

$$u_2 = \frac{1}{4}(0+0+u_1+u_3) = \frac{1}{4}(u_1, + u_3) \qquad \text{..... (ii)}$$

$$u_3 = \frac{1}{4}(0+0+u_2+50) = \frac{1}{4}(u_2+50) \qquad \text{..... (iii)}$$

Substituting (iii) and (i) in (ii) we get

$$u_2 = \frac{1}{4}\left[\frac{1}{4}u_2 + \frac{1}{4}(u_2+50)\right]$$

$\therefore \quad 16u_2 = 2u_2 + 50$

$\therefore \quad u_2 = 3.5714,\ u_1 = 0.89285\ u_3 = 13.39285.$

Applying (1) again at the mesh points u_4, u_5, u_6 we get

$$u_4 = \frac{1}{4}u_5 \qquad \therefore\ u_4 = 1.7857$$

$$u_5 = \frac{1}{4}(u_4+u_6) \qquad u_5 = 7.1429$$

$$u_6 = \frac{1}{4}(u_4+100) \qquad u_6 = 26.7857$$

7.9 HYPERBOLIC EQUATIONS

The example of hyperbolic partial differential equation is the wave equation.

$$\frac{\partial^2 u}{\partial t^2} = c^2\frac{\partial^2 u}{\partial x^2}$$

Its solution is the displacement function $U(x, t)$ defined for values of x from 0 to l and for t from o to ∞, satisfying the initial and boundary conditions. In the case of hyperbolic equations, we have two initial conditions and two boundary conditions. Such equations arise from convective type of problems in vibrations, wave mechanics and gas dynamics.

7.10 SOLUTION OF WAVE EQUATION

Consider the wave equation of one dimension

$$\frac{\partial^2 U}{\partial t^2} = c^2 \frac{\partial^2 U}{\partial x^2} \qquad ...(1)$$

Subject to the initial conditions $U = f(x)$, $\dfrac{\partial U}{\partial t} = g(x)$

$$0 \le x \le 1 \text{ at } t = 0 \qquad ...(2)$$

and the boundary conditions $u(0, t) = \varphi(t)$ $u(1,t) = \psi(t)$ ... (3)

Consider a rectangular mesh in the $x - t$ plane spacing h along x direction and k along time t direction. Denoting a mesh point $(x,t) = (ih, jk)$ or (i, j)

We have

$$\frac{\partial^2 u}{\partial x^2} = \frac{u_{i-1,j} - 2u_{i,j} + u_{i+1,j}}{h^2} \quad \text{and} \quad \frac{\partial^2 u}{\partial t^2} = \frac{u_{i,j-1} - 2u_{i,j} + u_{i,j+1}}{k^2}$$

Replacing the derivatives in (1) by their approximations we obtain

$$u_{i,j-1} - 2u_{i,j} + u_{i,j+1} = \frac{c^2k^2}{h^2}(u_{i-1,j} - 2u_{i,j} + u_{i+1,j})$$

$$\therefore \quad u_{i,j+1} = 2(1 - \alpha^2 c^2)\, u_{i,j} + \alpha^2 c^2 (u_{i-1,j} + u_{i+1,j}) - u_{i,j-1} \qquad ...(4)$$

where $\alpha = \dfrac{k}{h}$.

Now replacing the derivative in (2) by its central difference approximation, we get

$$\frac{u_{i,j+1} - u_{i,j-1}}{2k} = \frac{\partial u}{\partial t} = g(x)$$

or $u_{i,j+1} = u_{i,j-1} + 2\,kg\,(x)$ at $t = 0$

i.e., $u_{i,1} = u_{i,-1} + 2kg(x)$ for $j = 0$ ...(5)

Also initial condition $u = f(x)$ at $x = 0$ becomes $u_{i,-1} = f(x)$ ..... (6)

Combining (5) and (6) we have

$$u_{i,1} = f(x) + 2\text{kg}\,(x) \qquad .... (7)$$

Also (3) gives $u_{0,j} = \varphi(t)$ and $u_{1,j} = \psi(t)$

Hence the explicit form (4) gives the values of $u_{i,j+1}$ at the $(j + 1)$ level when the nodal values at $(j-1)^{th}$ and j^{th} levels are known from (6) and (7) as shown in the following figure.

A special case : The coefficient of $u_{i,j}$ in (4) will vanish if $\alpha = \frac{1}{c}$ or $k = \frac{h}{c}$.

Then (4) reduces to the simple form

$$u_{i,j+1} = u_{i-1,j} + u_{i+1,j} - u_{i,j-1} \qquad \text{....(8)}$$

This result provides an **explicit scheme** for the solution of the wave equation.

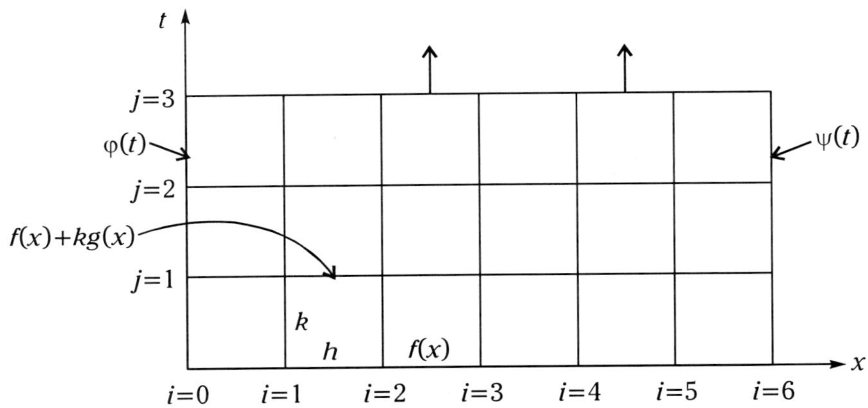

Obs 1: For $\alpha = \frac{1}{c}$ the solution of (4) is stable and coincides with the solution of (1)

For $\alpha < \frac{1}{c}$, the solution is stable but inaccurate.

For $\alpha > \frac{1}{c}$, the solution is unstable.

2. The formula (4) converges for $\alpha \le 1$ i.e., for $k \le h$.

Example 1: Evaluate the pivotal values of the equation $u_{tt} = 16\, u_{xx}$ taking $h = 1$ up to $t = 1.25$. The boundary conditions are $u(0,t) = u(5, t) = 0$, $u_t(x,0) = 0$ and $u(x,0) = x^2\,(5 - x)$.

Solution: Here $c^2 = 16$.

$\therefore$ The difference equation for the given equation is

$$u_{i,\,j+1} = 2(1 - 16\alpha^2)\, u_{i\,j} + 16\alpha^2\, (u_{i-1,\,j} + u_{i+1,\,j}) - u_{i,\,j-1} \text{ where } \alpha = \frac{k}{h}$$

Taking $h = 1$ and choose k so that the co-efficient of $u_{i,\,j}$ vanishes, we have $16\alpha^2 = 1$ i.e., $k = \frac{h}{4} = \frac{1}{4}$.

$\therefore$ (1) reduces to

$$u_{i,\,j+1} = u_{i-1,\,j} + u_{i+1,\,j} - u_{i,\,j-1} \quad ....(2)$$

Which gives a convergent solution (since $\frac{k}{h} < 1$). Its solution coincides with the solutions of the given differential equation.

Now since $u\,(0,\,t) = u\,(5,t) = 0$

$\therefore \quad u_{0,\,j} = 0$ and $u_{5,\,j} = 0$ for all values of j

i.e., the entries in the first and last columns are zero.

Since $u(x,0) = x^2\,(5 - x)$

$$u_{i,\,0} = i^2\,(5 - i) = 4,\ 12,\ 18,\ 16 \text{ for } i = 1,\ 2,\ 3,\ 4, \text{ at } t = 0.$$

These are the entries of the first row.

Finally, the initial condition $u_t\,(x,\,0) = 0$ becomes

$$\frac{u_{i,j+1} - u_{i,j-1}}{2k} = 0 \text{ when } j = 0 \text{ gives } u_{i,\,1} = u_{i,\,-1} \quad ...(3)$$

Putting $j = 0$ in (2)

$$u_{i,\,1} = u_{i-1,\,0} + u_{i+1,\,0} - u_{i,\,-1}$$

$$= u_{i-1,\,0} + u_{i+1,\,0} - u_{i,\,1} \quad \text{using (3)}$$

$$\therefore \quad u_{i,\,1} = \frac{1}{2}\left[u_{i-1,0} + u_{i+1,0}\right] \quad ....(4)$$

Taking $i = 1,2,3,4$ successively we get

$$u_{11} = \frac{1}{2}\,(u_{0,0} + u_{2,0}) = \frac{1}{2}\,(0 + 12) = 6$$

$$u_{2,1} = \frac{1}{2}\,(u_{1,0} + u_{3,0}) = \frac{1}{2}\,(4 + 18) = 11$$

$$u_{3,1} = \frac{1}{2}(u_{2,0} + u_{4,0}) = \frac{1}{2}(12 + 16) = 14$$

$$u_{4,1} = \frac{1}{2}(u_{3,0} + u_{5,0}) = \frac{1}{2}(18 + 0) = 9$$

These are the entries of the second row.

Putting $j = 1$ in (2) we get

$$u_{i,2} = u_{i-1,1} + u_{i+1,1} - u_{i,0}.$$

Taking $i = 1, 2, 3, 4$ successively we get

$$u_{1,2} = u_{0,1} + u_{2,1} - u_{1,0} = 0 + 11 - 4 = 7$$

$$u_{2,2} = u_{1,1} + u_{3,1} - u_{2,0} = 6 + 14 - 12 = 8$$

$$u_{3,2} = u_{2,1} + u_{4,1} - u_{3,0} = 11 + 9 - 18 = 2$$

$$u_{4,2} = u_{3,1} + u_{5,1} - u_{4,0} = 14 + 0 - 16 = -2$$

These are the entries of the third row.

Similarly, putting $j = 2, 3, 4$ successively in (3) the entries of the 4th, 5th and 6th rows are obtained.

Hence the entries of $u_{i,j}$ are shown in the following table.

j \ i	0	1	2	3	4	5
0	0	4	12	18	16	0
1	0	6	11	14	9	0
2	0	7	8	2	– 2	0
3	0	2	– 2	– 8	– 7	0
4	0	– 9	– 14	– 11	– 6	0
5	0	– 16	– 18	– 12	– 4	0

Example 2: The transverse displacement U of a point at a distance x from one end and at any time t of a vibrating string satisfies the equation $\frac{\partial^2 U}{\partial t^2} = 4\frac{\partial^2 U}{\partial x^2}$ with boundary conditions $u = 0$ at $x = 0, t > 0$ and $u = 0$

at $x = 4$, $t > 0$ and initial conditions $u = x(4 - x)$ and $\dfrac{\partial u}{\partial t} = 0$ at $t = 0$, $0 \le x \le 4$. Solve this equation numerically for one half period of vibration taking $h = 1$ and $k = \dfrac{1}{2}$.

Solution: Here $\dfrac{h}{k} = 2 = c$

$\therefore$ The difference equation for the given equation is

$$u_{i,\,j+1} = u_{i-1,\,j} + u_{i+1,\,j} - u_{i,\,j-1} \qquad \text{....(1)}$$

which gives the convergent solution $\qquad (\because\ k < h)$.

Now, since $u(0, t) = u(4,t) = 0$.

$$u_{0,j} = 0 \text{ and } u_{4,j} = 0 \text{ for all values of } j.$$

i.e., the entries in the first and last columns are zero.

Since $u(x,0) = x(4 - x)$

$$u_{i,0} = i(4 - i) = 3, 4, 3 \text{ for } i = 1, 2, 3 \text{ at } t = 0.$$

These are the entries of the first row.

Also $u_t\,(x, 0) = 0$ becomes

$$\frac{u_{i,\,j+1} - u_{i,\,j-1}}{2k} = 0 \ \text{ when } j = 0 \text{ gives } u_{i,1} = u_{i,\,-1} \qquad \text{... (2)}$$

Putting $j = 0$ in (1)

$$\begin{aligned} u_{i,1} &= u_{i-1,0} + u_{i+1,0} - u_{i,-1} \\ &= u_{i-1,0} + u_{i+1,0} - u_{i,1} \qquad \text{using (2)} \end{aligned}$$

Taking $i = 1, 2, 3$ we get

$$u_{11} = \frac{1}{2}\,[u_{0,0} + u_{2,0}] = 2$$

$$u_{2,1} = \frac{1}{2}\,[u_{1,0} + u_{3,0}] = 3$$

$$u_{3,1} = \frac{1}{2}\,[u_{2,0} + u_{4,0}] = 2$$

These are the entries of the second row

Putting $j = 1$ in (1)

$$u_{i,2} = u_{i-1,1} + u_{i+1,1} - u_{i,0}$$

Taking $I = 1, 2, 3$ we get

$$u_{1,2} = u_{0,1} + u_{2,1} - u_{1,0} = 0 + 3 - 3 = 0$$

$$u_{22} = u_{1,1} + u_{3,1} - u_{2,0} = 2 + 2 - 4 = 0$$

$$u_{32} = u_{2,1} + u_{4,1} - u_{3,0} = 3 + 0 - 3 = 0$$

These are the entries of the third row and so on.

Now the equation of the vibrating string of length l is

$$u_{tt} = c^2 u_{xx} \qquad (\because \; l = 4, c = 2)$$

$\therefore$ its period of vibration $= \dfrac{2l}{c} = \dfrac{2 \times 4}{2} = 4$ sec.

This shows that we have to compute

$u(x, t)$ up to $t = 2$.

Similarly we obtain the values of $u_{i,2}$ and $u_{i,3}$.

The values of $u_{i,j}$ are shown in the following table.

j \ i	0	1	2	3	4
0	0	3	4	3	0
1	0	2	3	2	0
2	0	0	0	0	0
3	0	– 2	– 3	– 2	0
4	0	– 3	– 4	– 3	0

7.11 QUESTION BANK

1) Find the value of $U(x,y)$ satisfying the Laplace equation $\dfrac{\partial^2 U}{\partial x^2} + \dfrac{\partial^2 U}{\partial y^2} = 0$ at the pivotal points of a square region with boundary values shown in the following table.

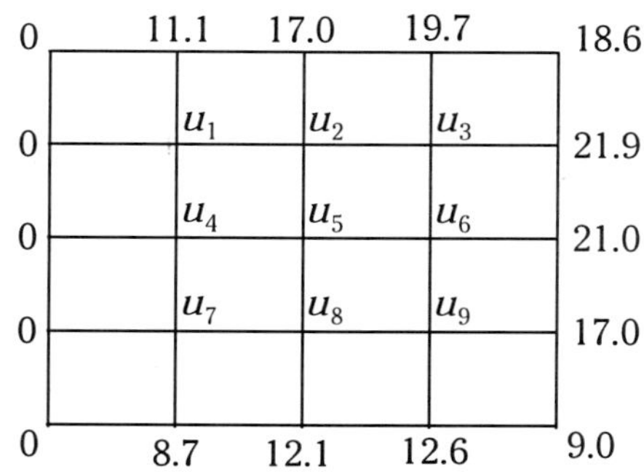

Ans.: $u_1 = 7.9$, $u_2 = 13.7$, $u_3 = 17.9$, $u_4 = 6.6$, $u_5 = 11.9$, $u_6 = 16.3$, $u_7 = 6.6$, $u_8 = 11.2$, $u_9 = 14.3$.

2) Using difference approximation solve $\nabla^2 u = 0$ at the nodal points of the square grid as shown in the following table using boundary values indicated.

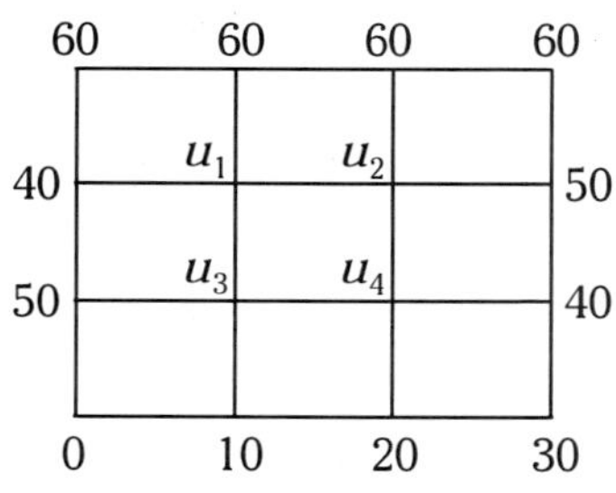

Ans.: $u_1 = 26.66$ $u_2 = 33.33$, $u_3 = 43.33$, $u_4 = 46.66$.

3) Solve $\nabla^2 u = 0$ under the conditions $(h = 1, k = 1)$

$$u(0,y) = 0,\ u(4,y) = 12 + y \text{ for o} \le x \le 4.$$

Ans.: $u_1 = 2.37$, $u_2 = 5.6$, $u_3 = 9.87$, $u_4 = 2.89$, $u_5 = 6.14$, $u_6 = 9.89$, $u_7 = 3.02$, $u_8 = 6.17$, $u_9 = 9.51$.

4) Solve $U_{xx} + U_{yy} = 0$ for the square mesh with boundary values as shown in the following table. Iterate till the mesh values are correct to two decimal places.

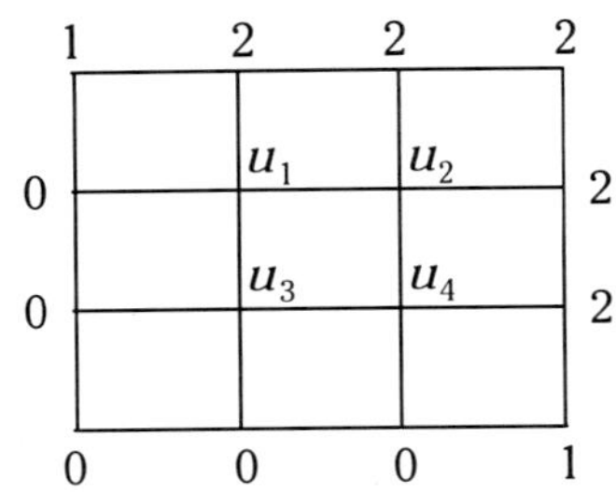

Ans.: $u_1 = 0.99$ $u_2 = 1.49$ $u_3 = 0.49$.

5) Solve $\dfrac{\partial^2 U}{\partial x^2} + \dfrac{\partial^2 U}{\partial y^2} = 8x^2 y^2$ for square mesh given $u = 0$ on the four boundaries dividing the square into 16 sub squares of length one unit.

Ans.: $-3, -2, -3, -2, -2, -2, -3, -2, -3$.

6) Find the solution of the parabolic equation $U_{xx} = 2u_t$ when $u(0, \text{t}) = u(4\ t) = 0$ and $u(x,0) = x(4 - x)$ taking $h = 1$ find the values upto $t = 5$.

Ans.:

j \ i	0	1	2	3	4
0	0	3	4	3	0
1	0	2	3	2	0
2	0	1.5	2	1.5	0
3	0	1	1.5	1	0
4	0	0.75	1	0.75	0
5	0	0.5	0.75	0.5	0

7) Solve the boundary value problem $U_t = U_{xx}$ under the conditions $u(0,t) = u(1,t) = 0$ and $u(x,0) = \sin \pi x$ $0 \le x \le 1$ using schmidt method. (Take $h = 0.2$ and $k = 0.02$)

Ans.:

j \ i	0	1	2	3	4	5
0	0	0.59	0.95	0.95	0.59	0
1	0	0.475	0.77	0.77	0.475	0
2	0	0.38	0.62	0.62	0.38	0
3	0	0.31	0.50	0.50	0.31	0
4	0	0.25	0.41	0.41	0.25	0
5	0	0.20	0.33	0.33	0.20	0

8) Solve the heat equation $\frac{\partial U}{\partial t} = \frac{\partial^2 U}{\partial x^2}$ subject to the conditions $u(0,t) = u(1,t) = 0$ and

$$u(x, 0) = 2x \text{ for } 0 \le x \le \frac{1}{2}$$

$$= 2(1 - x) \text{ for } \frac{1}{2} \le x \le 1.$$

Take $h = \frac{1}{4}$ and k according to Bender schmidt equation

Ans.:

j \ i	0	1	2	3	4
0	0	0.5	1	0.5	0
1	0	0.5	0.5	0.5	0
2	0	0.25	0.5	0.25	0
3	0	0.25	0.25	0.25	0

9) Solve the boundary value problem $U_{tt} = U_{xx}$ with the condition $u(0, t) = u(1, t) = 0$, $u(x,0) = \frac{1}{2}(1 - x)$ and $u_t(x,0) = 0$ taking $h = k = 0.1$ for $0 \le t \le 4$ compare your solution with the exact solution at $x = 0.5$ and $t = 0.3$.

Ans.:

$T = 0.3$, $x =$	0.1	0.2	0.3	0.4	0.5
Numerical Sol u :	0.02	0.04	0.06	0.075	0.08
Exact Sol. u:	0.02	0.04	0.06	0.075	0.08

10) Evaluate the pivotal values of the following equation taking $h = 1$ and upto one half of the period of vibration $16U_{xx} = U_{tt}$ given that $u(0,t) = u(5,t) = 0$

$u(x, 0) = x^2(x - 5)$ and $u_t(x,0) = 0$

j \ i	0	1	2	3	4	5
0	0	4	12	18	16	0
1	0	4	12	18	16	0
2	0	8	10	10	2	0
3	0	6	6	– 6	– 6	0
4	0	– 2	– 10	– 10	– 8	0
5	0	– 16	– 18	– 12	– 4	0

UNIT 8

DIFFERENCE EQUATIONS AND Z-TRANSFORMS

8.1 FINITE DIFFERENCES

Suppose that the function $y = f(x)$ is tabulated for the equally spaced values.

$$x : x_0, \quad x_0 + h, \ldots\ldots \quad x_0 + nh \text{ giving}$$
$$y : y_0, \quad y_1, \ldots\ldots \quad y_n.$$

To determine the values of $f(x)$ or $f^1(x)$ or some intermediate values of x, the following types of differences are found useful.

Forward differences

The differences $y_1 - y_0, y_2 - y_1, \ldots y_n - y_{n-1}$, when denoted by $\Delta y_0, \Delta y_1 \ldots$ $\ldots\Delta y_{n-1}$ respectively are called the first forward differences where Δ is the forward difference operator. Thus the first forward differences are $\Delta y_r = y_{r+1} - y_r$.

Similarly the second forward differences are defined by

$$\Delta^2 y_r = \Delta y_{r+1} - \Delta y_r.$$

In general

$$\Delta^n y_r = \Delta^{n-1} y_{r+1} - \Delta^{n-1} y_r$$

defines the nth forward differences.

In a difference table, x is called the argument and y the function or the entry. y_0 the first entry is called the leading term and $\Delta y_0, \Delta^2 y_0, \Delta^3 y_0$ etc are called the leading differences.

Any higher order forward differences can be expressed in terms of the entries.

We have

$$\Delta^2 y_0 = \Delta y_1 - \Delta y_0 = (y_2 - y_1) - (y_1 - y_0) = y_2 - 2y_1 + y_0.$$

$$\Delta^3 y_0 = \Delta^2 y_1 - \Delta^2 y_0 = (y_3 - 2y_2 + y_1) - (y_2 - 2y_1 + y_0)$$

$$= y_3 - 3y_2 + 3y_1 - y_0.$$

$$\Delta^4 y_0 = \Delta^3 y_1 - \Delta^3 y_0 = y_4 - 4y_3 + 6y_2 - 4y_1 + y_0$$

The co-efficients occuring on *r, h, s* being the binomial co-efficients, we have in general,

$$\Delta^n y_0 = y_n - {}^n c_1 y_{n-1} + {}^n c_2 y_{n-2} - \dots + (-1)^n y_0.$$

Backward differences

The differences $y_1 - y_0, y_2 - y_1 \dots y_n - y_{n-1}$ when denoted by ∇y_1, ∇y_2, ... ∇y_n respectively are called the first backward differences where ∇ is the backward difference operator. Similarly we define higher order backward differences. Thus we have

$$\nabla y_r = y_r - y_{r-1}$$

$$\nabla^2 y_r = \nabla y_r - \nabla y_{r-1}$$

$$\nabla^3 y_r = \nabla^2 y_r - \nabla^2 y_{r-1} \text{ etc.}$$

Central differences

Sometimes it is convenient to employ another system of differences known as central differences. In this system, the central difference operator δ is defined by the relations

$$y_1 - y_0 = \delta_{1/2},\ y_2 - y_1 = \delta y_{3/2} \dots\dots\ y_n - y_{n-1} = \delta y_{\frac{n-1}{2}}$$

Similarly higher order central differences are defined as

$$\delta y_{3/2} - \delta y_{1/2} = \delta^2 y_1$$

$$\delta y_{5/2} - \delta y_{3/2} = \delta^2 y_2$$

$$\delta^2 y_2 - \delta^2 y_1 = \delta^3 y_{3/2} \text{ and so on.}$$

Differences of a polynomial

The nth differences of a polynomial of the nth degree are constant and all higher order differences are zero.

i.e., $\Delta^n f(x) = a\, n!\, h^n$

$$\Delta^{n+1} f(x) = 0$$

Shift operator E

The shift operator E is the operator of increasing the argument x by h so that

$$E f(x) = f(x+h),\ E^2 f(x) = f(x+2h).\ E^3 f(x) = f(x+3h) \text{ etc.}$$

The inverse operator E^{-1} is defined by $E^{-1} f(x) = f(x-h)$

If y_x is the function $f(x)$ then $Ey_x = y_{x+h}$ and $E^{-1} y_x = y_{x-h}$.

Also $E^n y_x = x + nh$, and $Ey^{-n}{}_x = y_{x-nh}$.

Relation between the operators.

I) $\Delta = E - 1$ II) $\nabla = 1 - E^{-1}$

III) $\Delta = E\nabla = \nabla E$ IV) $E = e^{hd}$.

8.2 DIFFERENCE EQUATION

A difference equation is a relation between the differences of an unknown function at one or more general values of the argument.

Thus, $\Delta y_{(n+1)} + y_{(n)} = 2$ ... (1)

$\Delta y_{(n+1)} + \Delta^2 y_{(n-1)} = 1$ ... (2)

are the difference equations.

An alternative way of writing difference equation is as under.

Since $\Delta y_{(n+1)} = y_{(n+2)} - y_{(n+1)}$

Equation (1) takes the form ... $y_{(n+2)} - y_{(n+1)} + y_{(n)} = 2$ ... (3)

and $\Delta^2{}_{(yn-1)} = y_{(n+1)} - 2y_{(n)} + y_{(n-1)}$.

$\therefore$ Equation (2) takes the form

$y_{(n+2)} - 2y_{(n)} + y_{(n-1)} = 1$ ... (4)

Difference equations are met with under the name of recurrence relations.

Order of a difference equation

Order of a difference equation is the difference between the largest and the smallest arguments occurring in the difference equation divided by the unit of increment.

Thus (3) above is of the second order, for

$$\frac{\text{Largest argument} - \text{smallest argument}}{\text{Unit of increament}}$$

$$= \frac{(n+2)-n}{1} = 2$$

and (4) is of the third order, for

$$\frac{(n+2)-(n-1)}{1} = 3$$

Note: While finding the order of a difference equation it must always be expressed in a form tree from Δ's for the highest power of Δ does not give order of the difference equation.

Solution : Solution of a difference equation is an expression for $y_{(n)}$ which satisfies the given difference equation.

The general solution of a difference equation is that in which the number of arbitrary constants is equal to the order of the difference equation.

A particular solution (particular integral) is that solution which is obtained from the general solution by giving particular values to the constants.

8.3 FORMATION OF DIFFERENCE EQUATIONS

Example 1: Form the difference equation corresponding to the family of curves $y = ax + bx^2$.

Solution:

$$y = ax + bx^2 \qquad ...(1)$$

$$\therefore \Delta y = a\,\Delta(x) + b\,\Delta(x^2)$$

$$= a(x+1-x) + b[(x+1)^2 - x^2]$$

$$= a + b(2x+1)$$

$$\Delta^2 y = 2b[(x+1) - x] = 2b$$

$$\therefore \qquad b = \frac{1}{2}\Delta^2 y.$$

and

$$a = \Delta y - b(2x+1)$$

$$= \Delta y - \frac{1}{2}\Delta^2 y(2x+1)$$

Substituting these values of a and b in (1) we get

$$y = \left[\Delta y - \frac{1}{2}\Delta^2 y(2x+1)\right] x + \frac{1}{2}\Delta^2 y\, x^2.$$

$\therefore \quad (x^2 + x)\,\Delta^2 y - 2x\,\Delta\, y + 2y = 0.$

This is the difference equation which may be written in forms of E as

$$(x^2 + x)\, y_{x+2} - (2x^2 - 4x)\, y_{n+1} + (x^2 + 3x + 2)\, y_n = 0.$$

Example 2: From $y_n = a2^n + b\,(-2)^n$, derive a difference equation by eliminating the constants.

Solution:

$$y_n = a\,2^n + b\,(-2)^n$$
$$y_{n+1} = 2a\,2^n - 2b\,(-2)^n$$
$$y_{n+2} = 4a\,2^n + 4b\,(-2)^n$$

Eliminate a and b we get.

$$\begin{vmatrix} y_n & 1 & 1 \\ y_{n+1} & 2 & -2 \\ y_{n+2} & 4 & 4 \end{vmatrix} = 0$$

or $y_{n+2} - 4y_n = 0$. Which is the desired difference equation.

Example 3 : Form the difference equations generated by

i) $y_x = ax + b2^x$

ii) $y_n = (A + Bn)\,3^n$

iii) $y_n = A\,3^n + \mathrm{B}5^n$.

iv) $y = \dfrac{a}{x} + b$

v) $y_x = a\,2^x + b\,3^x + \mathrm{C}$.

8.4 LINEAR DIFFERENCE EQUATIONS

Definition: A linear difference equation is that in which y_{n+1}, y_{n+2} etc occur to the first degree only and are not multiplied together.

A linear difference equation with constant coefficients is of the form

$$y_{n+r} + a_1\, y_{n+r-1} + a_2\, y_{n+r-2} + \ldots. + a_n\, y_n = f(n).$$

Where $a_1, a_2, a_3 \ldots a_n$ are constants.

Now we deal with linear difference equations with constant coefficients.

Elementary properties : If $u_1\,(n), u_2(n) \ldots u_r\,(n)$ are r independent solutions of the equation.

$$y_{n+r} + a_1\, y_{n+r-1} + a_2\, y_{n+r-2} + \ldots. + a_r\, y_n = 0 \qquad \ldots(2)$$

then its complementary solution is

$$u_n = c_1 u_1 (n) + c_2 u_2 (n) + \text{.........} + c_r u_r (n)$$

where $c_1, c_2 \ldots c_r$ are arbitrary constants.

If v_n is a particular solution of (1), then the complete solution of (1) is

$$y_n = u_n + v_n.$$

The part u_n is called the complementary function (CF) and part v_n is called the particular function (PF) of (1)

Thus the complete solution (CS) of (1) is

$$y_n = \text{CF} + \text{PI}$$
$$= \text{complementary function} + \text{particular integral.}$$

8.5 RULES TO FIND COMPLEMENTARY FUNCTION

1) To begin with, consider the first order linear equation

$y_{n+1} - \lambda y_n = 0$ where λ is a constant.

Rewriting it as

$$\frac{y_{n+1}}{\lambda^{n+1}} - \frac{y_n}{\lambda_n} = 0 \text{ we have } \Delta\left(\frac{y_n}{\lambda_n}\right) = 0.$$

Which gives $\dfrac{y_n}{\lambda_n} = c$, a constant.

Thus the solution of $(E - \lambda) y_n = 0$ is $y_n = c\lambda^n$.

2) Now consider the second order linear equation

$y_{n+2} + ay_{n+1} + by_n = 0$ which in symbolic form is

$$(E^2 + aE + b) y_n = 0 \qquad \ldots (1)$$

Auxiliary equation is

$$E^2 + aE + b = 0$$

Let its root be λ_1, λ_2

Case I: If these roots are real and distinct then (1) is equivalent to

$$(E - \lambda_1)(E - \lambda_2) y_n = 0 \qquad \ldots(2)$$

or $$(E - \lambda_2)(E - \lambda_1) y_n = 0 \qquad \ldots(3)$$

If y_n satisfies the subsidiary equation $(E - \lambda_1) y_n = 0$ then it will also satisfy (3).

Similarly if y_n satisfies $(E - \lambda_2)\, y_n = 0$, then it will satisfy (2).

$\therefore$ We can derive two independent solutions of (1), by solving the two subsidiary equations.

$$(E - \lambda_1)\, y_n = 0 \text{ and } (E - \lambda_2)\, y_n = 0$$

Their solutions are $y_n = c_1\,(\lambda_1)^n$ and $y_2 = c_2\,(\lambda_2)^n$ respectively where c_1 and c_2 are arbitrary constants.

$\therefore$ $C.S$ of (1) is

$$y_n = c_1\,(\lambda_1)^n + c_2\,(\lambda_2)^n.$$

Case II : If the roots are real and equal (i.e., $\lambda_1 = \lambda_2$) then (2) becomes.

$$(E - \lambda_1)^2\, y_n = 0 \qquad \text{....(4)}$$

Let $y_n = (\lambda_1)^n\, Z_n$ here Z^n is a new dependent variable, then (4) takes the form

$$(\lambda_1)^{n+2}_{Zn+2} - 2\lambda_1\,(\lambda_1)^{n+1}_{Zn+1} + \lambda_1^2\,(\lambda_1)^{n}_{Zn} = 0$$

$\therefore$ $\quad Z_{n+2} - 2Z_{n+1} + Z_n = 0$

i.e., $\quad \Delta^2\, Z_n = 0$

$\therefore$ $\quad Z_n = c_1 + c_2 n$ where $c_{1,}\, c_2$ are constants.

Thus the solution of (1) becomes

$$y_n = (c_1 + c_2 n)\,(\lambda_1)^n$$

Case III: If the roots are imaginary (i.e., $\lambda_1 = \alpha + i\,\beta$, $\lambda_2 = \alpha - i\,\beta$) then the solution of (1) is

$$y_n = c_1\,(\alpha + i\beta)^n + c_2\,(\alpha - i\beta)^n$$

put $\quad \alpha = r\cos\theta, \quad \beta = r\sin\theta$

$$= r^n\,[c_1(\cos n\theta + i\sin n\theta) + c_2\,(\cos n\theta - i\sin n\theta)]$$

$$= r^n\,[A_1\cos n\theta + A_2\sin n\theta]$$

where A_1, A_2 are constants and $r = \sqrt{\alpha^2 + \beta^2}$, $\theta = \tan^{-1}\dfrac{\beta}{\alpha}$.

3) In general, to solve the equation

$$y_{n+r} + a_1\, y_{n+r-1} + a_2\, y_{n+r-2} + \ldots\ldots + a_r\, y_n = 0$$

where a's are constants.

i) Write the equation in the symbolic form

$$(E^r + a_1\, E^{r-1} + \ldots + a_r)\, y_n = 0.$$

ii) Write down the AE

$$E^r + a_1 E^{r-1} + \ldots + ar = 0 \text{ and solve it for } E.$$

iii) Write the solution as follows.

	Roots of A.E.	Solution i.e., CF.
1.	$\lambda_1, \lambda_2, \lambda_3$ ....(real and distinct)	$c_1(\lambda_1)^n + c_2(\lambda_2)^n + c_3(\lambda_3)^n + \ldots.$
2.	$\lambda_1, \lambda_2, \lambda_3$ ...(2 real & equal roots)	$(c_1 + c_2 n)(\lambda_1)^n + c_3(\lambda_3)^n + \ldots\ldots$
3.	$\lambda_1, \lambda_2, \lambda_3$ ...(3 real & equal roots)	$(c_1 + c_2 n + c_3 n^2)(\lambda_i)^n + \ldots\ldots.$
4.	$\alpha + i\beta, \alpha - i\beta$ (a pair of imaginary roots)	$r^n (c_1 \cos n\theta + c_2 \sin n\theta)$

Example 1: Solve the difference equation $u_{n+3} - 2\, u_{n+2} - 5u_{n+1} + 6\, u_n = 0.$

Solution: Given equation in symbolic form is

$$(E^3 - 2E^2 - 5E + 6)\, u_n = 0.$$

$\therefore$ A, E is $E^3 - 2E^2 - 5E + 6 = 0.$

$\therefore$ $E = 1 - 2, 3$

Thus the complete solution is

$$u_n = c_1 (1)^2 + c_2 (-2)^n + c_3 (3)^n.$$

Example 2 : Solve $u_{n+2} - 2u_{n+1} + u_n = 0.$

Given equation in symbolic form is

$(E^2 - 2E + 1)\, u_n = 0.$

$\therefore$ $E = 1.1.$

Thus the solution is

$$u_n = (c_1 + c_2 n)(1)^n. \text{ i.e., } u_n = c_1 + c_2 n.$$

Example 3: Solve $y_{n+1} - 2y_n \cos\alpha + y_{n-1} = 0.$

This is second order difference equation in y_{n-1}

Which in symbolic form is

$$(E^2 - 2E\cos\alpha + 1)\, y_{n-1} = 0.$$

$\therefore$ AE is $E^2 - 2E\cos\alpha + 1 + 0.$

$$\therefore \quad E = \frac{2\cos\alpha \pm \sqrt{4\cos^2\alpha - y}}{2}$$

$$= \cos\alpha \pm i\sin\alpha.$$

Thus the solution is

$$y_{n-1} = (1)^{n-1}\,[c_1 \cos (n-1)\,\alpha + c_2 \sin (n-1)\,\alpha]$$
$$= c_1 \cos n\alpha + c_2 \sin n\,\alpha$$

Example 1: Solve the following difference equations

1) $y_{n+2} - y_{n+1} - 2y_n = 0$
2) $(\Delta^2 - 3\,\Delta + 2)\,y_n = 0$
3) $u_{n+3} - 3u_{n+2} + 4u_n = 0$
4) $u_{n+3} - 3u_{n+1} + 2u_n = 0$
5) $u_{n+2} - 6u_{x+1} + 9u_n = 0.$

8.6 RULES TO FIND THE PARTICULAR INTEGRAL

Consider the equation

$$y_{n+r} + a_1\, y_{n+r-1} + a_2\, y_{n+r-2} + \ldots + a_r\, y_n = f(n)$$

Which in symbolic form is $\varphi(E)\, y_n = f(n)$

Where $\varphi(E) = E^r + a_1\, E^{r-1} + \ldots + ar.$

Then the particular integral is $\text{PI} = \dfrac{1}{\varphi(E)} f(n)$

Case I: When $f(n) = a^n$

$$\therefore \quad \text{PI} = \frac{1}{\varphi(E)} a^n, \text{ put } E = a.$$

$$= \frac{1}{\varphi(a)} a^n, \text{ provided } \varphi(a) \neq 0$$

If $\varphi\,(a) = 0$ then for the equation

i) $(E - a)\, y_n = a^n$, $\text{PI} = \dfrac{1}{E-a} a^n = n a^{n-1}$

ii) $(E - a)^2\, y_n = a^n$, $\text{PI} = \dfrac{1}{(E-a)^2} a^n = \dfrac{n(n-1)}{2!} a^{n-2}$

iii) $(E - a)^3\, y_n = a^n \quad \dfrac{1}{(E-a)^3} = \dfrac{n(n-1)\,(n-2)}{3!} a^{n-3}$ and so on.

Example 1: Solve $y_{n+2} - 4y_{n+1} + 3y_n = 5^n$.

Solution: Given equation in symbolic form is

$$(E^2 - 4E + 3)\, y_n = 5^n.$$

$\therefore$ AE is $E^2 - 4E + 3 = 0$ or $(E - 1)(E - 3) = 0$.

$\therefore$ $E = 1, E = 3$.

$\therefore$ $CF = C_1 (1)^n + c_2 (3)^n = c_1 + c_2 3^n$.

and
$$PI = \frac{1}{E^2 - 4E + 5} 5^n \quad \text{put} \quad E = 5.$$

$$= \frac{1}{25 - 4 \times 5 + 3} 5^n = \frac{1}{8} 5^n$$

Thus the complete solution is

$$y_n = c_1 + c_2\ 3^n + \frac{5^n}{8}$$

Example 2: Solve $u_{n+2} - 4u_{n+1} + 4u_n = 2^n$.

AE is $E^2 - 4E + 4 = 0$ $\therefore$ $E = 2, 2$.

$\therefore$ $CF = (c_1 + c_2 n)\, 2^n$.

$$PI = \frac{1}{(E-2)^2} 2^n = \frac{n(n-1)}{2!} 2^{n-2} = n\,(n - 1)\, 2^{n-3}$$

Hence the complete solution is

$$u_n = (c_1 + c_2 n)\, 2^n + n\,(n - 1)\, 2^{n-3}$$

Case II: When $f(n) = \sin kn$.

$$PI = \frac{1}{\varphi(E)} \sin kn = \frac{1}{\varphi(E)} \frac{(e^{ikn} - e^{-ikn})}{2i}$$

$$= \frac{1}{2!} \left[\frac{1}{\varphi(E)} a^n - \frac{1}{\varphi(E)} b^n \right]$$

Now we proceed as in case 1, where $a = e^{ik}$ and $b = e^{-ik}$

2) When $f(n) = \cos kn$

$$PI = \frac{1}{\varphi(E)} \cos kn = \frac{1}{\varphi(E)} \left(\frac{e^{ikn} + e^{ikn}}{2} \right)$$

$$= \frac{1}{2}\left[\frac{1}{\varphi(E)}a^n + \frac{1}{\varphi(E)}b^n\right] \quad \text{as before.}$$

Now proceed as in case 1.

Example 3: Solve $y_{n+2} - 2\cos\alpha \,.\, y_{n+1} + y_n = \cos\alpha n$

Solution: Given equation in symbolic form is

$$(E^2 - 2\cos\alpha\, E + 1)\, y_n = \cos\alpha n$$

AE is $E^2 - 2\cos\alpha\; E + 1 = 0$

$$\therefore \quad E = \frac{2\cos\alpha \pm \sqrt{4\cos^2\alpha - 4}}{2} = \cos\alpha \pm i\sin\alpha$$

$$\therefore \quad \text{CF} = (1)^n\,[c_i \cos\alpha\, n + c_2 \sin\alpha n]$$

$$= c_1 \cos\alpha n + c_2 \sin\alpha n$$

$$\text{PI} = \frac{1}{E^2 - 2E\cos\alpha + 1}\cos\alpha n$$

$$= \frac{1}{E^2 - E\,(e^{i\alpha} + e^{-i\alpha}) + 1}\left(\frac{e^{i\alpha n} + e^{-i\alpha n}}{2}\right)$$

$$= \frac{1}{2}\left[\frac{1}{(E - e^{i\alpha})\,(E - e^{-i\alpha})}e^{i\alpha n} + \frac{1}{(E - e^{i\alpha})(E - e^{-i\alpha})}e^{-i\alpha n}\right]$$

put $E = e^{i\alpha}$ put $E = e^{-i\alpha}$

$$= \frac{1}{2}\left[\frac{1}{(E - e^{i\alpha})}\,\frac{1}{(e^{i\alpha} - e^{-i\alpha})}e^{i\alpha n} + \frac{1}{(E - e^{-i\alpha})}\,\frac{1}{(e^{-i\alpha} - e^{i\alpha})}e^{-i\alpha n}\right]$$

$$= \frac{1}{4i\sin\alpha}\left[\frac{1}{E - e^{i\alpha}}e^{i\alpha n} - \frac{1}{E - e^{-i\alpha}}e^{-i\alpha n}\right]$$

$$\left(\because \frac{e^{i\alpha} - e^{i\alpha}}{2i} = \sin\alpha\right)$$

$$= \frac{1}{4i\sin\alpha}\left[ne^{i\alpha(n-1)} - ne^{-i\alpha(n-1)}\right]$$

$$= \frac{1}{4\sin\alpha}\left[\frac{ne^{i\alpha(n-1)} - e^{-i\alpha(n-1)}}{2i}\right]$$

$$= \frac{n}{2\sin\alpha}\left[\sin(n-1)\alpha\right]$$

$$= \frac{n\sin(n-1)\alpha}{2\sin\alpha}$$

Hence the complete solution is

$$y_n = c_1 \cos \alpha n + c_2 \sin \alpha n + \frac{n\sin(n-1)\alpha}{2\sin\alpha}$$

Case III : When $f(n) = n^p$

$$\text{PI} = \frac{1}{\varphi(E)} n^P = \frac{1}{\varphi(1+\Delta)} n^P$$

1) Expand $[\varphi(1 + \Delta)]^{-1}$ in ascending powers of Δ by the binomial theorem, as far as the term in Δ^p.

2) Express n^p in the factorial form and operate on it with each term of the expansion.

Example 1: Solve $y_{n+2} - 4y_n = n^2 + n - 1$

Given equation is $(E^2 - 4)\, y_n = n^2 + n - 1$

$\therefore$ AE is $E^2 - 4 = 0$. $\therefore\ E = \pm 2$

$\therefore$ CF $= c_1\,(2)^n + c_2\,(-2)^n$.

$$\text{PI} = \frac{1}{E^2-4}\,(n^2 + n - 1)$$

$$= \frac{1}{(1+\Delta)^2-4}\,[n\,(n-1) + 2n - 1]$$

$$= \frac{1}{\Delta^2+2\Delta-3}\left[(n)^2 + 2(n) - 1\right]$$

$$= -\frac{1}{3}\left[1-\left(\frac{2}{3}\Delta+\frac{\Delta^2}{3}\right)\right]^{-1}\left\{[n]^2+2[n]-1\right\}$$

$$= -\frac{1}{3}\left[1+\frac{2}{3}\Delta+\frac{7}{9}\Delta^2.........\right]\left[[n]^2+2[n]-1\right]$$

$$= -\frac{1}{3}\left[\left([n]^2+2[n]-1\right)+\frac{2}{3}(2(n))+2)+\frac{7}{9}\times 2\right]$$

$$= -\frac{1}{3}\left[[n]^2 + \frac{10}{3}[n] + \frac{17}{9}\right]$$

$$= -\frac{n^2}{3} - \frac{7}{9}n - \frac{17}{27}$$

Hence the complete solution is

$$y_n = c_1\, 2^n + c_2(-2)^n - \frac{n^2}{3} - \frac{7n}{9} - \frac{17}{27}$$

Case IV: When $f(n) = a^n\, F(n)$, $F(n)$ being a polynomial of finite degree in n.

$$\text{PI} = \frac{1}{\varphi(E)} a^n F(n) = a^n \frac{1}{\varphi(aE)} F(n)$$

Example 1: Solve $y_{n+2} - 2y_{n+1} + y_n = n^2\, 2^n$.

Given equation is $(E^2 - 2E + 1)\, y_n = n^2\, 2^n$.

Its $\quad \text{CF} = c_1 + c_2\, n.$

and $\quad \text{PI} = \frac{1}{(E-1)^2} 2^n n^2$

$$= 2^n \frac{1}{(2E-1)^2} n^2$$

$$= 2^n \frac{1}{(1+2\Delta)^2} n^2$$

$$= 2^n\, [\,1 + 2\,\Delta\,]^{-2}\, [n(n-1) + n]$$

$$= 2^n\, [\,1 - 4\Delta + 12\,\Delta^2 \ldots]\, ([n]^2 + [n])$$

$$= 2^n\, [n]^2 + [n] - 4\; 2[n] + 1 + 12 \times 2$$

$$= 2^n\, ([n]^2 - 7\,[n] + 20$$

$$= 2^n\, (n^2 - 8n + 20) \qquad \because [n]^2 = n(n\text{–}1) \text{ factorial}$$

Hence the complete solution is

$$y_n = c_1 + c_2\, n + 2^n\, (n^2 - 8n + 20).$$

Example 2: Solve the following difference equations.

1) $y_{n+2} - 5y_{n+1} - 6y_n = 2^n$.
2) $(E^2 - 5E + 6)y = x + 2^x$
3) $y_{n+2} - 3y_{n+1} + 2y_n = n^2 + 2n - 1$.

4) $y_{n+2} - 6y_{n+1} + 8y_n = 2^n + 6n.$

5) $y_{n+1} - 3y_n = 3^n (n + 2)$

6) $y_{n+2} + y_{n+1} - 56y_n = 2^n (n^2 - 3)$

7) $u_{n+2} - 4u_{n+1} + 4u_n = n^2 2^n.$

Example 3: Solve the difference equation $y_{k+2} - 4y_{k+1} + 4y_k = 0$ and find solution such that $y_0 = 1$ and $y_1 = 3$.

Solution: The difference equation can be written as

$$(E^2 - 4E + 4)y_k = 0.$$

$\therefore$ AE is $E^2 - 4E + 4 = 0$. $\therefore$ $E = 2,2$

$\therefore$ $y_k = (c_1 + c_2 k) 2^k$ is the general solution.

Putting $k = 0$ and $k = 1$ we get

$$1 = c_1 \text{ and } 3 = 2 (c_1 + c_2)$$

$$\therefore \quad c_1 = 1, c_2 = \frac{1}{2}$$

$$\therefore \quad y_k = 2^k \left(1 + \frac{k}{2}\right) = (k + 2) 2^{k-1}$$

Example 4: Solve $y_{k+3} + y_{k+2} - 8y_{k+1} - 12y_k = 2k^2 + 5.$

Solution: Given equation is $y_{k+3} + y_{k+2} - 8y_{k+1} - 12y_k = 2k^2 + 5.$

$\therefore$ Auxiliary equation is

$$(E^3 + E^2 - 8E - 12) = 0$$

$$(E - 3) (E + 2)^2 = 0 \qquad \therefore \; E = 3, E = -2, -2$$

$$\therefore \quad \text{CF} = c_1 (3)^k + (c_2 + c_3 k) (-2)^k$$

$$\text{and} \quad \text{PI} = [(E - 3) (E + 2)^2]^{-1} (2k^2 + 5)$$

$$= [(\Delta - 2) (\Delta + 3)^2]–1 (2k^2 + 5) \qquad \because E = 1 + \Delta$$

$$= [\Delta^3 + 4\Delta^2 + - 3\Delta - 18]^{-1} (2k^2 + 5)$$

$$= -\frac{1}{18}\left[1 + \frac{3\Delta - 4\Delta^2 - \Delta^3}{18}\right]^{-1} (2k^2 + 5)$$

$$= -\frac{1}{18}\left[1 - \frac{3\Delta - 4\Delta^2 - \Delta^3}{18} + \left(\frac{3\Delta - 4\Delta^2 - \Delta^3}{18}\right)^2 \cdots\right] (2k^2 + 5)$$

$$= -\frac{1}{18}\left[1-\frac{\Delta}{6}+\frac{\Delta^2}{4}........\right](2k^2+5)$$

$$= -\frac{1}{18}\left[24^2+5-\frac{4k+2}{6}+\frac{k}{4}\right]$$

$$= -\frac{k^2}{9}+\frac{1}{27}k-\frac{17}{54}$$

Hence the general solution is given by

$$y_k = c_1\, 3^k + (c_2 + c_3 k)\,(-2)^k - \frac{k^2}{9} + \frac{k}{27} - \frac{17}{54}$$

Example 3: Solve $y_{k+2} - 2y_{k+1} + 5y_k = 2.3^k - 4.7^k$.

Solution: The given equation in symbolic form is

$(E^2 - 2E + 5)\, y_k = 2.3^k - 4.7^k$.

AE is $E^2 - 2E + 5 = 0$ $\quad\therefore E = 1 + 2i$

$$\therefore \quad \text{CF} = 5^{k/2}\,(c_1 \cos k\theta + c_2 \sin k\theta) \qquad \left(\because r = \sqrt{1+2^2}+\sqrt{5} = 5^{1/2}\right)$$

$$\text{PI} = \frac{1}{E^2-2E+5}(2.3^k - 4.7^k)$$

$$= 2\,.\frac{1}{E^2-2E+5}3^k - 4\,\frac{1}{E^2-2E+5}7^k$$

$$= 2\,.\frac{1}{3^2-2(3)+5}3^k - 4\,\frac{1}{7^2-2(7)+5}7^k$$

$$\therefore \quad y_k = 5^{k/2}\left(c_1 \cos k\theta + c_2 \sin k\theta\right) + \frac{1}{4}3^k - \frac{1}{10}7^k$$

where $\theta = \tan^{-1} 2$

Example 4: Solve $y_{k+2} + y_{k+1} + y_k = k\, 2^k$.

Solution: $(E^2 + E + 1)\, y_k = k.\, 2^k$.

$$\text{PI} = \frac{k\,2^k}{(E^2+E+1)}$$

$$= 2^k\,[(2E)^2 + 2E + 1]^{-1}\,(k)$$

$$= 2^k\,[7 + 10\,\Delta + 4\,\Delta^2]^{-1}\,(k) \quad \because E = 1 + \Delta.$$

$$= \frac{2^k}{7}\left[1 - \frac{10\Delta + 4\Delta^2}{7} \ldots\ldots\right](k)$$

$$= \frac{2^k}{7}\left(k - \frac{10}{7}\right)$$

$\therefore \quad$ GS = CF + PI

CF $= c_1 \cos k\theta + c_2 \sin k\theta$ where $\theta = \tan^{-1}(-\sqrt{3})$

$$\therefore \quad y_k = c_1 \cos k\theta + c_2 \sin k\theta + \frac{2^k}{7}\left(k - \frac{10}{7}\right)$$

Example 5: Solve $y_{k+2} - 5y_{k+1} + 6y_k = 5^k$

$$\text{PI} = \frac{1}{E^2 - 5E + 6} 5^k \qquad \text{put } E = 5$$

$$= \frac{5^k}{6}$$

and CF $= c_1 2^k + c_2 3^k$

$\therefore \quad$ GS = CF + PI

$$\therefore \quad y_k = c_1 2^k + c_2 3^k + \frac{5^k}{6}$$

Example 6: $y_{n+2} - 2y_{n+1} + y_n = n^2 2^n.$

AE is $E^2 - 2E + 1 = 0 \qquad \therefore E = 0.\ E = 1$

CF $= (c_1 + c_2 n)(1)^n = c_1 + c_2 n.$

$$\text{PI} = \frac{1}{(E-1)^2} 2^n n^2$$

$= 2^n [1 + 2\Delta]^{-2} [n(n-1) + n]$

$= 2^n [1 - 4\Delta + 12\Delta^2 \ldots] [(n)^2 + (n)]$

$= 2^n [(n)^2 + (n) - 4(2(n) + 1) = 2^n (n^2 - 8n = 20)$
$+ 12 \times 2]$

$\therefore \quad$ GS = CF + PI

$y_n = c_1 + c_2 n + 2^n (n^2 - 8n + 20).$

8.7 THE Z-TRANSFORMS

The Z-transform plays an important role in the field of Communication Engineering and Control Engineering at the stage of analysis and presentation of discrete-time linear shift invariance systems. When continuous signals are sampled, discrete-time functions arise. The application of Z-transform in discrete time systems is similar to that of the Laplace transform in continuous time systems.

If the function u_n is defined for discrete values ($n = 0, 1, 2, 3 \ldots$) and $u_n = 0$ for $n < 0$ then its Z-transform is defined to be

$$Z(u_n) = U(z) = \sum_{n=0}^{\infty} u_n z^{-n}$$

Whenever the infinite series converges.

The inverse Z-Transform is written as

$$Z^{-1}[U(z)] = u_n.$$

Note : i) If $u_n = 0$ for $n < 0$ then u_n is called a casual sequence.

ii) If $Z(u_n) = \sum_{n=0}^{\infty} u_n z^{-n}$ then this is one sided Z-transform.

iii) If u_n is defined for $n = 0, \pm 1, \pm 2, \pm 3 \ldots$ then

$Z(u_n) = \sum_{n=-\infty}^{\infty} u_n z^{-n}$ is called two sided Z-transform.

If we insert a particular complex number Z into the power series (1) the resulting value of $Z(u_n)$ will be a complex number. Thus the Z-transform $U(z)$ is a complex valued function of a complex variable z.

8.8 Z-TRANSFORM OF SOME STANDARD SEQUENCES

The application of the definition gives the following useful results.

1) Prove that $Z(a^n) = \dfrac{z}{z-a}$

Proof: By definition

$$Z(a^n) = \sum_{n=0}^{\infty} a^n z^{-n}$$

$$= 1+\frac{a}{z}+\frac{a^2}{z^2}+\frac{a^3}{z^3}+ \ldots\ldots\ldots\ldots$$

$$= \left(1-\frac{a}{z}\right)^{-1} = \frac{1}{1-\frac{a}{z}} = \frac{z}{z-a}$$

$$\therefore \qquad \boxed{z(a^n) = \frac{z}{z-a}}$$

2) Prove that $Z(n^p) = -z\dfrac{d}{dz}Z(n^{p-1})$

where p is a positive integer.

Proof: By definition we have

$$Z(n^p) = \sum_{n=0}^{\infty} n^p z^{-n}$$

$$= z\sum_{n=0}^{\infty} n^{p-1} n z^{-(n+1)} \qquad \ldots(1)$$

Changing p to $p-1$ in the definition we get

$$Z(n^{p-1}) = \sum_{n=0}^{\infty} n^{p-1} z^{-n}$$

Differentiate w.r.t. z

$$\frac{d}{dz}\left[Z(n^{p-1})\right] = \sum_{n=0}^{\infty} n^{p-1}(-n)\, z^{-(n+1)} \qquad \ldots(2)$$

Substituting (2) in (1) we get

$$Z(n^p) = -z\frac{d}{dz}\left[Z(n^{p-1})\right]$$

Which is the desired **recurrence formula**.

Using the recurrence formula we deduce the following basic formulae.

3) Prove that $Z(1) = \dfrac{z}{Z-1}$

Proof : We have

$$Z(a^n) = \frac{z}{z-a}$$

Put $a = 1$

$$\therefore \quad Z(1) = \frac{z}{z-1}$$

4) Prove that $Z(n) = \dfrac{z}{(z-1)^2}$

Proof : By putting $p = 1$ in $Z(n^p) = -z \dfrac{d}{dz} Z(n^{p-1})$ we get

$$\therefore \quad Z(n^1) = -z \frac{d}{dz} Z(1)$$

$$= -z \frac{d}{dz}\left(\frac{z}{z-1}\right)$$

$$= -z_1 \left[\frac{z-1-z}{(z-1)^2}\right]$$

$$\therefore \quad Z(n) = \frac{z}{(z-1)^2}$$

5) Prove that $Z(n^2) = \dfrac{z^2+z}{(z-1)^3}$

Proof : We have

$$Z(n^2) = -z \frac{d}{dz} Z(n)$$

$$= -z \frac{d}{dz}\left[\frac{z}{(z-1)^2}\right]$$

$$= -z \left[\frac{(z-1)^2 1 - z\, 2(z-1)}{(z-1)^4}\right]$$

$$= \frac{z^2+z}{(z-1)^3}$$

6) Prove that $Z(n^3) = \dfrac{z^3 + 4z^2 + z}{(z-1)^4}$

Proof : We have

$$Z(n^3) = -z\frac{d}{dz}Z(n^2)$$

$$= -z\frac{d}{dz}\left[\frac{z^2+z}{(z-1)^3}\right]$$

$$= -z\left[\frac{(z-1)^3(2z+1)-(z^2+z)\,3(z-1)^2}{(z-1)^6}\right]$$

$$\therefore \quad Z(n^3) = \frac{z^3+4z^2+z}{(z-1)^4}$$

7) Prove that $Z(n^4) = \dfrac{z^4 + 11z^3 + 11z^2 + z}{(z-1)^5}$

Proof : As above

$$Z(n^4) = \frac{z^4+11z^3+11z^2+z}{(z-1)^5}$$

8.9 LINEARITY PROPERTY

Let a, b, c be any constants and u_n, v_n, w_n be any discrete functions then

$$Z[au_n + bv_n - cw_n] = aZ(u_n) + bZ(v_n) - CZ(w_n).$$

By definition

$$Z[au_n + bv_n - Cw_n] = \sum_{n=0}^{\infty} (au_n + bv_n - cw_n)\, z^{-n}$$

$$= a\sum_{n=0}^{\infty} u_n z^{-n} + b\sum_{n=0}^{\infty} v_n z^{-n} - c\sum_{n=0}^{\infty} w_n z^{-n}$$

$$= a\,Z(u_n) + bZ(v_n) - c\,Z(w_n)$$

8.10 DAMPING RULE

If $\quad Z(u_n) = U(z)$ then $Z(a^{-n} u_n) = U(az)$

By definition

$$Z(a^{-n} u_n) = \sum_{n=0}^{\infty} a^{-n} u_n z^{-n}$$

$$= \sum_{n=0}^{\infty} u_n (az)^{-n}$$

$$= U(az)$$

$$\therefore \quad Z(a^{-n} u_n) = U(az)$$

Cor.: $\quad Z(a^n u_n) = U\left(\frac{z}{a}\right)$

The geometric factor a^{-n} when $|a| < 1$, damps the function u_n, hence the name damping rule.

The application of the damping rule leads to the following standard results.

1) Prove that $Z(na^n) = \dfrac{az}{(z-a)^2}$

Proof : We know that

$$Z(n) = \frac{z}{(z-1)^2}$$

$$\therefore \quad Z(na^n) = u(a^{-1} z)$$

$$= \frac{a^{-1}z}{(a^{-1}z-1)^2}$$

$$= \frac{az}{(z-a)^2}$$

2) Prove that $Z(n^2 a^n) = \dfrac{az^2 + a^2 z}{(z-a)^3}$

Proof : We know that

$$Z(n^2) = \frac{z^2 + z}{(z-1)^3}$$

$$\therefore \quad Z(n^2 a^n) = u(a^{-1} z)$$

$$= \frac{(a^{-1}z)^2 + a^{-1}z}{(a^{-1}z-1)^3}$$

$$= \frac{az^2 + a^2 z}{(z-a)^3}$$

3) Prove that

i) $Z(\cos n\,\theta) = \dfrac{z(z-\cos\theta)}{z^2 - 2z\cos\theta + 1}$

ii) $Z(\sin n\,\theta) = \dfrac{z\sin\theta}{z^2 - 2z\cos\theta + 1}$

iii) $Z(a^n \sin n\theta) = \dfrac{z(z-a\cos\theta)}{z^2 - 2az\cos\theta + a^2}$

iv) $Z(a^n \sin n\theta) = \dfrac{az\sin\theta}{z^2 - 2az\cos + a^2}$

Proof : We know that

$$Z(1) = \frac{z}{z-1} \text{ and } e^{-in\,\theta} = \cos n\theta - i \sin n\theta$$

Consider

$$Z(e^{-in\,\theta}) = Z[(e^{-i\theta})^n \,.\, 1]$$

$$= \frac{ze^{i\theta}}{ze^{i\theta}-1} = \frac{z}{z-e^{-i\theta}}$$

Multiply and divide by $(z - e^{i\theta})$

$$= \frac{z(z-e^{i\theta})}{(z-e^{-i\theta})(z-e^{i\theta})}$$

$$= \frac{z[z-(\cos\theta + i\sin\theta)]}{z^2 - z_1(e^{i\theta}+e^{-i\theta})+1}$$

$$\therefore \quad Z(\cos\theta - i\sin n\theta) = \frac{z(z-\cos\theta) - iz\sin\theta}{z^2 - 2z\cos\theta + 1}$$

Equating real and imaginary parts we get

$$Z(\cos n\theta) = \frac{z(z-\cos\theta)}{z^2 - 2z\cos\theta + 1} \qquad \text{....(i)}$$

and
$$Z(\sin n\theta) = \frac{z\sin\theta}{z^2 - 2z\cos\theta + 1} \qquad \text{...(ii)}$$

Again from (1)

$$Z(a^n \cos n\theta) = \frac{a^{-1}z(a^{-1}z - \cos\theta)}{(a^{-1}z)^2 - 2(a^{-1}z)\cos\theta + 1}$$

$$= \frac{z(z - a\cos\theta)}{z^2 - 2az\cos\theta + a^2} \qquad \text{....(iii)}$$

and from (ii)

$$Z(a^n \sin\theta) = \frac{a^{-1}z\sin\theta}{(a^{-1}z)^2 - 2(a^{-1}z)\cos\theta + 1}$$

$$= \frac{az\sin\theta}{z^2 - 2az\cos\theta + a^2} \qquad \text{....(iv)}$$

Example 4: Find the $Z(n+1)^2$

Solution: $Z(n+1)^2 = Z[n^2 + 2n + 1]$

$$= Z(n^2) + 2Z(n) + Z(1)$$

$$= \frac{z^2 + z}{(z-1)^3} + 2 \cdot \frac{z}{(z-1)^2} + \frac{z}{(z-1)}$$

$$= \frac{z^3 + z^2}{(z-1)^3}$$

Example 5: Obtain the Z-transform of $\sin(3n+5)$

Solution: We have

$$Z[\sin(3n+5)] = Z[\sin 3n \cos 5 + \cos 3n \sin 5]$$

$$= \cos 5\, Z(\sin 3n) + \sin 5\, Z(\cos 3n).$$

We know that $Z(\cos n\theta) = \dfrac{z(z-\cos\theta)}{z^2 - 2z\cos\theta + 1}$

$$= \frac{\cos 5z \sin 3}{z^2 - 2z\cos 3 + 1} + \frac{\sin 5z(z - \cos 3)}{z^2 - 2z\cos 3 + 1}$$

$$= \frac{z(\sin 3\cos 5 - \cos 3\sin 5) + z^2 \sin 5}{z^2 - 2z\cos 3 + 1}$$

$$= \frac{z(z\sin 5 - \sin 2)}{z^2 - 2z\cos 3 + 1}$$

Example 6: Obtain the Z-transform of cosh $n\theta$ and sinh $n\theta$

Solution: We have

i) $$Z(\cosh n\theta) = \frac{1}{2}\left[Z(e^{n\theta}) + Z(e^{-n\theta})\right]$$

$$= \frac{1}{2}\left[Z\left\{(e^{-\theta})^{-n} . 1\right\} + Z\left\{(e^{\theta})^{-n} .1\right\}\right]$$

Taking $u_n = 1$ $$= \frac{1}{2}\left[\left(\frac{z}{z-1}\right)_{z \to ze^{-\theta}} + \left(\frac{z}{z-1}\right)_{z \to ze^{\theta}}\right]$$

$$= \frac{1}{2}\left[\frac{ze^{-\theta}}{ze^{-\theta} - 1} + \frac{ze^{\theta}}{ze^{\theta} - 1}\right]$$

$$= \frac{1}{2}\left[\frac{z^2 - ze^{-\theta} + z^2 - ze^{-\theta}}{z^2 - z(e^{\theta} + e^{-\theta}) + 1}\right]$$

$$= \frac{1}{2}\left[\frac{2z^2 - 2z\left(\frac{e^{\theta} + e^{-\theta}}{2}\right)}{z^2 - 2z\frac{(e^{\theta} + e^{-\theta})}{2} + 1}\right]$$

$$\therefore Z(\cosh n\theta) = \frac{z_1^2 - z\cosh\theta}{z^2 - 2z\cosh\theta + 1}$$

ii) $$\sinh n\theta = \frac{1}{2}(e^{n\theta} - e^{-n\theta})$$

$$\therefore Z(\sinh n\theta) = \frac{1}{2}\left[\frac{ze^{-\theta}}{ze^{-\theta} - 1} - \frac{ze^{\theta}}{ze^{\theta} - 1}\right]$$

$$= \frac{1}{2}\left[\frac{z^2 - ze^{-\theta} - z^2 + ze^{\theta}}{z^2 - 2z\cosh\theta + 1}\right]$$

$$= \frac{1}{2}\frac{\left[z(e^{\theta} - e^{-\theta})\right]}{z^2 - 2z\cosh\theta + 1}$$

$$\therefore\ Z(\sinh n\theta) = \frac{z\sinh\theta}{z^2 - 2z\cosh\theta + 1}$$

Example 7: Find Z-transform of

i) Unit impulse sequence $\delta(n) = \begin{cases} 1 & n = 0 \\ 0 & n \neq 1 \end{cases}$

ii) Unit step sequence $u(n) = \begin{cases} 0 & n < 0 \\ 1 & n \geq 0 \end{cases}$

Solution: We have

i) $$Z[\delta(n)] = \sum_{n=0}^{\infty} \delta(n)z^{-n}$$

$$= 1 + 0 + 0 \ldots\ldots = 1$$

ii) $$Z[u(n)] = \sum_{n=0}^{\infty} u(n)z^{-n}$$

$$= 1 + z^{-1} + z^{-2} + z^{-3} \ldots\ldots\ldots\ldots$$

$$= [1 - z^{-1}]^{-1} = \frac{z}{z-1}$$

Example 8: Find the Z-transform of

1) $3n - 4\sin\dfrac{n\pi}{4} - 5a^2$

Solution: We have $Z(u_n) = U(z)$

$$\therefore\quad Z\left[3n - 4\sin\frac{n\pi}{4} - 5a^2\right]$$

$$= 3Z(n) - 4Z\left(\sin\frac{n\pi}{4}\right) - 5a^2 Z(1)$$

$$= \frac{3z}{(z-1)^2} - \frac{4z}{\sqrt{2}\,(z^2 - \sqrt{2}z + 1)} - \frac{5a^2 z}{(z-1)}$$

Example 9: Find the Z-transform of

$$\cos\left(\frac{n\pi}{2} + \frac{\pi}{4}\right)$$

$$= \cos\frac{n\pi}{2}\cos\frac{\pi}{4} - \sin\frac{n\pi}{2}\sin\frac{\pi}{4}$$

$$\therefore \quad Z\left[\cos\left(\frac{n\pi}{2} + \frac{\pi}{4}\right)\right] = \cos\frac{\pi}{4}\, Z\left(\cos\frac{n\pi}{2}\right) - \sin\frac{\pi}{4}\, Z\left(\sin\frac{n\pi}{2}\right)$$

$$= \frac{1}{\sqrt{2}} \frac{z\left(z - \cos\frac{\pi}{2}\right)}{z^2 - 2z\cos\frac{\pi}{2} + 1} - \frac{1}{\sqrt{2}} \frac{z\sin\frac{\pi}{2}}{z^2 - 2z\cos\frac{\pi}{2} + 1}$$

$$= \frac{1}{\sqrt{2}}\left[\frac{z^2 - z}{z^2 + 1}\right]$$

Example 10: Find the Z-transform of

$$\cos\left(\frac{n\pi}{8} + \theta\right)$$

$$= \cos\frac{n\pi}{8}\cos\theta - \sin\frac{n\pi}{8}\sin\theta$$

<\> $$Z\left(\cos\frac{n\pi}{8} + \theta\right) = \cos\theta\, Z\left(\cos\frac{n\pi}{8}\right) - \sin\theta\, Z\left(\sin\frac{n\pi}{8}\right)$$

$$= \frac{\cos\theta \,.\, z\left(z - \cos\frac{\pi}{8}\right) - \sin\theta \,.\, z\sin\frac{\pi}{8}}{z^2 - 2z\cos\frac{\pi}{8} + 1}$$

$$= \frac{z^2 \cos\theta - z\left(\cos\theta\cos\frac{\pi}{8} + \sin\theta\sin\frac{\pi}{8}\right)}{z^2 - 2z\cos\frac{\pi}{8} + 1}$$

$$= \frac{z^2 \cos\theta - z\cos\left(\theta - \frac{\pi}{8}\right)}{z^2 - 2z\cos\frac{\pi}{8} + 1}$$

8.11 SHIFTING u_n TO THE RIGHT

If $Z(u_n) = U(z)$ then $Z(u_{n-k}) = z^{-k}\,U(z)$, $\quad k > 0$

By definition

$$Z(u_{n-k}) = \sum_{n=0}^{\infty} u_n - k\ z^{-n}$$

$$= u_0\, z^{-k} + u_1\, z^{-1} + u_2\, z^{-2} \ldots\ldots.. \qquad (\because u_n = 0,\ n < 0)$$

$$= z^{-k}\,(u_0 + u_1 z^{-1} + u_2 z^{-2} \ldots\ldots.)$$

$$= z^{-k} \sum_{n=0}^{\infty} u_n z^{-n}$$

$$\therefore \quad Z(u_n - k) = z^{-k}\,U(z)$$

Shifting u_n to the left

If $\quad Z(u_n) = U(z)$ then

$$Z(u_{n+k})\, z^k = z^k \left[U(z) - u_0 - u_1 z^{-1} - u_2 z^{-2} \ldots\ldots.. u_{n-k} z^{-(k-1)}\right]$$

By definition

$$Z(u_{n+k}) = \sum_{n=0}^{\infty} u_{n+k} z^{-n}$$

$$= z^k \sum_{n=0}^{\infty} u_{n+k} z^{-(n+k)}$$

$$= z^k \left[\sum_{n=0}^{\infty} u_n z^{-n} - \sum_{n=0}^{k-1} u_n z^{-n}\right]$$

$$\therefore \quad Z(u_n + k) = \left[U(z) - u_0 - u_1 z^{-1} - u_2 z^{-2} \ldots\ldots\ldots u_{k-1}\, z^{-(k-1)}\right]$$

Using this theorem, we deduce the following results.

1) $Z(u_{n+1}) = z\,[U(z) - u_0]$

2) $Z(u_{n+2}) = z^2\,[U(z) - u_0 - u_1 z^{-1}]$

3) $Z(u_{n+3}) = z^3\,[U(z) - u_0 - u_1 z^{-1} - u_2 z^{-2}]$

4) $Z(u_{n+4}) = z^4\,[U(z) - u_0 - u_1 z^{-1} - u_2 z^{-2} - u_3 z^{-3}]$

Example 11: Prove that $Z\left(\frac{1}{n!}\right) = e^{\frac{1}{z}}$ and hence evaluate

$Z\left(\frac{1}{(n+1)!}\right)$ and $Z\left(\frac{1}{(n+2)!}\right)$

Proof: We have by definition

$$Z\left(\frac{1}{n!}\right) = \sum_{n=0}^{\infty} \frac{1}{n!} z^{-n}$$

$$= 1 + \frac{z^{-1}}{1!} + \frac{z^{-2}}{2!} + \frac{z^{-3}}{3!} + \ldots\ldots\ldots\ldots$$

$$= e^{z^{-1}} = e^{1/z}$$

$$\therefore \quad Z\left(\frac{1}{n!}\right) = e^{1/z}$$

Shifting $\left(\frac{1}{n!}\right)$ one unit to the left gives

$$Z\left(\frac{1}{(n+1)!}\right) = z\left[Z\left(\frac{1}{n!}\right) - 1\right]$$

$$= z\left[e^{1/z} - 1\right]$$

Similarly, shifting $\left(\frac{1}{n!}\right)$ two units to the left

$$Z\left(\frac{1}{(n+2)!}\right) = z^2\left[e^{1/z} - 1 - z^{-1}\right]$$

Example 12: Find the *Z*-transform of $\frac{1}{2}(n+1)(n+2)$

Solution: $Z\left\{\frac{1}{2}(n+1)(n+2)\right\}$

$$= \frac{1}{2}\left[Z(n^2) + 3Z(n) + 2Z(1)\right]$$

$$= \frac{1}{2}\left[\frac{z^2+z}{(z-1)^3} + \frac{3z}{(z-1)^2} + \frac{2z}{(z-1)}\right], \text{ if } |z| > 1.$$

Example 13: Find the *Z*-transform of $Z(a^n \cos n\pi)$

Solution:

$$Z(a^n \cos n\pi) = \sum_{n=0}^{\infty} a^n \cos n\pi\, z^{-n}$$

$$= 1 - \frac{a}{z} + \left(\frac{a}{z}\right)^2 - \left(\frac{a}{z}\right)^3 + \ldots\ldots\ldots.\infty. \quad (\cos n\pi = (-1)^n)$$

$$= \left(1 + \frac{a}{z}\right)^{-1} \text{ if } \left|\frac{a}{z}\right| < 1 \quad \text{if } |z| > |a|$$

$$= \frac{z}{z+a}$$

8.12 INITIAL VALUE THEOREM

If $Z(u_n) = U(z)$ then $u_0 = \lim_{z\to\infty} U(z)$

Proof : We know that

$$U(z) = Z(u_n) = u_0 + u_1 z^{-1} + u_2 z^{-2} + \ldots\ldots\ldots$$

Taking limits as $z \to \infty$ we get

$$\lim_{z\to\infty} U(z) = u_0$$

Similarly, additional initial values can be found successively

i.e.,
$$u_1 = \lim_{z\to\infty}\left[z\{U(z)-u_0\}\right]$$
$$u_2 = \lim_{z\to\infty}\left[z^2\{U(z)-u_0-u_1 z^{-1}\}\right]$$
$$u_3 = \lim_{z\to\infty}\left[z^2\{U(z)-u_0-u_1 z^{-1}-u_2 z^{-2}\}\right]$$

8.13 FINAL VALUE THEOREM

If $Z(u_n) = U(z)$ then $\lim\limits_{z\to\infty}(u_n) = \lim\limits_{z\to 1}(z-1)\,U(z)$

Proof : By definition

$$Z(u_{n+1}-u_n) = \sum_{n=0}^{\infty}(u_{n+1}-u_n)z^{-n}$$

$$\therefore \quad Z(u_{n+1})-Z(u_n) = \sum_{n=0}^{\infty}(u_{n+1}-u_n)z^{-n}$$

$$\therefore \quad z[U(z)-u_0]-u(z) = \sum_{n=0}^{\infty}(u_{n+1}-u_n)z^{-n}$$

$$\therefore \quad (z-1)\,U(z)-u_0 z = \sum_{n=0}^{\infty}(u_{n+1}-u_n)z^{-n}$$

Taking the limits on both sides as $z \to 1$ we get

$$\lim_{z\to 1}\left[(z-1)U(z)\right]-u_0 = \sum_{n=0}^{\infty}(u_{n+1}-u_n)$$

$$= \lim_{n\to\infty}\left[(u_1-u_0)+(u_2-u_1)+\ldots\ldots+(u_{n+1}-u_n\right]$$

$$= \lim_{n\to\infty} u_{n+1}-u_0$$

$$\lim_{n\to\infty}(u_n) = \lim_{z\to 1}(z-1)\,U(z)$$

Example 14: If $U(z) = \dfrac{2z^2 + 5z + 14}{(z-1)^4}$ evaluate u_2 and u_3

Solution:

$$U(z) = \frac{2z^2 + 5z + 14}{(z-1)^4}$$

$$\therefore \quad U(z) = \frac{1}{z^2}\,\frac{2 + 5z^{-1} + 14z^{-2}}{(1-z^{-1})^4}$$

By initial value theorem

$$u_0 = \lim_{z\to\infty} U(z) = 0$$

Similarly

$$u_1 = \lim_{z\to\infty} \left[z\{U(z) - u_0\}\right] = 0$$

$$u_2 = \lim_{z\to\infty} \left[z^2\left\{U(z) - u_0 - u_1 z^{-1}\right\}\right]$$

$$= 2 - 0 - 0 = 2$$

and

$$u_3 = \lim_{z\to\infty} z^3\left[U(z) - u_0 - u_1 z^{-1} - u_2 z^{-2}\right]$$

$$= \lim_{z\to\infty} z^3\left[U(z) - 0 - 0 - 2z^{-2}\right] \qquad \because u_2 = z$$

$$= \lim_{z\to\infty} z^3\left[\frac{2z^2 + 5z + 14}{(z-1)^4} - \frac{2}{z^2}\right]$$

$$= \lim_{z\to\infty}\left[\frac{13z^2 + 2z^2 + 8z - 2}{z^2(z-1)^4}\right]$$

$$= 13$$

8.14 CONVOLUTION OF SEQUENCES

The convolution of two sequences $\{u_n\}$ and $\{v_n\}$ is defined as

$$w(n) = \sum_{m\to 0}^{\infty} u_m \cdot v_{n-m} = u_n\ v_n$$

Convolution Theorem

If $Z^{-1}[U(z)] = u_n$ and $Z^{-1}[V(z)] = v_n$ then

$$Z^{-1}[U(z)\,V(z)] = \sum_{m=0}^{n} u_m \cdot v_{n-m} = u_n\, v_n$$

Proof : We have

$$U(z) = \sum_{m=0}^{n} u_n\, z^{-n} \text{ and } V(z) = \sum_{n=0}^{\infty} v_n z^{-n}$$

$$\therefore \quad U(z), V(z) = (u_0 + u_1 z_1^{-1} + u_2 z^{-2} + \ldots + u_n z^{-n} + \ldots \infty)$$

$$X(v_0 + v_1 z_1^{-1} + v_2 z^{-2} + \ldots + v_n z^{-n} + \ldots \infty)$$

$$= \sum_{n=0}^{\infty} (u_0 v_n + u_1 v_{n-1} + \ldots + u_n v_0)$$

$$= (u_0 v_n + u_1 v_{n-1} + u_2 v_{n-2} \ldots + u_n v_0)$$

$$= u_n\, v_n$$

$$\therefore \quad Z^{-1}[U(z) \,.\, V(z)] = u_n * v_n$$

Hence the proof.

Example 15: Use convolution theorem to evaluate

$$Z^{-1}\left\{\frac{z^2}{(z-a)(z-b)}\right\}$$

Solution: We know that

$$Z^{-1}\left\{\frac{z}{z-a}\right\} = a^n \text{ and } z^{-1}\left\{\frac{z}{z-b}\right\} = b^n$$

$$\therefore \quad Z^{-1}\left\{\frac{z^2}{(z-a)\,(z-b)}\right\} = Z^{-1}\left\{\frac{z}{z-a} \cdot \frac{z}{z-b}\right\}$$

$$= a^n * b^n$$

$$= \sum_{m=0}^{n} a^m\, b^{n-m}$$

$$= b^n \sum_{m=0}^{n} \left(\frac{a}{b}\right)^m \quad \text{which is G.P.}$$

$$= b^n \left[\frac{\left(\frac{a}{b}\right)^{n+1} - 1}{\frac{a}{b} - 1} \right] = \frac{a^{n+1} - b^{n+1}}{a - b}$$

8.15 CONVERGENCE OF Z-TRANSFORMS

1) *Z*-transform operation is performed on a sequence u_n which may exist in the range of integers $-\infty < n < \infty$, we write

$$U(z) = \sum_{n=-\infty}^{\infty} u_n z^{-n} \qquad \text{....(1)}$$

Where u_n represents a number in the sequence for *n*, an integer. The region of the *Z*-plane in which (1) converges absolutely is known as the region of convergence (R O C) of *U*(*z*) we have discussed one-sided *Z*-transform only for which $n \geq 0$. Here the sequence is always right sided and R O C is always inside any prescribed contour $|z| < |b|$

2) Two sided Z-transform of u_n is defined by

$$U(z) = \sum_{n=-\infty}^{\infty} u_n z^{-n}$$

In this case, the sequence is two sided and the region of convergence for (2) is the annular region $|b| < |z| < |c|$. The inner circle bounds the terms in negative powers of *z* and outer circle bounds the terms in positive powers of *z*. The shaded annular of convergence is necessary for the two sided sequence and its *Z*-transform to exist.

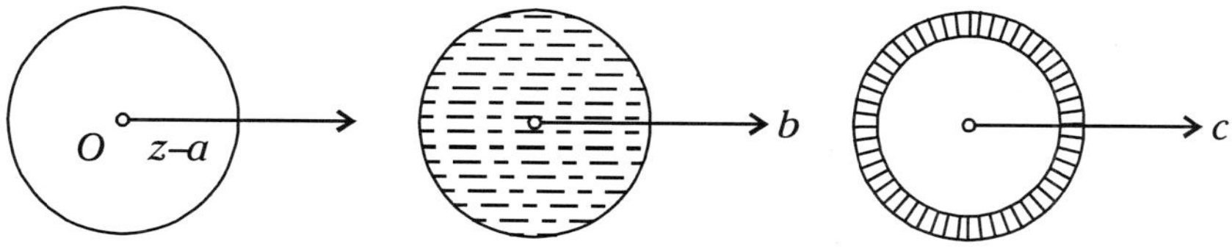

$|z| > |a|$ $\qquad$ $|z| < |b|$ $\qquad$ $|b| < |z| < |c|$

8.16 EVALUATION OF INVERSE Z-TRANSFORMS

We can obtain the inverse Z-transforms using any of the following three methods.

1) Partial fractions method

This method is similar to that of finding the inverse Laplace transforms using partial fractions. The method consists of decomposing $\frac{u(z)}{z}$ into partial fractions. Multiplying the resulting expression by z and then inverting the same.

Example 1: Find the inverse Z-transform of $\frac{2z^2+3z}{(z+2)(z-4)}$

Solution: Let $U(z) = \frac{2z^2+3z}{(z+2)(z-4)}$

$$\therefore \quad \frac{U(z)}{z} = \frac{2z+3}{(z+2)(z-4)} = \frac{A}{z+2} + \frac{B}{z-4}$$

$$\therefore \quad 2z+3 = A(z-4) + B(z+2)$$

Put $z = -2$ $\qquad A = \frac{1}{6}$

Put $z = 4$ $\qquad B = \frac{11}{6}$

$$\therefore \quad U(z) = \frac{1}{6}\frac{z}{z+2} + \frac{11}{6}\frac{z}{z-4}$$

Taking inverse Z-transform on both sides

$$Z^{-1}[U(z)] = \frac{1}{6}Z^{-1}\left\{\frac{z}{z+2}\right\} + \frac{11}{6}Z^{-1}\left\{\frac{z}{z-4}\right\}$$

$$\therefore \quad u_n = \frac{1}{6}(-2)^n + \frac{11}{6}(4)^n$$

Example 2 : Find the inverse Z-transform of $\frac{z}{(z-2)(z+3)^2}$

Solution: $U(z) = \dfrac{z}{(z-2)\,(z+3)^2}$

$$\therefore \quad \frac{U(z)}{z} = \frac{z}{(z-2)\,(z+3)^2} = \frac{A}{(z-2)} + \frac{B}{(z+3)} + \frac{C}{(z+3)^2}$$

$$\therefore \quad 1 = A\,(z+3)^2 + B\,(z-2)\,(z+3) + C\,(z-2)$$

Putting $z = 2$, $z = -3$ and $z = 0$ we get

$$A = \frac{1}{25}, \quad B = -\frac{1}{25} \quad C = \frac{1}{5}$$

Taking inverse Z-transform on both sides

$$Z^{-1}\{U(z)\} = \frac{1}{25}\left[Z^{-1}\left\{\frac{z}{z-2}\right\} - Z^{-1}\left\{\frac{z}{z+3}\right\} - 5\,Z^{-1}\left\{\frac{z}{(z+3)^2}\right\}\right]$$

$$u_n = \frac{1}{25}\left[(2)^n - (-3)^n + \frac{5}{3}n(-3)^n\right]$$

$$\because \; Z^{-1}\left\{\frac{az}{(z-a)^2}\right\} = na^n$$

Example 3 : Find the inverse Z-transform of $\dfrac{z^3 - 20z}{(z-2)^3(z-4)}$

Solution: Here $U(z) = \dfrac{z^3 - 20z}{(z-2)^3(z-4)}$

$$\therefore \quad \frac{U(z)}{z} = \frac{z^2 - 20}{(z-2)^3(z-4)} = \frac{Az^2 + Bz + C}{(z-2)^3} + \frac{C}{z-4}$$

$$\therefore \quad z^2 - 20 = (Az^2 + Bz + C)\,(z-4) + D\,(z-2)^3.$$

Putting $z = 4, 2, 0$ and 1

$D = -\dfrac{1}{2}$ $\quad 4A + 2B + C = 8 \quad$ Solving these equations we get

$C + 2D = 5 \qquad A = \dfrac{1}{2},\; B = 0,\; C = 6,\; D = -\dfrac{1}{2}$

$3(A + B + C) + D = 19$

$$\therefore \quad \frac{U(z)}{z} = \frac{1}{2}\left[\frac{z^2+12}{(z-2)^3} - \frac{1}{z-4}\right]$$

$$\therefore \quad U(z) = \frac{1}{2}\left[\frac{z^3+12z}{(z-2)^3}\right] - \frac{1}{2}\left[\frac{z}{z-4}\right]$$

$$z = \frac{1}{2}\left[z\frac{(z-2)^2+4z^2+8z}{(z-2)^3}\right] - \frac{1}{2}\left[\frac{z}{z-4}\right]$$

$$U(z) = \frac{1}{2}\ \frac{z}{z-2} + \frac{2z(z+2)}{(z-2)^3} - \frac{1}{2}\ \frac{z}{z-4}$$

On inversion

$$u_n = \frac{1}{2}\ (2)^n + n^2\ (2)^n - \frac{1}{2}\ (4)^n.$$

Example 4: Find the inverse Z-transform of $\dfrac{z}{(z+1)^2}$ by division method.

Solution: Given $U(z) = \dfrac{z}{(z+1)^2}$

$$= z^{-1} - \frac{2+z^{-1}}{z^2+2z+1} \quad \text{by actual division}$$

$$= z^{-1} - 2z^{-2} + 3z^{-3} - \frac{4z^{-2}+3z^{-3}}{z^2+2z+1}$$

Continuing this process of division we get an infinite series

$$U(z) = \sum_{n=0}^{\infty} (-1)^{n-1} nz^{-n}$$

Thus $u_n = (-1)^{n-1} n$.

Example 5: Find the inverse Z-transform of $\log\left(\dfrac{z}{z+1}\right)$ by power series method.

Solution: Given $U(z) = \log\left(\frac{z}{z+1}\right)$

Put $z = \frac{1}{y}$

$$\therefore \quad U(z) = \log \frac{\frac{1}{y}}{\frac{1}{y}+1} = \log\frac{1}{(1+y)} = \log 1 - \log(1+y)$$

$$= 0 - \log(1+y)$$

$$= -y + \frac{1}{2}y^2 - \frac{1}{3}y^3 + \ldots\ldots\ldots$$

$$= \quad z^{-1} + \frac{1}{2}z^{-2} - \frac{1}{3}z^{-3} + \ldots\ldots\ldots$$

Thus $u_n = 0$ for $n = 0$

$$= \frac{(-1)^n}{n} \qquad \text{otherwise.}$$

Example 6: Find the inverse Z-transform of

$$\frac{2(z^2 - 5z + 6.5)}{(z-2)\,(z-3)^2} \quad \text{for } 2 < |z| < 3$$

Solution: Given $U(z) = \frac{2(z^2 - 5z + 6.5)}{(z-2)\,(z-3)^2}$

$$= \frac{A}{z-2} + \frac{B}{z-3} + \frac{C}{(z-3)^2}$$

Solving by partial fractions A = 1, B = 1 and C = 1

$$\therefore \quad U(z) = \frac{1}{z-2} + \frac{1}{z-3} + \frac{1}{(z-3)^2}$$

$$= \frac{1}{z}\left(1 - \frac{2}{z}\right)^{-1} - \frac{1}{3}\left(1 - \frac{z}{3}\right)^{-1} + \frac{1}{9}\left(1 - \frac{z}{3}\right)^{-2}$$

$$= \frac{1}{z}\left[1+\frac{2}{z}+\frac{4}{z^2}+\frac{8}{z^3}+........\right]$$

$$-\frac{1}{3}\left[1+\frac{z}{3}+\frac{z^2}{9}+\frac{z^3}{27}+.........\right]$$

$$+\frac{1}{9}\left[1+\frac{2z}{3}+\frac{3z^2}{9}+\frac{4z^3}{27}+..................\right]$$

$$= \left[\frac{1}{z}+\frac{2}{z^2}+\frac{2^2}{z^3}+\frac{2^3}{z^4}+...........\right]$$

$$-\left[\frac{1}{3}+\frac{z}{3^2}+\frac{z^2}{3^3}+\frac{z^4}{3^4}+.............\right]$$

$$+\left[\frac{1}{3^2}+\frac{2z}{3^3}+\frac{3z^2}{3^4}+\frac{4z^3}{3^5}+..............\right]$$

$$= \sum_{n=1}^{\infty} 2^{n-1} z^{-n} - \sum_{n=0}^{\infty} \left(\frac{1}{3}\right)^{n+1} z^n + \sum_{n=0}^{\infty} (n+1)\left(\frac{1}{3}\right)^{n+2} z^n$$

On inversion we get

$$u_n = 2^{n-1}\ n \geq 1 \text{and } u_n = -(n+2)\,3^{n-2}, \qquad n \leq 0.$$

Example 7: Find the inverse Z-transform of $\dfrac{8z}{(2z-1)(4z-1)}$

Solution: Given $u(z) = \dfrac{8z}{(2z-1)(4z-1)}$

$$= \frac{z}{\left(z-\frac{1}{2}\right)\left(z-\frac{1}{4}\right)}$$

$\therefore \quad \dfrac{U(z)}{z} = \dfrac{A}{z-\frac{1}{2}}+\dfrac{B}{z-\frac{1}{4}}$ by Partial fractions we get $A = 2,\ B = -1$

$$= \frac{2}{z-\frac{1}{2}} - \frac{1}{z-\frac{1}{4}}$$

$$\therefore \quad U(z) = \frac{2z}{z-\frac{1}{2}} - \frac{z}{z-\frac{1}{4}}$$

On inversion

$$u_n = 2Z^{-1}\left\{\frac{z}{z-\frac{1}{2}}\right\} - Z^{-1}\left\{\frac{z}{z-\frac{1}{4}}\right\}$$

$$= 2\left(\frac{1}{2}\right)^n - \left(\frac{1}{4}\right)^n, \quad n = 0, 1, 2, \ldots\ldots\ldots$$

Example 8: Find the inverse Z-transform of $\dfrac{z}{z^2+7z+10}$

Solution: Given $U(z) = \dfrac{z}{z^2+7z+10}$

$$\therefore \quad \frac{U(z)}{z} = \frac{1}{(z+2)(z+5)} = \frac{A}{z+2} + \frac{B}{z+5}$$

$$= \frac{1}{3}\frac{1}{z+2} - \frac{1}{3}\cdot\frac{1}{z+5} \qquad \text{Solving } \left(A = \frac{1}{3},\ B = -\frac{1}{3}\right)$$

$$\therefore \quad U(z) = \frac{1}{3}\frac{z}{z+2} - \frac{1}{3}\frac{z}{z+5}$$

On inversion

$$u_n = \frac{1}{3}(-2)^n - \frac{1}{3}(-5)^n \qquad \left(\because\ Z(a^n) = \frac{z}{z-a}\right)$$

8.17 APPLICATION TO DIFFERENCE EQUATIONS

Working procedure

i) Take the Z-transform of both sides of the difference equations, using the formulae of $Z - \text{T}$ and the given conditions.

ii) Transpose all the terms without $U(z)$ to the right

iii) Divide by the coefficient of $U(z)$ getting $U(z)$ as a function of z.

iv) Express this function in terms of the Z-transforms of known functions and then take the inverse Z-transform on both sides.

This gives u_n as a function of n, which is the desired solution.

v) Note the following results

$$Z(u_n) = U(z)$$
$$Z(u_{n+1}) = z\,[(U(z) - u_0]$$
$$Z(u_{n+2}) = z^2\,[U(z) - u_0 - u_1 z^{-1}]$$
$$Z(u_{n+3}) = z^3\,[U(z) - u_0 - u_1 z^{-1}\; u_2\, z^{-2}] \text{ etc.}$$

Example 1: Using Z-transform solve the difference equation

$$u_{n+2} + 4u_{n+1} + 3u_n = 3^n \text{ given } u_0 = 0 \text{ and } u_1 = 1$$

Solution: The difference equation is

$$u_{n+2} + 4u_{n+1} + 3u_n = 3^n \qquad \ldots (1)$$

Taking Z-transform both sides of the equation

$$Z(u_{n+2}) + 4Z(u_{n+1}) + 3z(u_n) = Z(3^n).$$

$$\therefore \quad z^2\,[\,U(z) - u_0 - u_1\, z^{-1}] + 4z\,[U(z) - u_0] + 3U(z) = \frac{z}{z-3}$$

$$\therefore \quad (z^2 + 4z + 3)\,U(z) = z + \frac{z^2 - 2z}{z-3}$$

$$\therefore \quad U(z) = \frac{z^2 - 2z}{(z+1)(z+3)(z-3)}$$

$$\therefore \quad \frac{U(z)}{z} = \frac{z-2}{(z+1)(z+3)(z-3)}$$

$$= \frac{A}{z+1} + \frac{B}{z+3} + \frac{C}{z-3} \text{ by partial fraction}$$

Solving $A = \frac{3}{8},\; B = \frac{-5}{12},\; C = \frac{1}{24}$.

$$= \frac{3}{8}\frac{1}{z+1} - \frac{5}{12}\frac{1}{z+3} + \frac{1}{24}\frac{1}{z-3}$$

$$\therefore \quad U(z) = \frac{3}{8}\frac{z}{z+1} - \frac{5}{12}\frac{z}{z+3} + \frac{1}{24}\frac{z}{z-3}$$

On inversion we get

$$u_n = \frac{3}{8}(-1)^n - \frac{5}{12}(-3)^n + \frac{1}{24}(3)^n.$$

Example 2: Solve $y_{n+2} + 6y_{n+1} + 9y_n = 2^n$ with the given conditions $y_0 = y_1 = 0$ by using Z – transform.

Solution: Given difference equation is

$$y_{n+2} + 6y_{n+1} + 9y_n = 2^n \qquad \text{....(1)}$$

Taking Z-transform on both sides of the equation (1)

$$Z(y_{n+2}) + 6Z(y_{n+1}) + 9Z(y_n) = Z(2^n)$$

$$\therefore \quad z^2 [\mathrm{Y}(z) - y_0 - y_1 z^{-1}] + 6z [\mathrm{Y}(z) - y_0] + 9\mathrm{Y}(Z) = \frac{z}{z-2}$$

Since $y_0 = 0$, $y_1 = 0$ we have

$$(z^2 + 6z + 9)\, y(z) = \frac{z}{z-2}$$

$$\therefore \quad Y(z) = \frac{z}{(z-2)(z+3)^2}$$

$$\therefore \quad \frac{Y(z)}{z} = \frac{1}{(z-2)(z+3)^2} = \frac{A}{z-2} + \frac{B}{z+3} + \frac{C}{(z+3)^2}$$

$$\therefore \quad 1 = A(z+3)^2 + B(z-2)(z+3) + C(z-2)$$

Put $z = 2$, $z = -3$ and $z = 0$ we get $A = \frac{1}{25}$, $B = \frac{1}{25}$ and $C = -\frac{1}{5}$

$$\therefore \quad \frac{Y(z)}{z} = \frac{1}{25}\left[\frac{1}{z-2} - \frac{1}{z+3} - \frac{5}{(z+3)^2}\right]$$

$$\therefore \quad Y(z) = \frac{1}{25}\left[\frac{z}{z-2} - \frac{z}{z+3} - 5\frac{z}{(z+3)^2}\right]$$

On inversion

$$y_n = \frac{1}{25}\left[Z^{-1}\left(\frac{z}{z-2}\right) - Z^{-1}\left(\frac{z}{z+3}\right) - \frac{5}{-3}Z^{-1}\left(\frac{-3z}{(z+3)^2}\right)\right]$$

$$= \frac{1}{25}\left[2^n - (-3)^n + \frac{5}{3}n(-3)^n\right]$$

$$\left(\because Z^{-1}\left(\frac{az}{(z-a)^2}\right) = na^n\right)$$

$$\therefore \quad y_n = \frac{1}{25}\left[2^n - (-3)^n + \frac{5}{3}n(-3)^n\right]$$

Example 3: Find the response of the system $y_{n+2} - 5y_{n+1} + 6y_n = u_n$ with $y_0 = 0, y_1 = 1$ and $u_n = 1$ for $n = 0, 1, 2, 3$ by Z-transform method.

Solution: Given the response of the system is

$$y_{n+2} - 5y_{n+1} + 6u_n = u_n \quad \ldots (1)$$

Taking Z-transform on both sides of equation (1)

$$Z(y_{n+2}) - 5Z(y_{n+1}) + 6Z(y_n) = Z(1) \quad \because u_n = 1.$$

$$z^2[Y(z) - y_0 - y_1 z^{-1}] - 5z[Y(z) - y_0] + 6Y(z) = \frac{z}{z-1}$$

substituting $y_0 = 0, y_1 = 1$ we get

$$(z^2 - 5z + 6)\,Y(z) = \frac{z}{z-1} + z = \frac{z^2}{z-1}$$

$$\therefore \quad \frac{Y(z)}{z} = \frac{z}{(z-1)(z-2)(z-3)}$$

$$= \frac{A}{z-1} + \frac{B}{z-2} + \frac{C}{z-3}$$

Solving by partial fractions we get $A = \frac{1}{2},\ B = -2,\ C = \frac{3}{2}$

$$\frac{Y(z)}{z} = \frac{1}{2}\frac{1}{z-1} - 2\frac{1}{z-2} + \frac{3}{2}\frac{1}{z-3}$$

$$Y(z) = \frac{1}{2}\frac{z}{z-1} - 2\frac{z}{z-2} + \frac{3}{2}\frac{z}{z-3}$$

On inversion we get

$$y_n = \frac{1}{2}(1)^n - 2(2)^n + \frac{3}{2}(3)^n$$

Example 4: Using Z-transform solve the difference equation

$$u_{n+2} - 2u_{n+1} + u_n = 3n + 5$$

Solution: Given equation is

$$u_{n+2} - 2u_{n+1} + u_n = 3n + 5$$

Taking Z-transform on both sides of the equation

$$Z[u_{n+2}] - 2Z[u_{n+2}] + Z[u_n] = Z[3n+5]$$

$$z^2[U(z) - u_0 - u_1] - 2z[U(z) - u_0] + U(z) = 3\frac{z}{(z-1)^2} + 5\frac{z}{(z-1)}$$

$$\therefore \quad (z^2 - 2z + 1)\,U(z) = \frac{5z^2 - 2z}{(z-1)^2} + u_0(z^2 - 2z) + u_1 z$$

$$\therefore \quad U(z) = \frac{5z^2 - 2z}{(z-1)^4} + u_0\frac{z^2 - 2z}{(z-1)^2} + u_1\frac{z}{(z-1)^2}$$

on inversion

$$u_n = Z^{-1}\left[\frac{5z^2 - 2z}{(z-1)^4}\right] + u_0 Z^{-1}\left[\frac{z^2 - 2z}{(z-1)^2}\right] + u_1 Z^{-1}\left[\frac{z}{(z-1)^2}\right] \quad ....(1)$$

We know that $Z(1) = \frac{z}{z-1}$, $Z(n) = \frac{z}{(z-1)^2}$ $\quad Z(n^2) = \frac{z^2 + z}{(z-1)^3}$

and $\quad Z(n^3) = \frac{z^3 + 4z^2 + z}{(z-1)^4}$

Consider

$$\frac{5z^2-2z}{(z-1)^4} = A\frac{z^3+4z^2+z}{(z-1)^4}+B\frac{(z^2+z)}{(z-1)^3}+\frac{z}{(z-1)^2}+\frac{z}{z-1}$$

Equating coefficients of like powers of z we get

$$A=\frac{1}{2}, \quad B=1, \quad C=-\frac{3}{2} \quad D=0$$

$$\therefore \qquad Z^{-1}\left[\frac{5z^2-2z}{(z-1)^4}\right]=\frac{1}{2}(n)^3+(n)^2-\frac{3}{2}n$$

Next,

$$Z^{-1}\left[\frac{z^2-2z}{(z-1)^2}\right]=Z^{-1}\left[\frac{z}{z-1}\right]-Z^{-1}\left[\frac{z}{(z-1)^2}\right]$$

$$= 1 - n.$$

and
$$Z^{-1}\left[\frac{z}{z-1}\right]=n$$

Substituting in (1)

$$u_n = \frac{1}{2}\ n\,(n-1)\,(n+3)+u_0\,(1-n)+u_1\,n$$

$$= \frac{1}{2}\ n(n-1)\,(n+3)+u_0\,(1-n)+u_1\,n$$

$$u_n = \frac{1}{2}\ n\,(n-1)\,(n+3)+u_0+(u_1+u_0)n$$

Example 5: Solve the difference equation

$u_{n+2} - 4u_n = n - 1$ with $u_0 = 1$ and $u_1 = 2$ by using Z-transform.

Solution: Given difference equation is

$$u_{n+2} - 4u_n = n - 1$$

Taking the Z-transform on both sides of the equation

$$Z\,(u_{n+2}) - 4Z\,(u_n) = Z\,(n) - Z\,(1)$$

$$\therefore \quad z^2 [U(z) - u_0 - u_1 z^{-1}] - 4\, U(z) = \frac{z}{(z-1)^2} - \frac{z}{z-1}$$

$$\therefore \quad (z^2 - 4)\, U(z) = u_0 z^2 + u_1 z + \frac{z(2-z)}{(z-1)^2}$$

$$= z(z+2) - \frac{z(z-2)}{(z-1)^2}$$

$$U(z) = \frac{z(z+2)}{z^2-4} - z\frac{(z-2)}{(z^2-4)\,(z-1)^2}$$

$$\frac{U(z)}{z} = \frac{1}{z-2} - \frac{1}{(z-1)^2\,(z+2)}$$

$$= \frac{1}{z-2} - \left[\frac{A}{z+2} + \frac{B}{z-1} + \frac{C}{(z-1)^2}\right] \text{ by P.F.s}$$

$$A = -\frac{1}{9},\ B = \frac{1}{9},\ C = -\frac{1}{3}$$

$$= \frac{1}{z-2} - \frac{1}{9}\,\frac{1}{z+2} + \frac{1}{9}\,\frac{1}{z-2} - \frac{1}{3}\,\frac{1}{(z-1)^2}$$

$$\therefore \quad U(z) = \frac{z}{z-2} - \frac{1}{9}\,\frac{z}{z+2} + \frac{1}{9}\,\frac{z}{z-1} - \frac{1}{3}\,\frac{z}{(z-1)^2}$$

On inversion

$$u_n = (2)^n - \frac{1}{9}(-2)^n + \frac{1}{9}(1) - \frac{1}{3}n(1)^n$$

$$= 2^n - \frac{1}{9}(-2)^n + \frac{1}{9} - \frac{1}{3}n$$

8.18 QUESTION BANK

1) Evaluate the followings

i) $Z(n^2)$ ii) $Z(n^3)$

iii) $Z(n\,a^n)$ iv) $Z(n^2\,a^n)$

v) $Z(a^n \cos n\theta)$ vi) $Z(a^n \sin n\theta)$

vii) $Z(\cosh n\theta)$ viii) $Z(\sinh n\theta)$

ix) $Z(\sin(3n+5))$ x) $Z\left(\frac{1}{(n+2)!}\right)$

Ans:

i) $\frac{z^2+z}{(z-1)^3}$ ii) $\frac{z^3+4z^2+z}{(z-1)^4}$

iii) $\frac{az}{(z-a)^2}$ iv) $\frac{az^2+a^2z}{(z-a)^3}$

v) $\frac{z(z-a\cos\theta)}{z^2-2az\cos\theta+a^2}$ vi) $\frac{az\cos\theta}{z^2-2az\cos\theta+a^2}$

vii) $\frac{z(z-\cosh\theta)}{z^2-2z\cosh\theta+1}$ viii) $\frac{z\sinh\theta}{z^2-2z\cosh\theta+1}$

ix) $\frac{z^2\sin 5 - z\sin 2}{z^2-2z\cos 3+1}$ x) $z^2\left(e^{1/z}-1-z^{-1}\right)$

2) Show that $Z\left(\frac{1}{n+1}\right) = z\log\frac{z}{z+1}$

3) Prove that $Z[n\,f(n)] = -z\frac{d}{dz}[zf(n)]$

4) Using $Z(n) = \frac{z}{(z-1)^2}$ show that $Z(n\cos n\theta) = \frac{(z^3+z)\cos\theta-2z^2}{(z^2-2z\cos\theta+1)^2}$

5) Using $Z(n^2) = \frac{z^2+z}{(z-1)^3}$ show that $Z(n+1)^2 = \frac{z^3+z^2}{(z-1)^3}$.

6) If $U(z) = \dfrac{2z^2 + 5z + 14}{(z-1)^4}$ evaluate u_2 *and* u_3

Ans: $u_2 = 2,$ $u_3 = 13.$

7) Find the inverse Z-transform of $\dfrac{5z}{(2-z)(3z-1)}$

Ans: $\left(\dfrac{1}{3}\right)^n - 2^n$

8) Find the inverse Z-transform of $\dfrac{4z^2 - 2z}{z^3 - 5z^2 + 8z - 4}$

Ans.: $2 = (2)^n + 3\,(n-1)\,2^n,\ n \geq 1.$

9. Obtain $Z^{-1}\left\{\dfrac{1}{(z-2)(z-3)}\right\}$ for $|z| < 2$.

Ans.: $\left(-\dfrac{1}{3} - \dfrac{z}{3^2} - \dfrac{z^2}{3^3} - \dfrac{z^3}{3^4} \ldots\ldots\ldots\right) + \left(\dfrac{1}{2} + \dfrac{z}{2^2} + \dfrac{z^2}{2^3} + \dfrac{z^3}{2^4} + \ldots\ldots\ldots\right)$

10. Solve the difference equation.

$$y_{n+2} - 4y_{n+1} + 3y_n = 5^n$$

Ans.: $y_n = c_1 + c_2\,3^n + \dfrac{1}{8}\,5^n.$

11. Solve $u_{n+2} - 4u_n = n - 1$ by using Z-transform

Ans.: $u_n = c_1\,2^n + c_2\,(-2)^n - \dfrac{n}{3} + \dfrac{1}{9}.$

12. Solve $y_{n+2} + 2y_{n+1} + y_n = n$ with $y_0 = y_1 = 0$ by using Z-transform

Ans.: $y_n = \dfrac{3}{4}n$

13. Solve $u_{k+2} - 2u_{k+1} + u_k = 2^k$ with $u_0 = 2$, $u_1 = 1$ by using Z-transform.

Ans.: $u_k = 1 - 2k + 2^k$.

14. Solve $y_{n+2} - 6y_{n+1} + 8y_n = 2^n + 6n$ using Z-transform,

Ans.: $y_n = c_1 4^n + \left(c_2 - \frac{n}{4}\right) 2^n + 2n - \frac{8}{3}$

15. Find the impulse response of a system described by

$$y_{(n+1)} + 2y_{(n)} = \delta_n \qquad y_0 = 0.$$

Ans.: $y_n = (-2)^{n-1}, \quad n \geq 1$

Model Question Paper

Third Semester B.E. Degree Examination

Engineering Mathematics – III

Time: 3 Hrs Max. Marks : 100

Note : Answer any Five full questions choosing at least two from each part.

PART – A

Q1. (a) Obtain the Fourier series to represent the function $f(x)$ defined as follows:

$$f(x) = \begin{cases} \pi x & 0 \le x \le 1 \\ \pi(2-x) & 1 < x \le 2 \end{cases}$$

and hence deduce that

$$\frac{1}{1^2} + \frac{1}{3^2} + \frac{1}{5^2} + \ldots\ldots = \frac{\pi^2}{8}$$

(b) Find a series of cosines of multiple of x which will represent $x \sin x$ in $(0, \pi)$

(c) Express y as Fourier series upto second harmonic given

x:	0	1	2	3	4	5
y:	4	8	15	7	6	2

(7 + 7 + 6 = 20 Marks)

Q2. (a) Express the function $f(x) = \begin{cases} 1 & \text{for } |x| \le 1 \\ 0 & \text{for } |x| > 1 \end{cases}$ as a Fourier integral.

Hence evaluate $\int_0^\infty \frac{\sin \lambda \cos x\lambda}{\lambda} d\lambda$.

(b) Find the Fourier sine transform of $f(x) = e^{-|x|}$ and hence evaluate

$$\int_0^\infty \frac{x \sin mx}{1+x^2} dx, \quad m > 0.$$

(c) Find the inverse Fourier sine transform of $\frac{u}{1+u^2}$

(7 + 7 + 6 = 20 Marks)

Q3. (a) Derive one dimensional wave equation in the form $\frac{\partial^2 y}{\partial t^2} = c^2 \frac{\partial^2 y}{\partial x^2}$

(b) Solve $\frac{\partial^2 u}{\partial x^2} + \frac{\partial^2 u}{\partial y^2} = 0$ which satisfies the conditions

$u(0, y) = u(l, y) = u(x, 0) = 0$ and $u(x, a) = \sin \frac{\pi x}{l}$.

(c) Obtain the D'Alembert's solution of the wave equation $\frac{\partial^2 u}{\partial t^2} = c^2 \frac{\partial^2 u}{\partial x^2}$. (7 + 7 + 6 = 20 Marks)

Q4. (a) Fit a parabola to the following data

x:	1	2	3	4	5	6	7	8	9
y:	2	6	7	8	10	11	11	10	9

Estimate y when $x = 4.5$.

(b) Solve the LPP graphically maximize $z = 3x_1 + 4x_2$ subject to $4x_1 + 2x_2 \leq 80$; $2x_1 + 5x_2 \leq 180$ and $x_1 \geq 0$, $x_2 \geq 0$.

(c) Using simplex method maximize $z = 5x_1 + 3x_2$ subject to $x_1 + x_2 \leq 2$, $5x_1 + 2x_2 \leq 10$, $3x_1 + 8x_2 \leq 12$ and $x_1, x_2 \geq 0$.

(7 + 6 + 7 = 20 marks)

Part – B

Q5. (a) Find the real root of $3x - \cos x - 1 = 0$ by Newton's-Raphson Method. (x in radians)

(b) Solve the system of equations

$8x - y + z = 18$; $2x + 5y - 2z = 3$, $x + 2y - 3z = -6$

by using Gauss-Seidal iterative method.

(c) Find the largest eigenvalue and the corresponding eigenvector of the matrix $\begin{bmatrix} 1 & -3 & 2 \\ 4 & 4 & -1 \\ 6 & 3 & 5 \end{bmatrix}$ by power method. (7 + 7 + 6 = 20 Marks)

Q6. (a) The following table gives the distances in nautical miles of the visible horizon for the given heights in feet above the earth's surface.

Height (x) :	100	150	200	250	300	400
Distance (y) :	10.63	13.03	15.04	16.81	18.42	21.27

Estimate the value of y when $x = 410$ ft.

(b) Using Newton's divided difference formula evaluate $f(8)$ and $f(15)$ given that

x :	4	5	7	10	11	13
$f(x)$:	48	100	294	900	1210	2028

(c) Evaluate $\int_4^{5.2} \log_e^x \, dx$ by using Weddle's rule taking seven ordinates. (7 + 7 + 6 = 20 Marks)

Q7. (a) Derive the standard 5 point formula for the Laplace equation $\frac{\partial^2 u}{\partial x^2} + \frac{\partial^2 u}{\partial y^2} = 0$.

(b) Solve the Laplace equation $u_{xx} + u_{yy} = 0$ for the square mess given in the following figure with boundary values.

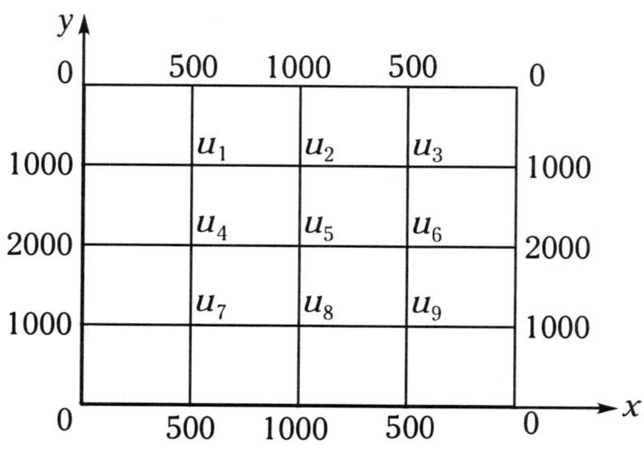

(c) Solve the equation $\frac{\partial u}{\partial t} = \frac{\partial^2 u}{\partial x^2}$ subject to the conditions $u(x, 0) = \sin \pi x$, $0 \le x \le 1$, $u(0, t) = u(1, t) = 0$ carry-out computations for two levels taking $h = \frac{1}{3}$, $k = \frac{1}{36}$. (4 + 8 + 8 = 20 Marks)

Q8. (a) Evaluate i) $Z(n^2 a^n)$ ii) $Z(a^n \cos n\theta)$ iii) $Z(\sin(4n+5))$

(b) Find the inverse Z-transform of $\dfrac{5z}{(2-z)(3z-1)}$

(c) Find the response of the system

$y_{n+2} - 5y_{n+1} + 6y_n = u_n$ with $y_0 = 0, y_1 = 1$ and $u_n = 1$ for $n = 0, 1, 2, 3,$ ...... by Z-transform method.

Index